Euclid Beach Park
CLEVELAND, O.
YEARBOOK

By Blake LeDuc Sr.

From The Collections
of
Angelo Datellis
and
Blake LeDuc Sr.

EMPLOYEE ROSTERS

Photographs • Postcards • Trivia

FIRST EDITION
September 1996

LIMITED EDITION
NO. 400

ISBN
0-9654588-0-6

Copyright © 1996 Duke Graphics Inc.
All Rights Reserved

Duke Graphics Inc.
Eastlake, Ohio 44095

No part of this book may
be copied without written
permission of
Duke Graphics Inc.

Printed in U.S.A.

DEDICATION

I would like to dedicate this book to Doris Humphrey Mackley, who gave me the strength and courage to believe in myself and in my personal growth. She touched my life in such a special way that words can never describe.

And also to my friends David Humphrey Scott and Georgene Scott, for not only being my friends but also my family during the past few years. Your love for your father, Dudley Humphrey Scott, and for Euclid Beach Park has kept it alive, and a lot would not be possible if not for you.

And also a special thank you to Blake LeDuc, Sr., for making this book possible. Had it not been for your hard work, determination and efforts, this book never would be possible. And most importantly, thank you for being a friend.

Angelo Datellis

SPECIAL GRATITUDE

A special gratitude and thanks should go to Angelo Datellis from everyone that has any interest in Euclid Beach Park.

Without his love for Euclid Beach and his dedication to preserving it, most of the material in this book would have been lost forever.

Thanks Angelo

Dedicated to the love of my life,
My wife...Donna

Blake LeDuc Sr.

I would also like to thank the following people for their
support and interest in this project:

Marian R. Nathan	Attorney
Pam J. Darnell	Webb Communications
David Humphrey Scott	Preface
Frank Brodnick	Euclid Beach Nuts
Dudley S. Humphrey Jr.	The Humphrey Co.
Bob Murphy	The News-Herald

...as well as my family and all of the employees of
Duke Printing and Mailing Inc., especially:

Pauline Schneider	Data Entry
Ron Ostrow	Digital Graphics and Layout
Robert Nozling	Image Assembly
Barb Nadock	Fulfillment
Lynette Cogley	Order Information

Blake LeDuc Sr.

Preface

In My Opinion
by David Humphrey Scott

It is a well-established fact that Dudley S. Humphrey II and his two brothers, David and Harlow, took the amusement park known as Euclid Beach Park from bankruptcy to absolute success in a single year's time. The Humphrey family accomplished this extreme turnaround when they took direct control of the park prior to the summer of 1901.

Dudley Sherman Humphrey I and his brother William came to Ohio in 1836 and settled in what is now Parma, in the area southwest of Cleveland. Here they prospered as farmers, lumbermen and clock makers. Humphrey family members were among the early trustees of Parma Township, and later served as postmasters.

While residents were shooting bears and wolves on what is now Pearl Road, the Humphrey were busy renaissance men, taking on new challenges at every turn. By the time they moved from Parma in 1851, they had built a five bedroom log cabin.

The family left Parma for Wakeman, Ohio, where they bought a 400-acre farm, established another lumberyard, built the first corduroy road to Vermilion on Lake Erie, and operated an oil barrel factory. The family prospered financially until 1890, when a series of disasters struck.

First, two successive years of drought caused the complete loss of crops; then the lumberyard was totally destroyed by fire; and finally, they were victims of the financial collapse of the oil industry. These problems forced the Humphrey family into a sherrif's sale, leaving them with but the clothes on their backs and $20,000 still in debt.

With the aid of a friendly neighbor who gave them a horse and wagon, the family was able to move to the Cleveland area, where they established themselves in the retail popcorn business from horse-drawn carts.

After Euclid Beach Park was started in 1895 by Mr. Ryan and his friends, the Humphreys sold popcorn under their own name at great profit for only the first three years. They refused to return for the next two years of the Ryan ownership of the park because they did not approve of many of the activities permitted in the park. Some of their primary objections included shills and barkers, gambling, and the beer garden. Worst of all, the park featured a bowling alley where people were permitted to bowl on Sunday!

Yes, the family was square as bricks, and to that I can only say "God bless them." Later, when the Humphrey family took over the park, a tall, see-through fence was installed with a sign that read, "Private property, public welcome." All of the aspects of the park that they disagreed with under the previous ownership were removed.

Advertising flyers were sent to all church, union, business, ethnic and area groups promising "Nothing to depress or demoralize." These changes to the park were so successful that by the end of the 1901 season, the Humphreys controlled 80 percent of all the stock in Euclid Beach Park.

The Park Family

The Park Family includes all employees who have ever worked at Euclid Beach Park. This group includes managers, ride operators, kitchen aides, maintenance, grounds keepers, ticket sellers, security police, campground personnel and many others. All of the Humphrey family ideals would *never have been successful were it not for the park family's ability to work under and maintain the very strict rules that were now in place.*

Those employees who lived on the park property were limited to the number of house guests allowed at any given time. Consumption of alcohol was strictly limited in the home, and not allowed at all in public view. Park residents were also limited to the number of guest cars allowed on the property, and those cars were assigned designated spots. All cars of park employees living on the property had license plates issued by the park office, running from AH 101 to AH 149. This system allowed park security to pass cars through the gates during off-park hours without a hassle.

For many years, ride operators, dance hall employees and kitchen aides were not permitted to smoke. The dress code for employees was even more strict than it was for customers. Neither men nor women were allows to wear shorts; women's skirts must be below the knee; halters for women and men going shirtless were not permitted. People living on park property were not allowed to walk through the park in bathing suits, even if they were going straight to the swimming beach!

It was through the ingenuity and determination of the park family that many previously unknown devices, machines and processes were developed and built within the confines of the park. One of these machines was the huge power sweeper that was used to clean sidewalks every day. Another was a continuous concrete mixing machine that allowed things like the concrete head of the pier and the cement cottages to be poured in a single pour, leaving no joints.

At the time of the development of the Surprise House in the 1930s, an electrical proximity switch was developed by Harvey Humphrey and George Lister in the park's electrical lab. The "electric eye" had just been invented, the first one installed in the entrance to Chin's Restaurant in downtown Cleveland. People would drive for miles to walk through a door that would open automatically! The electric eye installed in Euclid Beach Park was one step better— it did not require any holes in the wall or a visible light beam.

I also can recall the early recording system installed in the dance hall so that bands could record their music. The park system used a diamond stylus to cut the grooves in 12 inch aluminum disks, played at 78 rpm.

With World War II's shortage of steel, the pattern for replacing the tracks of the roller coasters was switched. Track from the bottom of the hills, where the wear preface was greatest, was removed and exchanged with track from the top, where the wear was minimal. In this way, the roller coasters continued to run during the war.

During the Great Depression, when the Humphreys could not borrow money from the bank to meet the winter payroll, park maintenance employees continued work without pay so the park would be ready come spring. Their spirit was rewarded by the Humphreys, who did not charge rent on the homes used by park employees. The Humphreys regularly brought in truckloads of produce from the company-owned farm to distribute to park employees at no charge.

All in all, had it not been for the excellent work ethics of the park family, the principles established by the Humphrey family never would have been maintained.

To this I can only say, "God bless you all."

TABLE OF CONTENTS

Dedications	i
Special Thanks	ii
Preface	iii
Contents	v
History	vi
Pre-Humphrey	1
Pre-1901	2
1901	4
1903	5
1910	6
Gate	7
Numbers Without Names	8
1938 Roster	9
1939	13
1940	17
1941	23
1942	29
1943	37
1944	45
1945	51
1946	57
1947	65
1948	71
1949	77
1950	81
1951	87
1952	93
1953	99
1954	105
1955	111
1956	117
1957-58	121
1959-60	129
1961-62	137
1963-64	141
1965-66	145
Wages	148
Concession Operators	150
Tickets	151
Ride Count	154
Ride Cost	159
Picnics	161
School Days	169
Special Shows	172
Maps and Mileage	173
Supplies and Suppliers	176
Camp Trivia	181

This book is not meant to be a history of Euclid Beach, but another look at the park and it's employee's through rare and mostly unpublished material.

Two excellent books are available on the history of the park from Amusement Park Books Inc.

**Euclid Beach Park
"Closed for the Season"**

**Euclid Beach Park
"A Second Look"**

I.

Many illustrations in this book were proofed from original metal printers engravings and have captions marked by a large roman numeral, such as above.

II. Pre Humphrey brochure cover. ca. 1898.

III. Main Gate. ca. 1899.

IV. First Humphrey stand on Public Square. ca. 1895.

V. Harvey Humphrey demonstrates corn popper invention of The Humphrey Co. ca. Pre-1900.

An original Humphrey popcorn wagon.

First Humphrey stand at the beach. ca. 1898.

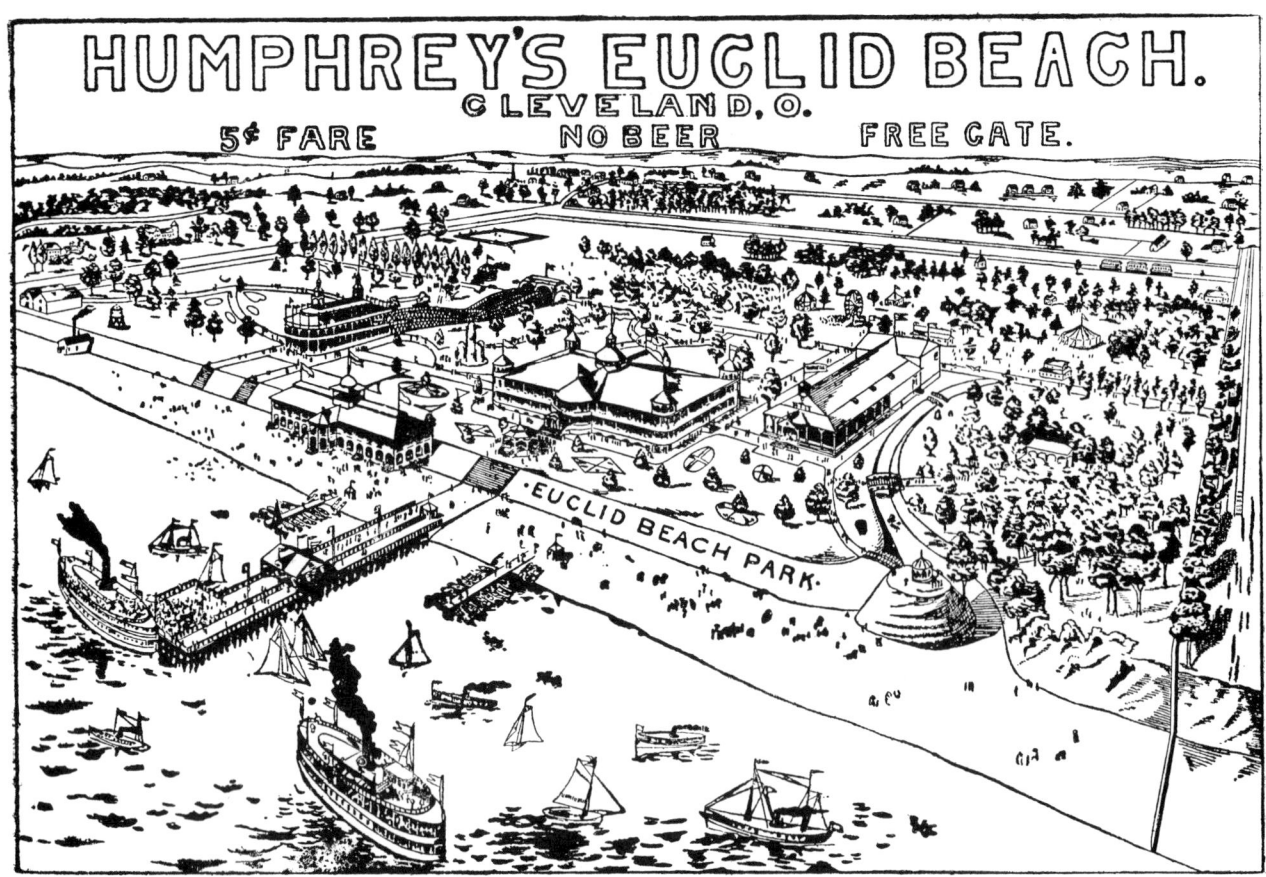
VI. The Humphrey era at Euclid Beach Park begins. 1901.

The original Humphrey main gate. ca. 1901.

On patrol in front of the World Theatre.

V. Dance pavillion and swing. ca. 1902.

Newspaper advertisement. ca. 1910.

HUMPHREY CO. MANAGERS AT EUCLID BEACH PARK.

The top line reading from left to right:
1. Robt. Ryle.
2. Peter Brugo.
4. Mrs. Ida A. Magner.
5. Mrs. Frank England.
6. Mrs. L. M. Hormel.
7. Miss Lottie Lesher.

Second row, left to right:
1. T. B. Rutherford.
2. A. W. Fritz.
3. Geo. Snyder.
4. H. D. Walker.
5. W S. Walker
6. Frank Sherwood.
7. S. Haines.
8. Harris C. Shannon.
9. John Anderson.
10. Ed. Beamer.
11. L. W. Scutt.
12. R. A. Pease.

Third row, sitting down, reading from left to right:
1. F. R. Forgason.
2. L. W. Dill.
3. C. R. Ramsey.
4. Dudley H. Scott.
5. J. C. Bright.
6. Ed. Austin.
7. E. S. Moore.
8. L. J. Currier.
9. Frank Kilby.
10. Clif B. Johnson.
11. R. Z. Owen.
12. Harvey Lawson.

Bottom row, from left to right:
1. Kenneth N. McClintock.
2. Al. Abby.
3. Ralph Trunkey.
4. Paul Mercer.

Following are the names of eight other department heads that do not appear in the group above:

Mr. W. J. McLevie, constructional manager for The Humphrey Company for the past eight years.

Mr. Harry W. Stroud, manager of Euclid Beach Camp Grounds, 11 years.

Mr. John McDonald, captain of Euclid Beach police for for ten years.

Mr. John E. Walter, manager Main Lunch for eight years.

Doc. McIlrath, manager fishing department for nine years.

Mr. Orr Brown, head porter for six years.

AT EUCLID BEACH.

Following are some of the people employed at the places of business at Euclid Beach Park:

SKATING RINK.
F. E. Kilby, Manager.
J. E. Tennant, Floor Manager.
F. R. Kreger, Floor Director.
Norman Poland, Director of Music.
E. Romanis Equipment, Manager.
Miss N. Sullivan, Head Check Lady.
Miss F. Hogue, Assistant.
Mrs. A. Ryel, Cashier.
John Kobie, Officer.
William O'Neil.
Norman Tegun.

SKATE MEN.
W. Atkins.
W. Watkiss.
W. Huyat.
E. Murphy.
E. Goldburg.
R. Sampson.
H. Stanbury.
C. Ferrel.
H. Skelley.

DANCE HALL.
L. W. Scutt, Manager.
Joseph Squires and Peter Wood, Floor Managers.
David Swank, Chester Richards, E. Hipkins, Charles Britton, Ticket Takers.
Roy Hickox, Charles Persell, on Rope.
Ed Tisher, R. Kagin, Floor Guards.
Extra Men:—
E. J. Woolmington.
M. Downing.
A. Seiquist.
J. Whitehead.
R. Paton.
H. Steller.
H. Caldwell.
J. Jones.

AERIAL SWING.
Wm. Richards, Manager.
Jack Brown, Electrician.
Henry Fussner, Ticket Taker.

FLYING PONIES.
Ralph Trunkey.
W. B. Carter, Helper.

CARROUSEL.
Howard Bales, Manager.
William Fussner.
Russel Davies.
George Weber.
James Johnson.

SUPERVISION OF GROUNDS.
H. W. Lawson, Superintendent;
Chist Holtz, Harvey Hoffmer, Teamsters.
George Fox, Dick Muckel, Arthur ich,
Lylie, Joe Lylie, John Pappaleo, John Murrachak, John Murphey, James Hussy, General Work.

CIGAR STAND.
Mary Stringer, Manager.
Alice H. Heidt, Relief.

CHESTNUT GROVE STAND.
Clifford Smith.
Clarence Hunter.
Scott Hunt.
O'Dell Dean.
Billie Richards.
Clarence Brennen.
Cliff B. Johnston, Manager.

THE SCENIC RAILWAY.
R. Hagedorn.
G. Krause.
Carl Zanzig.
C. Herrick.
O. Hanchett.
A. Muller.
E. Hickock.
D. Bowie.
R. McCoy.
George Reinhard, Manager.

POP CORN STAND.
T. J. Denman.
W. Braidey.
J. V. Kelley.
James Murray.
Lloyd Denman.
R. E. Murray.
J. M. Heid.
E. Monroe.
George Webber.
M. Davie.
E. Barney.
George Wilson.
F. L. Price, Manager.

BOWLING ALLEYS.
J. C. Bright, Mgr.
Ed Fisher.
Tom Conry.

BOWLING ALLEY LUNCH.
Chas. Smith.
Thomas Skelly.
Wm. Brennan.
Extra—I. Vaughn, Dan Brennan.
R. Z. Owen, Mgr.

SOUVENIR STAND.
Miss May Treter, Manager.
Gertrude Hardy, Assistant.

PHOTO STUDIO.
S. B. Johnson.
S. I. Riblet.
C. F. Riblet, Manager.

LADIES' DEPARTMENT.
Miss Payne, Manager.
Miss Littlejchn, Assistant.

FIGURE EIGHT.
Manager, Frank Sherwood.
Roland Barhyte, Brakeman.
A. Lance.
Harry Stellar, Ticket Taker.
Clint Storms, Ticket Taker.

MAIN LUNCH.
Gilbert Rider.
Wm. Keegan.
Ralph Snedden.
Lloyd Collier.
Erle Patchin.
B. Perkins.
Asp, Stanley.
Patrick Burke.
Carlton Truman.
Lester Marshall.
Arthur Faulkner.
Ray Buell.
Robert Leetch.
Thomas Carson.
J. E. Walter, Manager.

BATH HOUSE.
J. A. Anderson, Manager.
Mrs. Dwyer, Assistant.
Mrs. J. A. Anderson.
Life Guard, John Elish.
W. Davis, Assistant.
Gale Kirts, Special Guard and Instructor.
Attendants, G. Freeman, Harry Steller, Charlie Koons, Mrs. Hildreth.

MOVING PICTURE THEATER.
Dick Mullen, Operator.
Bertha De Voe, Pianist.
Geo. Wilcox, Trombone.
H. Poland, Manager.
Tom Poland, Assistant.

PENNY ARCADE.
Earl Fernal.
Frank Baldwin.
Norman Freeman.
Avery Finger.
Miss O. Redmond, Cashier.
Miss R. Clegg, Assistant Cashier.
C. Gent, Manager.

PORTERS.
Elmer Land.
Elmer Fletcher.
William Williams.
James Johnson.
Orr Brown, Manager.

BOX BALL ALLEY.
Sylvester Dwyer, Manager.

PAINTING AND DECORATING.
Ed Austin, Manager.
Frank Richie, Assistant.

CASHIERS.
Booth A, Mrs. Flynn.
Mrs. B. Murray, Booth B.
Booth C, Mrs. Austin.
Booth D, Miss B. Bayne.
Booth E, Mrs. J. E. Walters.
Booth F, Mrs. Terrel.
Booth G, Miss Alice Ryel.
Booth I, Mrs. Franklin.
Booth J, Mrs. R. Z. Owen.
Booth M, Mrs. A. Percell.
Mrs. Reddy, Pony Track.
Relief, Miss Florence Smith, Mrs. J. T. Barry, Miss Josie Smith.

MECHANICAL DEPARTMENT.
Dudley H. Scott.
Frank Wood, Plumber.
Charles McHale, Machinist.
Joe Baldwin, Assistant Plumber.
Henry Hess, Blacksmith.

CASTLE INN.
Ida L. Magner, Manager.

EUCLID BEACH DINING HALL.
W. S. Walker, Manager.

BUILDING DEPARTMENT.
W. J. McLevie, Constructional Manager.
Walter Briggs, Assistant.
Jim Walker.
Jim Robinson.

THE GARAGE.
W. R. Ryel, Manager.
O. P. Sutton, Mechanic.
John Palmer, Night Attendant.

CHAUFFEURS.
C. H. Hallwood.
Fred M. Kushen.
Fred Furze.

SIGHT-SEEING CAR.
W. R. Ryel, Manager.
Julian E. Carlisle, Announcer.

DANCE HALL MUSICIANS.
Prof. L. J. Currier, Leader.
Ed. Persell, Piano; Geo. Wilcox, Trombone; James Hamley, Violin; Roy Brady, Violin; John Polak, Cornet; Otto Uher, Clarinet; Fred Greenwald, Flute; E. M. Orpin, Cello; Sam Unger, Bass; Frank Shole, Drums.

OFFICE DEPARTMENT.
Harvey P. Shannon, Manager.
Geo. Harris, Assistant.

ELECTRICAL DEPARTMENT.
Ed S. Moore, Head Electrician.
Al Abbey, Assistant.

ICE CREAM CONE STAND.
Ida Maier, Manager.
Don Baldwin, Assistant.
Joe Baldwin, Extra.

BALLOON STAND.
Mrs. S. England, Manager.
Alice Heidt, Assistant.

POSTOFFICE.
A. W. Fritz, Manager.
Miss Grace R. Curtis, Assistant.
Dorothy Ewald, Extra.
Helen Fitzgerald, Extra.

POLICE FORCE.
Frank England.
J. J. Yeager.
Geo. E. Dennison.
John F. Kobie.
John Nalbach.
Harry Barnhardt.
Geo. Joseph.
Harry Kidwell.
John Guthrie.
John Mainer.
John MacDonald, Captain.

PONY TRACK.
Archie Lerick.
Dick Muckle, Extra.
Herman Smith.
Harry Lerick.
Geo. Fox, Manager.

CAMPING GROUNDS.
W. Sasager.
Homer Ward.
Walter Ward.
W. Kuhlman.
Ray Thomas.
Ed. Seila.
Fred Nelson.
Harry Watson.
Henry Barnhardt.
Nettie McGregor.
Harry W. Stroud, Manager.

OUTSIDE LIGHTING.
T. J. Skelley.
R. A. Pease, Manager.
A. MacDonald, Assistant.

TONSORIAL DEPARTMENT.
Ed. Beamer, Manager.
Harvey Decker, Assistant.

PUBLISHING DEPARTMENT.
A. W. Fritz, Manager and Editor.
Euclid Beach News.

CARE OF BOATS.
Ollie MacElrath, Manager.

FISHING TACKLE.
Dr. MacElrath, Manager.

SURVEYING DEPARTMENT.
P. M. Killaly, Manager.

CHECKING ROOM AT THE DANCE PAVILION.
Mrs. L. Hormel, Manager.
Mrs. F. Storms, Assistant.
Luther Martin, Extra.

STAND AT PUBLIC SQUARE.

DAY FORCE.
C. Henry.
L. Wilson.
T. McComb.
S. Smith.
R. Draband.
A. Holden.
Miss N. Hayes.
Mrs. W. A. Moffatt.
Miss M. Battersby.
Miss E. Dowdell.
Miss C. Farrell.
Miss N. Welch.
Miss M. Moran.
Miss G. Osborn.
W. A. Moffatt, Manager.

NIGHT FORCE.
H. Haase.
H. Schweitzer.
J. Hill.
S. Louck.
F. Russell.
Miss K. Otterman.

L. F. Price is the manager of the popcorn and candy stand at the Beach, one of the most important places to manage in the summer season. With the exception of one year Mr. Price has been with the Humphrey Co. since 1903, filling several different positions during that time, giving the best satisfaction in them all. He was this spring given charge of this important place and his success has been very satisfactory to the company.

Musical Department.—Peter Brugo, manager. Mr. Brugo is a man of wonderful musical talent and we doubt if his equal can be found in this country. The large organ that is used in the roller skating rink and in the other places where the company uses organs were mostly constructed by him, also most of the music used in them is written by Mr. Brugo.

(Above) Main Gate shown here built in wood. Later covered in Permastone.

1917 - 1937
MONTHLY EMPLOYMENT TOTALS
NUMBERS WITHOUT NAMES

	JAN	FEB	MAR	APR	MAY	JUN	JUL	AUG	SEP	OCT	NOV	DEC
1917	32	42	66	94	148	216	252	243	173	49	31	30
1918	35	35	38	57	115	200	227	230	126	51	34	33
1919	17	20	33	52	121	208	253	250	153	51	39	35
1920	31	33	34	55	124	204	262	261	163	63	71	59
1921	67	71	106	129	203	266	289	279	140	57	54	46
1922	31	31	36	73	121	292	257	251	160	60	52	48
1923	44	43	48	81	141	258	300	275	180	71	62	58
1924	59	57	67	100	176	252	325	317	198	95	74	70
1925	60	62	72	131	166	207	353	326	220	75	61	59
1926	52	50	56	97	175	265	324	321	192	46	55	50
1927	49	48	55	98	152	277	325	307	161	58	45	46
1928	42	47	54	88	141	257	310	305	171	55	48	51
1929	47	48	62	102	147	285	321	307	226	109	53	48
1930	47	57	69	109	159	258	310	307	134	21	38	36
1931	36	38	40	93	128	200	255	255	131	63	25	24
1932	12	13	22	60	87	196	243	197	96	26	31	28
1933	31	32	33	83	91	177	193	192	115	59	32	32
1934	33	33	34	74	76	205	213	213	118	34	35	35
1935	35	36	37	73	73	208	221	221	111	35	36	36
1936	36	37	40	93	128	209	236	233	138	52	43	42
1937	42	42	46	159	214	351	387	328	242	54	47	49

Early Humphrey Co. Employee picnic.

1938
EMPLOYEE ROSTER

LAST	FIRST	MIDDLE
ANDERSON	EMMA	KING
ANDERSON	JOHN	ANDREW
ARTELJ	JOHN	ANTON
BABER	ALLAN	W
BAIRD	MAE	ELEANOR
BAIRD	WILSON	EDWARD
BAKER	CILLIUS	MOSE
BAKER	LUCION	
BARAGA	FRANCES	K
BARRETT	CHARLES	DENNIS
BARRETT	JACK	M
BARRY	KENNETH	
BARTO	ALOYSIUS	FRANK
BASS	ISABEL	LOUISE
BATCHLET	ROBERT	LOUIS
BEADLE	CHARLES	ALBERT
BECHER	EMMA	SWANK
BECK	NEVITT	SIDMAN
BENEDICT	HOWARD	L
BENSON JR	PETER	
BERTRAM	THOMAS	EDWARD
BETZ	RALPH	KENNETH
BILLENS	WILLIAM	FREDRICK
BISSELL	HOWARD	LOREN
BIZJAK	FRANK	
BLANK	ROBERT	EUGENE
BLEAM	LULA	ELSIE
BLECKER	HARRY	
BLOEDE	WILLIAM	CARL
BLOW	ARLINE	BEIL
BOCK	WILLIAM	PAUL
BOGGESS	ELIZABETH	SUE
BOWDEN	FRANCES	MARIAN
BOWHALL	ELMER	F
BOZA	OLIVER	
BRAIDICH	ALBINA	
BRAIDICH	NICK	ALBERT
BREYLEY	ELEANOR	MAY
BREYLEY	HAZEL	MAY
BRICKLEY	ROBERT	ANSON
BRIMLOW	ROYCE	EARL
BROADBENT	ALBERT	HENRY
BROADBENT	JOSEPH	
BROUGHTON	JAMES	LEWIS
BUCHWALD	ROBERT	FREDRICK
BUERKEL	JOHN	M
BURBRIDGE	CARL	
BURROWS	CHARLES	FREDERICK
BURTON	JAMES	
BURYA	JOSEPH	JOHN
BUSH	EVERETT	HAMILTON
BUTLER	EVA	ENGEL
CADA	PHILIP	
CALLAGHAN	BEATRICE	BERNICE
CALLAGHAN	GEORGE	EDWARD
CAMERON JR	DONALD	DOUGLAS
CAMPBELL	ALEXANDER	ROBERT
CARROLL	CLARA	LEGGON
CASE	DONALD	LEW
CENTA	JOSEPHINE	ESTHER
CERCEK	JOHN	LOUIS
CETINA	EDWARD	
CETINA	WILLIAM	FREDRICK
CHAMBERLAIN	ELIZABETH	JANE
CHAMBERLAIN	PAULINE	MARIE
CHEMICK	BETTY	
CHINN	CYRIL	HARRIS
CINCO	RUDY	JOE
CIRINO	ROCCO	
CLARKE	EDWARD	JACK
CLOUSER	LESTER	JAMES
COLLINS	LAYTON	T
CONSLA	JACK	LOFTUS
CONTENTO	GUISEPPE	
COOK	LLOYD	HAYES
COOLIDGE	RALPH	
COTMAN	JOHN	JOSEPH
COWHARD	HATTIE	MAY
COX	DAVID	OLIVER
CRAIN	WILLIS	
CRITZER	RAYMOND	DUNNING
CROTTY	LAURENCE	ALOYSIUS
CRUZ	SANTIAGO	
DADEY	ROBERT	SCOTT
DANEWITZ	JOHN	A
DAVIES	RALPH	ERNEST
DAVIO	MILDRED	BERQUIN
DAVIS JR	ROSS	ESHER
DAYKIN	OTTO	
DEGERIO	LOUIS	
DELAMBO	FRANK	FRANCIS
DENNISON	JAMES	E
DETTMAN	LESTER	
DICKEY	HERBERT	ROBERT
DICKEY	RICHARD	ALBERT
DLUGOLESKI	STANLEY	WILLIAM
DOLES	FRANK	LEE
DONAHUE	FRANK	LEO
DONAHUE	LAURA	GRAY
DONAHUE JR	FRANKLIN	LEO
DOUGAN	JERRY	PENROSE
DOWDELL	E	
DUDECK	EDWARD	JOSEPH
DUNCAN	BERNARD	
DUNN	WALTER	
DUTCHCOT JR	W	J
ECKERMAN	RUSSELL	LEROY
ELLIS	HARRY	EUGENE
EPAVES	JACK	ROBERT
FARLEY	JACK	CEDRIC
FEDELE	ANTHONY	
FEITEN	EARL	ARTHUR
FIFOLT	MILDRED	ALICE
FINDLAY	ALEX	
FINERAN	MARY	BALL
FISH	WALTER	EDWARD
FISHER	ELDA	
FISHER	HARLEY	DEE
FITZGERALD	PETER	MICHAEL
FLATING	ROBERT	ALFRED
FLETCHER	ELLIS	
FLETCHER	ELMER	
FORD	HAROLD	
FOX	GEORGE	UPTON
FOX	JOHN	FRED
FRANK	HARRY	FRED
FRANLEY	HAROLD	CHARLES
FREDERICKS	FRANK	JOSEPH
FREDERICKS	NEIL	FRANCIS
FUHRMAN	FRED	FRANK
GARDINER	ROBERT	ALLEN
GAST	WILLIAM	HOEPP
GERM	JOHN	A
GILOY	ARTHUR	LAMBERT

LAST	FIRST	MIDDLE
GILROY	JACK	ANTHONY
GLASS	LEONARD	
GLAVIC	JOE	
GOGLIN	ALFRED	GEORGE
GRAF	ROBERT	FREDERICK
GRAHAM	HARRY	CLIFFORD
GREENWAY	FRED	STUDER
GREGG JR	EARLE	COVINGTON
GREIG	WILLIAM	C
GRIEBEL	RUSSELL	JOHN
GRIFFITH	WILLIAM	JOHN
GRIGOR	ALEX	PANTON
GROSSMAN	FRANK	TONY
GUNTON	WINIFRED	RISTOW
HAGERMAN	JACK	WILLIAM
HALL	WARREN	WILLIAM
HANNA	LORETTA	HANNA
HANSEN	ALFRED	IRVING
HARDY	HENRY	FRANCIS
HARROLD	THADDEUS	JAMES
HARTMAN	CARL	
HARTMAN	PETER	CARL
HARVEY	JOHN	WILLIAM
HAWKINS	JAMES	FRANKLIN
HAWKINS	THELMA	
HAYDL	CARL	WILLIAM
HAYES	WILLIS	JAMES
HEATON	MAY	
HECKMAN	EDWARD	W
HELLER	EUGENE	JOSEPH
HENCK	FRANK	HAMILTON
HENSEL	ROBERT	LEWIS
HERGENROEDER	CHARLES	WILLIAM
HIMEBAUGH	NORMAN	STEWART
HIRSCH	HAROLD	
HOEGLER	ALBERT	J
HOEGLER	WALTER	JOHN
HOGG	JACK	
HOHL	WILBUR	
HOLLMEYER	ROBERT	NORMAN
HOOD	EMILY	AUGUSTA
HOOPER	ARCHIE	
HORA	SARAH	
HOWE	RICHARD	HEYD
HUEBER	DONALD	FRANCIS
HULING	FRANCIS	SHEPARD
HUNSBERGER	EDYTHE	MAY
HUNT	BLANCHE	LEAH
HUNTER	ROY	CHARLES
INGALLS	LAIRD	
JABLONSKI	LEONARD	M
JACKLITZ	ALBERT	
JACKSON	ALLAN	WALTER
JACKSON	HAROLD	
JACKSON	NORMAN	KARL
JAMIESON	CLARENCE	BARCLEY
JAMIESON	DAVID	
JAMIESON	MARGARET	EILEEN
JANZ	JOHN	
JENKINS	CLIFF	
JOHNSON	JAMES	CARLYLE
JOHNSTON	CLIFFORD	BASIL
JOHNSTON	DONALD	HUMPHREY
JOHNSTON	WILLIAM	CLIFFORD
JUDD	CLARENCE	HADEN
JUDD	ESTIL	LEE
KAPPEL	MARGARET	
KAPPY	WILLIAM	FRED
KASTELIC	FRANK	MICHAEL
KAUFFMAN	RALPH	
KECK JR	MAX	A
KELLEY	CECIL	HOLMES
KELLEY	THOMAS	
KELLEY	THOMAS	
KELLY	DAISY	CURTIS
KENJERSKI	MERESLOW	
KESSLER	AGNES	CECELIA
KIKOLI	FRANK	
KILBY	FRANK	EVERETT
KIRSCH	NELSON	P
KIRSCH	PAUL	C
KLEBENGOT	ANNE	MARIE
KLINE	RICHARD	DAVID
KNAUFF	CHARLES	LOUIS
KOBIE	JOHN	
KOJAN	JOHN	S
KOPFSTEIN	VINCENT	A
KOSKY	JOSEPH	JOHN
KOSKY	STANLEY	A
KOSS	PATRICIA	VERA
KOSS	WILLIAM	HARRY
KOSTYK	STEVE	
KOZIKOWSKI	EDWARD	STANLEY
KRAUSE	GEORGE	
KREEGER	FRED	CARL
KRONBERGER	BEN	J
KRUSELL	BEN	FRED
KRUSELL	ED	FRANK
KUBELAVIC	MARY	ANN
KUBELAVIC	MICHAEL	JOHN
KUBELAVIC	PETER	PAUL
KUBELAVIC	SAM	PETE
KUNZ	ROBERT	
KUSHLAN	LOUIS	C
LAMBIE JR	JOHN	EDWARD
LANE	MELVIN	
LANGE	JOHN	FERDINAND
LEACH	DONALD	EVERETT
LECHOWICZ	BENEDICT	
LEONE	GEORGE	
LEONE	LOUISE	
LEVSTIK	EMIL	LOUIS
LEWIS	CHARLES	
LINGIS	PETER	WALTER
LISTER	GEORGE	HAMPTON
LISTER	GEORGE	W
LITTLE	PHILLIP	FREDERICK
LLOYD	FREDERIC	MOORE
LOGAN	LUCILLE	MATHILDA
LONG	WILLIAM	E
LONGFIELD	TESS	EVANS
LONGWELL	JOSEPH	B
LOVE	FRED	JAMES
LOVE	ROBERT	KELLOGG
LOVETT	ARTHUR	JOHN
LUKAS	BEN	
MAGILL	THOMAS	JOSEPH
MALAVASIC	MAX	ADOLPHE
MALONEY	THOMAS	J
MANLEY	JOSEPH	WILLIAM
MARASH	EMIL	

Euclid Beach.

LAST	FIRST	MIDDLE
MARKENS	EDNA	BROWNING
MAXIM	JOSEPH	JOHN
MAY	DICK	OTTO
MAY	LYDIA	
MAY	OSWALD CARL	FREDERICK
MAY	WALTER	EMIL
MCBRIDE	HARRY	ELDEN
MCCAHAN	BERNARD	RUSSELL
MCCAULEY	HENRY	ORMSLEY
MCCLINTICK	WILLIAM	R
MCCORMICK	KATHERYN	RITA
MCFARLAND	PATRICIA	
MCKNIGHT	DAVID	K
MCMASTER	WILLIAM	GORDON
MCNIECE	GRACE	LILLIAN
MEANS	JOHN	WILLIAM
MEGLAN	ALBERT	CHARLES
METZGER	FRANK	
MEYER	LOUIS	
MIHALINEC	FABIAN	JOHN
MIHALINEC	LOUIS	J
MILLER	CARL	EDWARD
MILLER	CLARENCE	DARCY
MILLER	JOHANNA	
MILLER	MARGARET	ELIZABETH
MILLER	MARGARET	MARY
MILLER	MARY	L
MILNER	JOSEPH	LOUIS
MOONEY	JOHN	PATRICK
MORGAN	WILLIAM	F
MORRIS	ALICE	MABEL
MORRIS	WILLIAM	CARLOS
MORRIS	WILLIAM	G
MORRIS SR	WILLIAM	
MORSE	DONALD	L
MORSE	KENNETH	CHARLES
MOWER	MARK	
MOZINA	ANGELA	
MULVIHILL	THOMAS	
MURPHY	JOHN	FRANCIS
MURRAY	ROBERT	ALLEN
MUSSER	DON	LEWIS
MYERS	GEORGE	
MYLOR	JAMES	FRANK
NACHTIGAL	FRANK	P
NEED	KATHLEEN	CAREY
NEUBAUER	JOSEPH	MELVIN
NEUBAUER	KATHRYN	MELVIN
NEWTON	SOPHIA	PLANTSCH
NICKLES	CLARENCE	
NICKLES	HAROLD	RAYMOND
NIGGLE	ETHEL	COOK
OAKLEY	ADDIE	ELLEN
OCONNOR	KATHLEEN	M
OCONNOR	NORA	FLYNN
OHLSSON	NILS	B
OKONSKI	EDWARD	WILBUR
OLESKI	JOHN	THOMAS
OMERSA	JOHN	
OPALK JR	MARTIN	
OXER	ORLANDO	MONROE
PALECHEK	ROBERT	
PALMER	NORMAN	L
PARKER	WILLIAM	
PARR	WILLARD	FRANCIS
PAVLISKA	PAUL	
PELL	KELSO	
PERRY	WILLIAM	
PETERMAN	MARTIN	H
PETERS	ROSE	
PETRI	EDWIN	A
PHILLIPS	ROY	H
PICOZZI	WILLIAM	RALPH
PINHEIRO	ALBERT	JOSEPH
PLANTSCH	MICHAEL	
PLAYFORD	HAROLD	WILSON
PLAYFORD	WILLIAM	HOWARD
POJE	VINCE	JOHN
PRICE	FLOYD	L
PRICE	NEWELL	MCCLOYD
PRICE	OTTO	ORVILL
PUCHALLA	MOLLIE	GLIEBE
RAUCH	CARL	
RAY	TINA	
READING	GEORGE	
REARDON	ROGER	SAMUEL
REESE	HARRY	CARBIS
REINHARD	GEORGE	ANDREW
REINHARD	GEORGE	MARTIN
REITHOFFER	AL	
REITHOFFER	WILLIAM	F
RESSLER	DONALD	
REYNOLDS	HERSHEL	
ROAWDEN	GENEVIEVE	DOYLE
ROBEDA	JOHN	
ROBERTSON	MYRTLE	EVATER
ROBINSON	RALPH	
ROBISON	PAUL	EDGAR
ROGERS	WILLIAM	FRASER
ROTH	PAUL	
RUMMEL	RUDOLPH	PETER
RUTTER	EDWARD	SAMUEL
RYDER	GEORGE	WATKINS
SAEFKOW	HOWARD	
SATOR	ALICE	MARGARET
SAXTON	CLIFFORD	ELWOOD
SAXTON	CLINTON	KAWOOD
SAXTON	HOWARD	NILE
SAXTON	JOHN	NELSON
SAXTON	STANLEY	
SCANLON	WILLIAM	
SCARLATELLA	MICHEL	O
SCARLATELLI	FRANK	
SCARLATELLI	PHYLLIS	MARY
SCHELLENTRAGER C		C
SCHMIDT	TARAS	NICHOLAS
SCHNEERER	JOHN	EDWARD
SCHNEIDER	ELDEN	EVERETT
SCHOENBERG JR	ALFRED	K
SCHOONMAKER	WILLIAM	JAHNE
SCHROLL	WILLIAM	CLOYD
SCHULZE	EMDEN	CHARLES
SCHUMACHER	HERBERT	LEO
SCHUMACHER	RALPH	M
SCOTT	DUDLEY	HUMPHREY
SHANNON	HARRIS	COOPER
SHASBERGER	WILLIAM	M
SHEA	ARNOLD	EDWARD
SHISLER	ORVILLE	CHARLES
SHIVELY	EDNA	ELIZABETH
SHIVELY	GEORGE	ALBERT
SIDEWAND	HARRY	H

11

LAST	FIRST	MIDDLE
SIEMEN	VIOLET	FRANCES
SIMMERMACHER	WILLIAM	
SIMONCIC	OTTO	
SLUSSER	EUGENE	G
SLUSSER	WILLIAM	RAYMOND
SMART	ROWELL	VICTOR
SMITH	ARTHUR	BICKERTON
SMITH	RICHARD	CLAYTON
SMITH	RICHMOND	EDWARD
SMITH	ROLAND	WOODROW
SNIDER	WILLIAM	FRANKLIN
SOLESKE	PHILLIP	
SOUCHAK	NICK	JULIUS
SPENCE	RICHARD	SLATER
SPRAGUE	CLAIR	CLIFTON
STAMBERGER	ROBERT	FREDRICK
STANDISH	SHIRLEY	ROSEMARY
STEFANCIC	EMIL	JOSEPH
STEIGERWALD	EDWARD	FRANCIS
STELTER	HARRY	
STEVENSON	JIM	
STOLLMAYER	JOSEPH	CONRAD
STONE	ALBERT	JAMES
STONE	R	D
STONEBACK	HOWARD	DETWEILER
STRICKLAND	ROBERT	ROYAL
STRUMBLE	TONY	WILLIAM
STUART	EUGENE	MILLER
STUBER	OLGA	
STUCKER	WINFIELD	AUGUST
SULLIVAN	DAN	RICHARD
SUNAGEL	EDWARD	R
SUNAGEL	ROBERT	DALE
SUPANCE	EMIL	WILLIAM
SUPANCE	JOHN	OSCAR
SUSTARIC	STEVE	EDWARD
SWANK	DAVID	
TANN	THOMAS	EDWARD
TAURMAN	CHARLES	
TAYLOR	ARTHUR	
TAYLOR	CALLOWAY	
TEAL	JOSEPH	EDWARD
THEIL	ERNEST	ALFRED
THRASHER	WILLIAM	HENRY
TIMPERIO	NICK	JACK
TUCKERMAN	WILLIAM	DAVID
TURK	FRANCIS	JEROME
TURNER	CHARLOTTE	MAY
ULMER	HERBERT	
VADNAL	ANTHONY	STEVE
VAN COVE	ALEX	
VAN NUIS	ALFONSO	COVAS
VAN SYCKLE	LYNN	
VAN TASSEL	HELEN	L
VAN TASSEL	JAMES	MYRON
VANCE	HOBART	CLYDE
VOGEL	ROBERT	
VRANEKOVIC	JOSEPH	DAVID
WADE	GENEVA	
WALKER	CHARLES	KENNETH
WALSER	ROBERT	
WALTERS	CHARLES	MARION
WARWICK	HALLIE	EMALINE
WATERSON	WILLIAM	HENRY
WATERUBRY	KENNETH	R
WEBER	ARCHIBALD	AUGUST
WEBER	ARCHIE	WILLIAM
WEBER	HARRIET	ETHEL
WEBER	ROBERT	WILLIAM
WEBER	TOM	
WEBER	WILLIAM	
WEISS	HENRY	JOHN
WHITE	PHILIP	B
WHITEHOUSE	FLORENCE	CONKLE
WHITNEY	HAZEL	MAE
WICHERT	LEONARD	JOHN
WIELAND	GEORGE	SHERMAN
WIGGINS	EDWIN	RICHARD
WILLIAMS	EDISON	GATES
WILLIAMS	EDNA	
WILLIAMS	MILDRED	AUGUSTA
WILLIAMS	WALTER	DEWEY
WILLIAMS	WILLIAM	
WILMOT	HARDIE	LANGDON
WILSON	LLOYD	PAUL
WINKLE	DONALD	EUGENE
WINTER	DONALD	
WOODBURN	EARL	HARRISON
WOODS	FRANK	HENRY
WURTZ	BETTY	GEORGIA
YEAGER	ROBERT	P
YELITZ	ANDREW	
YOUNG	LLOYD	WESLEY
YOUNG	MOSES	
YOUNG	ROBERTA	
ZALNERATIS	CHARLES	RAYMOND
ZANDER	KATHERINE	REINERT
ZIMMERMAN	ORVILLE	

PLAY GOLF
at Euclid Beach

18 Acres devoted to brand new idea in Golf.
41 Tees, 9 Greens; arranged at driving distances of from 40 to 275 yards.
Players under cover permitting play Rain or Shine.
Two 9 hole putting greens.
Absolute privacy for beginners.
Professional instructors on the grounds at all times.
No appointments necessary.
Balls, Clubs and Caddies furnished.
New players may learn the game.
Old players may improve their game.
Plenty of parking space.
Open daily from 8:30 a. m. until dark.

National Golf Course Company

Lake Shore Boulevard at Euclid Beach
Opposite Tourists' Camp

1939 EMPLOYEE ROSTER

LAST	FIRST	MIDDLE
ALDRICH	RUTH	MARIE
ANDERSON	CECIL	ROLAND
ANDERSON	EMMA	KIND
ANDERSON	JOHN	ANDREW
ARTELJ	ANTON	
AUGUST	ROBERT	OLIN
BABER	ALLAN	W
BAIRD	MAE	ELEANOR
BAKER	CILLIUS	MOSE
BAKER	LUCION	
BARAGA	FRANCES	K
BARTO	ALOYSIUS	FRANK
BARTY	JOSEPH	JOHN
BASS	ISABEL	LOUISE
BATCHLET	ROBERT	LOUIS
BEACH	ROBERT	ALLEN
BECHER	EMMA	SWANK
BECK	NEVITT	SIDMAN
BENEDICT	FRANCIS	
BENEDICT	HOWARD	L
BERGACH	FRANK	
BERTRAN	THOMAS	EDWARD
BESSAI	HERBERT	JOHN
BESTGEN JR	OSCAR	K
BETZ	RALPH	KENNETH
BILLENS	WILLIAM	FREDRICK
BIRD	FRANKLIN	L
BLACKMORE	ROBERT	WILLIAM
BLANK	OBERT	EUGENE
BLEAM	LULA	ELSIE
BLECKER	HARRY	
BLOEDE	WILLIAM	CARL
BLOW	ARLINE	BEIL
BOCK	WILLIAM	PAUL
BODE	WILLIAM	HENRY
BOGGESS	ELIZABETH	SUE
BOLDIN	LAWRENCE	RUDOLPH
BORGES	WAYNE	H
BORING	EMERY	C
BORING	FLORENCE	LISTER
BOULTON	ALBERT	
BRAIDICH	ALBINA	
BRAIDICH	NICK	ALBERT
BREHM JR	HOWARD	E
BREYLEY	ELEANOR	MAY
BREYLEY	HAZEL	MAY
BROADBENT	ALBERT	HENRY
BROUGHTON	JAMES	LEWIS
BRUNETTI	GEORGE	
BUCHOLTZ	DOROTHY	MILLER
BUCHWALD	ROBERT	FREDRICK
BUERKEL	JOHN	M
BUESS	JACK	PAYNE
BURROWS	ROBERT	HARRISON
BURTON	JAMES	
CADA	PHILIP	STANLEY
CALDWELL JR	CHARLES	GROVER
CALLAGHAN	BEATRICE	BERNICE
CALLAGHAN	GEORGE	EDWARD
CALLAGHAN	ROBERT	E
CAMERON JR	DONALD	DOUGLAS
CAMPBELL	ALEXANDER	ROBERT
CARROLL	CLARA	LEGGON
CASE	DONALD	LEW
CENTA	JOSEPHINE	ESTHER
CERCEK	JOHN	LOUIS
CETINA	EDWARD	
CETINA	WILLIAM	FREDRICK
CHAMBERLAIN	ELIZABETH	JANE
CHAMBERLAIN	PAULINE	MARIE
CHEMICK	BETTY	
CHINN	CYRIL	HARRIS
CHUBB JR	THOMAS	FRANCIS
CIEHANOWICZ	ALEX	FRED
CINCO	RUDY	JOE
CIRINO	ROCCO	
CLARKE	EDWARD	JACK
CLAYTON	RAYMOND	
CLEARY	FLORENCE	SCOTT
CLOUSER	LESTER	JAMES
COCKRELL	EDWIN	EMERSON
COLLINS	LAYTON	T
COMERFORD	JOHN	RAYMOND
CONNOR	ROBERT	EDWARD
CONTENTO	GUISEPPE	
COOK	LLOYD	HAYES
COOLIDGE	RICHARD	
COTMAN	JOHN	JOSEPH
COWHARD	HATTIE	MAY
COX	VERDON	LUTHER
CRAIN	WILLIS	
CRAPNELL	HAROLD	
CRITZER	RAYMOND	DUNNING
CROWLEY	DANIEL	
CRUZ	SANTIAGO	
DADEY	ROBERT	SCOTT
DAGLEY	WILLIAM	EDWARD
DANHAUSER	DONALD	WILLIAM
DAVIDSON	WALTER	MILLS
DAVIES	WILLIAM	BENNETT
DAVIO CAMPBELL	MILDRED	BERQUIN
DAVIS	ELEANOR	
DAVIS JR	ROSS	ESHER
DAYKIN	OTTO	
DEEMS	RICHARD	DAVID
DEGERIO	LOUIS	
DELAMBO	FRANK	FRANCIS
DENNISON	JAMES	E
DENTON	ROBERT	HAROLD
DETTMAN	LESTER	WILLIAM
DEVLIN	FRANK	JOHN
DICKEY	HERBERT	ROBERT
DICKEY	RICHARD	ALBERT
DICKHEISER	ALFRED	GUSTAV
DLUGOLESKI	STANLEY	WILLIAM
DONAHUE	FRANK	LEO
DONAHUE	LAURA	GRAY
DONAHUE JR	FRANKLIN	LEO
DOUGAN	JERRY	PENROSE
DOWDELL	EMMETT	JOSEPH
DUNAWAY	MOSE	
DUNCAN	BERNARD	KEITH
DUNN	WALTER	
DUTCHCOT	EDWARD	OSCAR
ELLIOTT	ROBERT	P C
EPAVES	JACK	ROBERT
FANKHAUSER	ALBERT	
FARLEY	JOHN	CEDRIC
FEDELE	ANTHONY	
FEITEN	EARL	ARTHUR
FERRERI	GEORGE	
FIDDES JR	NORMAN	
FIFOLT	MILDRED	ALICE
FINDLAY	ALEX	

13

LAST	FIRST	MIDDLE
FINERAN	MARY	BALL
FISH	WALTER	EDWARD
FISHER	ELDA	
FISHER	HARLEY	DEE
FITZGERALD	PETER	MICHAEL
FLETCHER	ELLIS	
FLETCHER	ELMER	
FORD	HAROLD	
FOX	JOHN	FRED
FOYE	WILLIAM	HENRY
FRANK	ELMER	PAUL
FRANK	HARRY	FRED
FRANLEY	HAROLD	CHARLES
FREDERICKS	FRANK	JOSEPH
FREDERICKS	NEIL	FRANCIS
FRY	FRED	
FUHRMAN	FRED	FRANK
GARBIN	ALBERT	JOHN
GARDINER	ROBERT	ALLEN
GAST	FREDERICK	CARL
GAST	GEORGE	LOUIS
GAST	WILLIAM	HOEPP
GAY	CHARLES	
GERM	JOHN	A
GETZ	STUART	HAYNSWORTH
GILOY	ARTHUR	LAMBERT
GILROY	JACK	ANTHONY
GOINS	EDGAR	M
GRAF	ROBERT	FREDERICK
GRAHAM	HARRY	CLIFFORD
GRAHAM	RAYMOND	ROBERT
GRANT	LAURA	WILLIAMS
GRAY	WALTER	SCOTT
GREENWAY	FRED	STUDER
GRIEBEL	RUSSELL	JOHN
GRIFFITH	WILLIAM	JOHN
GRISWOLD	RAYMOND	ARNOLD
GRISWOLD	WILLIAM	WAYNE
GUNTON	WINIFRED	RISTOW
HAGERMAN	JACK	WILLIAM
HALL	WARREN	WILLIAM
HANLEY	DANIEL	F
HANNA	LORETTA	HANNA
HANSEN	ALFRED	IRVING
HARROLD	THADDEUS	JAMES
HARTMAN	CARL	
HARTMAN	PETER	CARL
HARVEY	JOHN	WILIAM
HAWKINS	JAMES	FRANKLIN
HAYES	WILLIS	JAMES
HEATON	MAY	
HENCK	FRANK	HAMILTON
HERGENROEDER	CHARLES	WILLIAM
HERTER	PHILIP	FREDERICK
HIMEBAUGH	NORMAN	STEWART
HIRTER	WARNER	WILLIAM

HOBAN	JACK	EDWARD
HOEGLER	ALBERT	J
HOEGLER	WALTER	JOHN
HOOD	EMILY	AUGUSTA
HORA	SARAH	
HORN	ROBERT	JOHN
HOTCHKISS	LEW	R
HOWE	RICHARD	HEYD
HUEBER	DONALD	FRANCIS
HULING	FRANCIS	SHEPARD
HUNSBERGER	EDYTHE	MAY
HUNT	BLANCHE	LEAH
HUNTER	ROY	CHARLES
HYLAND	WILLIAM	
INGALLS	LAIRD	
IRWIN	MURRAY	
JABLONSKI	LEONARD	M
JACKLITZ	ALBERT	
JACKSON	ALLAN	WALTER
JACKSON	NORMAN	KARL
JAMIESON	CLARENCE	BARCLEY
JAMIESON	DAVID	
JAMIESON	MARGARET	EILEEN
JANKOWSKI	STANLEY	
JANZ	JOHN	
JENKINS	CLIFF	
JOHANNSEN	R	C
JOHNSON	JAMES	CARLYLE
JOHNSTON	CLIFFORD	BASIL
JOHNSTON	DONALD	HUMPHREY
JOHNSTON	WILLIAM	CLIFFORD
JONES JR	JAMES	CHESTER
JUDD	CLARENCE	HADEN
JUDD	ESTIL	LEE
KANALLY	MARGARET	
KANE	JOHN	JOSEPH
KASTELIC	FRANK	MICHAEL
KECK JR	MAX	A
KELLEY	CECIL	HOLMES
KELLY	DAISY	CURTIS
KENEALY	JAMES	EDWARD
KERSHAW	RICHARD	ALLEN
KESSLER	AGNES	CECELIA
KESSLER	H	
KESSLER	HERMAN A	LOIS
KIKOLI	FRANK	
KILBY	FRANK	EVERETT
KIRSCH	PAUL	C
KLEINHENZ	RY	E
KNEPPER	RAY	
KOCHEVER	FRED	A
KOJAN	JOHN	S
KOSKY	STANLEY	A
KOSS	WILLIAM	HARRY
KOSTYK	STEVE	MIKE
KOSTYK	STEVE	MIKE
KOZEL	ROBERT	EDWARD
KOZIKOWSKI	EDWARD	STANLEY
KRAUSE	GEORGE	
KRAUSE	HARRY	
KREEGER	FRED	CARL
KRUSELL	BEN	FRED
KRUSELL	ED	FRANK
KUBELAVIC	MICHAEL	JOHN
KUBELAVIC	PETER	PAUL
KUNZ	ROBERT	
LAMBIE JR	JOHN	EDWARD
LANE	RANK	GODFRED
LANGE	JOHN	FERDINAND
LAWS	WALTER	HARRY
LECHOWICZ	BENEDICT	
LEONARD	ROBERT	LORIMER
LEONE	LOUISE	
LEVSTIK	EMIL	LOUIS
LEWIS	CHARLES	

FLYING PONIES, EUCLID BEACH, CLEVELAND, SIXTH CITY

LAST	FIRST	MIDDLE
LINGIS	PETER	WALTER
LISTER	GEORGE	HAMPTON
LISTER	GEORGE	W
LISTER	ROBERT	NORRIS
LISTER	RUTH	
LOGAN	LUCILLE	MATHILDA
LONGFIELD	TESS	EVANS
LONGWELL	JOSEPH	B
LOTTERBURY	WATSON	
LOVE	FRED	JAMES
LOVE	ROBERT	KELLOGG
LOVETT	ARTHUR	JOHN
LUKAS	BEN	
MACDONALD	JOHN	
MAGILL	THOMAS	JOSEPH
MALAVASIC	MAX	ADOLPHE
MALONEY	THOMAS	J
MANLEY	JOSEPH	WILLIAM
MANNING	LAWRENCE	
MARKENS	EDNA	BROWNING
MAXIM	JOSEPH	JOHN
MAY	DICK	OTTO
MAY	LYDIA	
MAY	OSWALD CARL	FREDERICK
MAY	WALTER	EMIL
MCBRIDE	HARRY	ELDEN
MCCARTHY	WARNER	B
MCCAULEY	HENRY	ORMSLEY
MCCLINTICK	WILLIAM	R
MCCLINTOCK	NEAL	
MCCORKLE	CHARLES	EDWARD
MCCORMICK	KATHRYN	RITA
MCFARLAND	PATRICIA	
MCGEE	VIRGINIA	MARIE
MCGRATH	EDWARD	J
MCKNIGHT	DAVID	K
MCLEAN	MAXWELL	D
MCMASTER	WILLIAM	GORDON
MCNAMARA	FRANK	PERSHING
MCNIECE	GRACE	LILLIAN
MEANS	JOHN	WILLIAM
MEYER	LOUIS	
MIHALINEC	FABIAN	JOHN
MILLER	CARL	JOHN
MILLER	CLARENCE	DARCY
MILLER	FRANK	PHILIP
MILLER	MARGARET	ELIZABETH
MILLER	MARGARET	MARY
MILNER	JOSEPH	LOUIS
MOONEY	JOHN	PATRICK
MORAN	CARL	ALEXANDER
MORGAN	WILLIAM	F
MORRIS	WILLIAM	CARLOS
MORRIS SR	WILLIAM	GEORGE
MOWER	MARK	
MULVIHILL	THOMAS	
MURPHY	ALFRED	JOHN
MURPHY	JOHN	FRANCIS
MURRAY	MELVIN	ALLEN
MUSSER	DON	LEWIS
MUZZIO	CARL	RAY
NACHTIGAL	FRANK	P
NACHTIGAL	JIM	JOHN
NADERER	GEORGE	R
NEUBAUER	JOSEPH	MELVIN
NEUBAUER	KATHRYN	MELVIN
NEWKIRK	NORMA	FINERAN
NICKLES	CLARENCE	MICHAEL
NULTY	JAMES	B
OAKLEY	ADDIE	ELLEN
OCONNOR	NORA	FLYNN
OHLSSON	NILS	B
OKONSKI	EDWARD	WILBUR
OLSEN	RAYMOND	GEORGE
OMERSA	JOHN	
ONEILL	CYRIL	STANLEY
OPALK JR	MARTIN	
ORR	ROBERT	DON
OVANIN	GEORGE	JOHN
OXER	ORLANDO	MONROE
PALECHEK	ROBERT	
PALMER	NORMAN	L
PARKER	WILLIAM	
PELL	KELSO	
PERKO	ROBERT	CHARLES
PERLIN	JAMES	ANTHONY
PETERS	ROSE	
PETRI	EDWIN	A
PHILPOTT	GEORGE	
PIZEM	ANNA	KUHEL
PLANTSCH	MICHAEL	
POJE	VINCE	JOHN
PRICE	ETHEL	PHILLIPS
PRICE	FLOYD	L
PRICE	OTTO	ORVILL
PRY	RALPH	LESLIE
PUCHALLA	MOLLIE	GLIEBE
PUGH	WILLARD	JEROME
RAKOVEC	JOSEPH	C
RAY	TINA	
RAYER	LOUIS	FRANCIS
REESE	HARRY	CARBIS
REICH	WILLIAM	WALTER
REICHLE JR	WILLIAM	JOHN
REINER	ROY	PHILLIPS
REINHARD	GEORGE	ANDREW
REINHARD	GEORGE	MARTIN
RESSLER	DONALD	
RICHKO	MAX	
ROBEDA	JOHN	
ROBERTSON	MYRTLE	EVATER
ROBINSON	HERBERT	DONALD
ROBINSON	RALPH	
ROSE	GEORGE	JOSEPH
ROSE	HARRY	THOMAS
ROSE	JACK	ARTHUR
RUTTER	EDWARD	SAMUEL
RYDER	GEORGE	WATKINS
SAEFKOW	HOWARD	
SANDERS	JOSEPH	JOHN
SASILEWSKI	THADDEUS	ANDREW
SATOR	ALICE	MARGARET
SAXTON	CLINTON	KAWOOD
SAXTON	HOWARD	NILE
SAXTON	JOHN	NELSON
SAXTON	STANLEY	
SCANLON	WILLIAM	
SCARLATELLA	MICHELO	
SCARLATELLI	FRANK	
SCARLATELLI	PHYLLIS	MARY
SCHLAPPUL	ALFRED	FRANK

15

LAST	FIRST	MIDDLE
SCHMIDT	TARAS	NICHOLAS
SCHNEERER	JOHN	EDWARD
SCHNEERER	WILLIAM	FRENCH
SCHNEIDER	ELDEN	EVERETT
SCHOENBERG JR	ALFRED	K
SCHROLL	WILLIAM	CLOYD
SCHUMACHER	FRED	G
SCHUMACHER	HERBERT	LEO
SCHUMACHER	RALPH	M
SHANDLE	ERNEST	JAMES
SHANNON	HARRIS	COOPER
SHANNON	HARVEY	PAGE
SHASBERGER	WILLIAM	M
SHAVE	A	R
SHAW	WILLIAM	
SHEA	ARNOLD	EDWARD
SHIVELY	EDNA	ELIZABETH
SHIVELY	GEORGE	ALBERT
SIDEWAND	HARRY	H
SILVEROLI	JOHN	JAMES
SIMMERMACHER	WILLIAM	
SIMONCIC	OTTO	
SKOK	HEDVI	LUDMILLA
SLUSSER	WILLIAM	RAYMOND
SMALL	ROBERT	JOHN
SMITH	ARTHUR	BICKERTON
SMITH	GILBERT	A
SMITH	RICHARD	CLAYTON
SMITH	RICHMOND	EDWARD
SMITH	ROBERT	HARRY
SMITH	ROLAND	WOODROW
SMITH	WILLIAM	G
SMITH JR	RICHARD	C
SMRDEL	STANLEY	JOSEPH
SNIDER	WILLIAM	FRANKLIN
SOLESKE	PHILLIP	
SOUCHAK	NICK	JULIUS
SPINDLE	PAUL	
STAMBERGER	ROBERT	FREDRICK
STANDISH	SHIRLEY	ROSEMARY
STEFANCIC	EMIL	JOSEPH
STEIGERWALD	EDWARD	FRANCIS
STEPHENS	CARL	RAYMOND
STOLLMAYER	JOSEPH	CONRAD
STONE	ALBERT	JAMES
STONEBACK	HOWARD	DETWEILER
STREIFENDER	JOSEPH	
STREIFENDER	ROBERT	
STRICKLAND	ROBERT	ROYAL
STRICKLAND	W	
STRICKLAND	WILLIAM	THOMAS
STUART	EUGENE	MILLER
STUBER	OLGA	
SUNAGEL	EDWARD	R
SUNAGEL	ROBERT	DALE
SUPANCE	EMIL	WILLIAM
SUPANCE	JOHN	OSCAR
SUSTARIC	STEVE	EDWARD
SWANK	DAVID	
TAGGART	CHARLES	EDWIN
TAURMAN	CHARLES	LEE
TAYLOR	ALBERT	
TAYLOR	ARTHUR	
TAYLOR	CALLOWAY	
TAYLOR	MATILDA	MARGARET
TEAL	ARTHUR	
THRASHER	WILLIAM	HENRY
TIMPERIO	NICK	JACK
TINDALL	ROBERT	EDWIN
TITMAN	HARVEY	JOSEPH
TOWNHILL	HARRY	
TRAUGHBER JR	CHARLES	
TUCKERMAN	WILLIAM	DAVID
TURNER	CHARLOTTE	MAY
TWEED	MILTON	IVAN
ULM	ROGER	CLARENCE
ULMER	HERBERT	
URANKAR	ALBIN	FRANCIS
VADNAL	ANTHONY	STEVE
VAN COVE	ALEX	
VAN GERVE	MARTIN	
VAN NUIS	ALFONSO	COVAS
VAN SYCKLE	LYNN	VAN LIEU
VAN TASSEL	HELEN	L
VAN TASSEL	JAMES	MYRON
VANCE	HOBART	CLYDE
VOGEL	ROBERT	
WADE	GENEVA	MORGAN
WAHL	RALPH	JOHN
WALTERS	GEORGE	RAYMOND
WARDEN	HERBERT	EDGAR
WARGOFCHIK	M	C
WARNER JR	ROSS	OLDROYD
WARWICK	HALLIE	EMALINE
WASILEWSKI	THADDEUS	ANDREW
WATERBURY	KENNETH	R
WATKINS	WILLIAM	SIDNEY
WEBER	ARCHIBALD	AUGUST
WEBER	ARCHIE	WILLIAM
WEBER	HARRIET	EHTEL
WEBER	ROBERT	WILLIAM
WEBER	TOM	RICHARD
WEISS	HENRY	JOHN
WHEATCROFT	ROBERT	WILLIAM
WHITE	PHILIP	B
WHITEHOUSE	FLORENCE	CONKLE
WICKES	GLENN	FRENCH
WIELAND	GEORGE	SHERMAN
WIGGINS	EDWIN	RICHARD
WILLIAMS	EDISON	GATES
WILLIAMS	EDNA	
WILLIAMS	MILDRED	AUGUSTA
WILLIAMS	WALTER	DEWEY
WILLIAMS	WILLIAM	
WILMOT	DONALD	CHAS
WILMOT	HARDIE	LANGDON
WILSON	LLOYD	PAUL
WILSON	LUMLEY	F
WILSON	WILLIAM	SETH
WINKLE	DONALD	EUGENE
WOODBURN	EARL	HARRISON
WOODS	FRANK	HENRY
WURTZ	BETTY	GEORGIA
YOUNG	LLOYD	WESLEY
YOUNG	MOSES	
YOUNG	ROBERTA	
ZALNERATIS	CHARLES	RAYMOND
ZIVIC	JOHN	A

THE HUMPHREY CO., BEACH PARK HOTEL. W. S. Walker, Manager, Cleveland, Ohio.

16

1940
EMPLOYEE ROSTER

LAST	FIRST	MIDDLE
ABBY	JONES	
ALDEN	ALLEN	ELMER
ALDRICH	RUTH	MARIE
ANDERSON	EMMA	KING
ANDERSON	JOHN	ANDREW
ARTELJ	ANTON	
ARTELJ	VALERIA	
ASHMUS	RICHARD	ALLEN
AUGUST	ROBERT	OLIN
BABER	ALLAN	W
BADKE	FRANK	CHARLES
BADKE	KENNETH	KARL
BAHNER	JOHN	MCLEAN
BAILEY	HUGH	MCVEY
BAIRD	HOWARD	JAMES
BAIRD	MAE	ELEANOR
BAKER	CILLIUS	MOSE
BAKER	KENNETH	DUNCAN
BAKER	LUCION	
BALKANY	ERNEST	
BARAGA	FRANCES	K
BARTO	ALOYSIUS	FRANK
BASS	ISABEL	LOUISE
BATCHLET	ROBERT	LOUIS
BEACH	ROBERT	ALLEN
BECHER	EMMA	SWANK
BECK	NEVITT	SIDMAN
BERGACH	FRANK	
BESTGEN JR	OSCAR	K
BETZ	RALPH	KENNETH
BILLENS	WILLIAM	FREDRICK
BIRD	FRANKLIN	L
BLANK	ROBERT	EUGENE
BLEAM	LULA	ELSIE
BLECKER	HARRY	
BLOEDE	WILLIAM	CARL
BLOW	ARLINE	BEIL
BODE	HAROLD	RAYMOND
BODE	WILLIAM	HENRY
BOEHN	HOWARD	CARL
BORING	EMERY	C
BORING	FLORENCE	LISTER
BORK	CARL	RICHARD
BOWDEN	SHELDON	EUGENE
BOWHALL	ELMER	F
BRAIDICH	NICK	ALBERT
BRAIDICH	ALBINA	
BREHM JR	HOWARDE	
BREW	ALFRED	LLOYD
BREYLEY	ELEANOR	MAY
BREYLEY	HAZEL	MAY
BRITTAIN	AMELIA	F
BRITTAIN	GEORGE	H
BROADBENT	ALBERT	HENRY
BROUGHTON	JAMES	LEWIS
BROWNING	PRINTCES	
BRUNETTI	GEORGE	
BUCHOLTZ	DOROTHY	MILLER
BUDNAR	JOHN	LOUIS
BULLARD	FRANCIS	RICHARD
BURNS	PATRICIA	LEA
BURRIDGE	WILLIAM	AUSTIN
BURROWS	ROBERT	HARRISON
BURTON	JAMES	
CABOT	NATHAN	
CALAVAN	HARRY	M
CALEB	ROBERT	A
CALLAGHAN	BEATRICE	BERNICE
CALLAGHAN	GEORGE	EDWARD
CALLAGHAN	ROBERT	E
CAMERON JR	DONALD	DOUGLAS
CAMPBELL	ALEXANDER	ROBERT
CAMPBELL	FRED	
CARROLL	CLARA	LEGGON
CASE	DONALD	LEW
CENTA	JOSEPH	
CENTA	JOSEPHINE	ESTHER
CERCEK	JOHN	LOUIS
CHAMBERLAIN	ELIZABETH	JANE
CHAMBERLAIN	PAULINE	MARIE
CHANEY JR	MORRIS	M
CHEMICK	BETTY	
CHINN	CYRIL	HARRIS
CHUBB JR	THOMAS	FRANCIS
CIRINO	MARY	CONSTANCE
CIRINO	ROCCO	
CLARK	CURTIS	WAYNE
CLARKE	EDWARD	JACK
CLAYTON	RAYMOND	
CLEARY	FLORENCE	SCOTT
CLOUD	HENRY	GRADY
COLLINS	ETTICE	
COLLINS	LAYTON	T
CONTENTO	GUISEPPE	
CONWAY	HARRY	F
COOK	LLOYD	HAYES
CORBETT	RICHARD	MARTIN
CORLETT	ALBERT	GARRETT
CORNELL	ARTHUR	SAMUEL
COWHARD	HATTIE	MAY
CRAIN	WILLIS	
CRAPNELL	HAROLD	G
CRITZER	RAYMOND	DUNNING
CROWLEY	DANIEL	
CRUZ	SANTIAGO	
CULL	WILLIAM	JUSTIN
CURRY	ANDREW	WILLIAM
DAGLEY	WILLIAM	EDWARD
DAUGHERTY	GEORGE	DANIEL
DAVIES	BRITTANIA	
DAVIES	WILLIAM	BENNETT
DAVIO CAMPBELL	MILDRED	BERQUIN
DAVIS	ELEANOR	
DAVIS JR	ROSS	ESHER
DAYKIN	OTTO	
DE LAMBO	FRANK	FRANCIS
DEACON	WALTER	EDGAR
DEARTH	ROBERT	ALDRED
DEEDS	FRANK	EDGERTON
DEGERIO	LOUIS	
DETTMER	RICHARD	
DEVLIN	FRANK	
DICKEY	HERBERT	ROBERT
DICKEY	RICHARD	ALBERT
DIEDERICH	DONALD	LAWRENCE
DIETRICH	FRED	CHARLES
DIXON	JOHN	DAVID
DLUGOLESKI	STANLEY	WILLIAM
DONNELL	EDWARD	S
DOUGAN	JERRY	PENROSE
DOUGHERTY	EDWARD	RUSSELL
DOW	JOHN	A
DOWDELL	EMMETT	JOSEPH
DRESSLER	ROBERT	
DUERR	ROBERT	ALLEN

LAST	FIRST	MIDDLE
DUNCAN	DONALD	E
DUNN	WALTER	
DUTCHCOT	EDWARD	OSCAR
ELLIOTT	ROBERT	P C
EPAVES	FLORENCE	IRENE
EPAVES	JACK	ROBERT
FEDELE	ANTHONY	
FEDELE	HARRY	
FERRERI	GEORGE	
FIDDES	WILLIAM	GORDON
FIDDES JR	NORMAN	GERALD
FIFOLT	MILDRED	ALICE
FINDLAY	ALEX	
FINERAN	MARY	BALL
FISHER	ELDA	
FISHER	HARLEY	DEE
FITZGERALD	PETER	MICHAEL
FIX JR	OLIVER	L
FLEMING	HAZEL	ANN
FLEMING	JAMES	R
FLENNIKEN	HUGH	WILEY
FLETCHER	ELLIS	
FLETCHER	ELMER	
FLYNN	HARRY	KENNETH
FOGARTY	GENEVIEVE	MALADY
FORD	HAROLD	
FORD	THERESA	BAKER
FOX	JOHN	FRED
FOX	OTTO	ROGERS
FOX JR	RALPH	LEWIS
FOYE	WILLIAM	HENRY
FRANEY	MARTIN	FRANCIS
FRANLEY	HAROLD	CHARLES
FRANTZ	ADOLF	
FRANZEN	JOHN	MATHEW
FREDERICKS	FRANK	JOSEPH
FREDERICKS	NEIL	FRANCIS
FRY	FRED	
FRY	JOHN	GERALD
GARBIN	ALBERT	JOHN
GARDINER	CHARLES	
GASKINS	THERESA	KINDLER
GAST	FREDERICK	CARL
GAST	GEORGE	LOUIS
GAST	WILLIAM	HOEPP
GEERTSON	CHARLES	
GERM	JOHN	A
GETZ	STUART	HAYNSWORTH
GIFFIN	NELLIE	J
GILOY	ARTHUR	LAMBERT
GILROY	JACK	ANTHONY
GOINS	EDGAR	M
GORNIK	LEONARD	
GOWER JR	WILBERT	N
GRAHAM	HAROLD	CLIFFORD

GRAHAM	KATHERINE	OSWALD
GRAHAM	RAYMOND	ROBERT
GRANT JR	ALEXANDER	
GREEN	C	L
GREEN	CLAUDE	ELLSWORTH
GREEN	WILLIAM	FRANCIS
GREENWAY	FRED	STUDER
GRIEBEL	RUSSELL	JOHN
GRIFFITH	WILLIAM	JOHN
GRISWOLD	WILLIAM	WAYNE
GRUTTADAURIS	DOMINICK	
GUNTON	WINIFRED	RISTOW
GUSTAFSON	RALPH	ARTHUR
HAGERMAN	JACK	WILLIAM
HAMILTON	WILLIAM	L
HANLEY	DANIEL	F
HANNA	LORETTA	
HARPER	RALPH	ALLEN
HARROLD	THADDEUS	JAMES
HARTMAN	CARL	
HARTMAN	PETER	CARL
HARVEY	JOHN	WILLIAM
HATTON	EDWARD	LOUIS
HAWKINS	JAMES	FRANKLIN
HAWKINS	THELMA	
HAYES	WILLIS	JAMES
HEATON	MAY	
HEIDELOFF	GLENN	FRANKLIN
HENCK	FRANK	HAMILTON
HENGELY	JOHN	
HENSEL	ROBERT	LEWIS
HERGENROEDER	CHARLES	WILLIAM
HERTER	PHILIP	FREDERICK
HILBRINK	WILL	RAY
HILE	WILLIAM	HENRY
HIMEBAUGH	NORMAN	STEWART
HOEGLER	ALBERT	J
HOEGLER	WALTER	JOHN
HOIER	H	
HOOD	EMILY	AUGUSTA
HOPKINS	J	ROGER
HORA	SARAH	
HOWE	RICHARD	HEYD
HOWELL	BOB	
HUDDLESON	ARTHUR	EUGENE
HUDEC	JOHN	ALFRED
HUEBER	DONALD	FRANCIS
HULING	ENID	
HULING	FRANCIS	SHEPARD
HUMPHREY	DORIS	ELIZABETH
HUMPHREY III	DUDLEY	SHERMAN
HUNSBERGER	EDYTHE	MAY
HUNT	BLANCHE	LEAH
HUNTER	ROY	CHARLES
HYLAND	WILLIAM	
INGALLS	LAIRD	
JABLONSKI	LEONARD	M
JACKLITZ	ALBERT	
JACKSON	NORMAN	KARL
JAMIESON	CLARENCE	BARCLEY
JAMIESON	DAVID	
JAMIESON	MARGARET	EILEEN
JANZ	JOHN	
JAZBINSKI	ALBERT	
JENKINS	CLIFF	
JOHNSON	GEORGE	E
JOHNSON	JAMES	CARLYLE
JOHNSTON	CLIFFORD	BASIL
JOHNSTON	DONALD	HUMPHREY
JOHNSTON	ROBERT	HAMILTON
JOHNSTON	WILLIAM	CLIFFORD
JOHNSTONE	JAMES	EDWARD
JONES JR	JAMES	CHESTER
JUDD	CLARENCE	HADEN
JUDD	ESTIL	LEE
JUDD	WALLACE	CHARLES

LAST	FIRST	MIDDLE
KANALLY	MARGARET	
KASTELIC	FRANK	MICHAEL
KECK JR	MAX	A
KELLAN	ALVIN	GUY
KELLEY	CECIL	HOLMES
KELLY	DAISY	CURTIS
KENEALY	JAMES	EDWARD
KENNEDY	ALFRED	
KENT	NORRIS	JOHN
KERSHAW	RICHARD	ALLEN
KESSLER	AGNES	CECELIA
KESSLER	HERMAN	ALOIS
KESSLER JR	HERMAN	ALOIS
KIDDEY	PAUL	EUGENE
KIKOLI	FRANK	
KILBY	FRANK	EVERETT
KNIGHT	PARKS	JAMES
KOCHEVER JR	FRED	A
KOJAN	JOHN	S
KOSICKY	HARRY	FRANK
KOSKY	JOSEPH	JOHN
KOSTYK	STEVE	MIKE
KOZEL	ROBERT	EDWARD
KOZIKOWSKI	EDWARD	STANLEY
KRALL	RUDOLPH	LOUIS
KRALY	CLIFFORD	
KRAUSE	HARRY	
KREEGER	FRED	CARL
KRIVENKI	JOHN	CARL
KRUSELL	BEN	FRED
KRUSELL	ED	FRANK
KUBELAVIC	PETER	PAUL
KUEPPER	RAY	
KUNZ	ROBERT	
LAMBIE JR	JOHN	EDWARD
LANGE	JOHN	FERDINAND
LAPINAS	JOSEPH	PETER
LAWS	WALTER	HARRY
LEACH	LARRY	A
LECHOWICZ	BENEDIC	T
LECHOWICZ	RALPH	ANTHONY
LEONE	LOUISE	
LEWIS	CHARLES	
LEWIS	JOHN	ROLAND
LEWIS	WILLIAM	BENJAMIN
LINGIS	PETER	WALTER
LISTER	GEORGE	HAMPTON
LISTER	GEORGE	W
LISTER	RUTH	
LIVINGSTON	SAM	SHELDON
LOGAN	LUCILLE	MATHILDA
LONGFIELD	TESS	EVANS
LOVE	ROBERT	KELLOGG
LOVETT	ARTHUR	JOHN
LUCIANO	ANTHONY	CARMEN
LUKAS	BEN	
MAC DONALD	JOHN	
MACK	ROBERT	ANTHONY
MAHON	PATRICIA	ANN
MALADY	JOHN	PATRICK
MALAVASIC	MAX	ADOLPHE
MALEK	LOUIS	CARL
MALONEY	THOMAS	J
MANLEY	JOSEPH	WILLIAM
MANNING	C	B
MARKENS	EDNA	BROWNING
MARSH	ROBERT	
MARTINECK JR	JULIUS	ALVA
MAXIM	JOSEPH	JOHN
MAY	DICK	OTTO
MAY	LOUIS	WILLIAM
MAY	LYDIA	
MAY	OSWALD	CARL FREDERICK
MAY	WALTER	EMIL
MCCAULEY	HENRY	ORMSLY
MCCLEARY	JOHN	ELWOOD
MCCORMICK	KATHRYN	RITA
MCFARLAND	PATRICIA	
MCGRATH	EDWARD	J
MCKNIGHT	DAVID	K
MCLEAN	MAXWELL	D
MCMASTER	WILLIAM	GORDON
MCMILLAN	WILIAM	
MCNAMARA	FRANK	PERSHING
MCNIECE	GRACE	LILLIAN
MEAKER	ROBERT	THOMAS
MEANS	JOHN	WILLIAM
MEDHURST	WILLIAM	HOWARD
MEGLICH	MATHEW	
MERRELL	MURIEL	WHITE
MEYER	LOUIS	
MIHALINEC	FABIAN	JOHN
MILLER	CARL	JOHN
MILLER	CLARENCE	DARCY
MILLER	JOHN	R
MILLER	MARGARET	ELIZABETH
MILLER	MARGARET	MARY
MILNER	JOSEPH	LOUIS
MOONEY	FRANK	EUGENE
MOONEY	JOHN	PATRICK
MOORE	FENTON	LEO
MORAN	CARL	ALEXANDER
MORGAN	WILLIAM	
MORGAN	WILLIAM	F
MORGAN JR	LEROY	
MORRIS	WILLIAM	CARLOS
MORRIS SR	WILLIAM	GEORGE
MOSES	HOMER	
MOWER	MARK	
MULVIHILL	THOMAS	
MURPHY	ALFRED	JOHN
MURPHY	JOHN	FRANCIS
MURRAY	MARY	
NACHTIGAL	FRANK	P
NACHTIGAL	JIM	JOHN
NADERER	GEORGE	
NAGEL	RALPH	ROBERT
NEAL	ROBERT	MYERS
NEUBAUER	KATHRYN	MELVIN
NEUBERT	RAYMOND	AUGUST
NEVILLE	WILLIAM	THOMAS
NEWKIRK	NORMA	FINERAN
NICKLES	C	
NOLAN	JOSEPH	MICKEY
OAKLEY	ADDIE	ELLEN
OCONNOR	NORA	FLYNN
OHLSSON	NILS	B
OLESKI	JOHN	THOMAS
OLSEN	RAYMOND	GEORGE
OMERSA	JOHN	
ONISI	RICHARD	EARL
OPALICH	DANIEL	

LAST	FIRST	MIDDLE
OPALK JR	MARTIN	
OXER	ORLANDO	MONROE
PADDEN	EUGENE	JOSEPH
PARKER	WILLIAM	
PATALINO	ANTHONY	PETER
PATRONITE	SAM	JOE
PELL	KELSO	
PENKO	THERESA	ROSE
PENOVICH	ROBERT	
PERKIN	ELLA	L
PERKO	ROBERT	CHARLES
PETERS	ROSE	
PETERSEN	GILBERT	ALLAN
PETERSILGE	CARL	LORENZ
PETERSON	KEITH	AMBROSE
PETRI	EDWIN	A
PETTI	ROBERT	DONALD
PFENNIGER	CARL	
PHILPOTT	GEORGE	
PICOZZI	WILLIAM	RALPH
PIRAINO	ANTHONY	FRANK
PLANTSCH	ANNA	METTER
PLANTSCH	MICHAEL	
POJE	VINCE	JOHN
PRICE	ETHEL	PHILLIPS
PRICE	FLOYD	L
PRICE	GEORGE	ORVILL
PRICE	OTTO	ORVILL
PRYMMER	GEORGE	ALLEN
PUCEL	FRANK	
PUGH	WILLARD	JEROME
PULSE	ROBERT	A
RAKOVEC	JOSEPH	C
RAU	ALBERT	MYRON
RAY	TINA	
REA	GILBERT	WILLIAM
REARICH	STANLEY	B
REEL	CHESTER	GEORGE
REESE	HARRY	CARBIS
REINER	RALPH	VINCENT
REINER	ROY	PHILLIPS
REINHARD	GEORGE	ANDREW
REINHARD	GEORGE	MARTIN
REINHARD	R	R
RESSLER	DONALD	
ROBBINS	KATHLYN	TEEPLE
ROBEDA	JOHN	
ROBINSON	JACK	WOODROW
ROBINSON	RALPH	
ROE	SUSIE	PARNELL
ROSE	GEORGE	JOSEPH
ROSE	JACK	ARTHUR
RUETH	EDWARD	
RUMERY	JOHN	WAYNE
RUTTER	EDWARD	SAMUEL
RYAVEC	ALBERT	PETER

SAEFKOW	HOWARD	
SANDERS	JOSEPH	JOHN
SATOR	ALICE	MARGARET
SAUERBRUN	JACK	E
SAUERBRUN	RICHARD	ADAM
SAXTON	CLIFFORD	ELWOOD
SAXTON	CLINTON	KAWOOD
SAXTON	HOWARD	NILE
SAXTON	JOHN	NELSON
SCARLATELLA	MICHELO	
SCARLATELLI	FRANK	
SCARLATELLI	PHYLLIS	MARY
SCHMIDT	TARAS	NICHOLAS
SCHNEERER	JOHN	EDWARD
SCHNEERER	WILLIAM	FRENCH
SCHNEIDER	ELDEN	EVERETT
SCHULTZ	CARL	AUGUSTUS
SCHUMACHER	FRED	G
SCHUMACHER	RALPH	M
SHANDLE	ERNEST	JAMES
SHANNON	HARRIS	COOPER
SHANNON	HARVEY	PAGE
SHASBERGER	WILLIAM	M
SHAVE	A	R
SHEA	ARNOLD	EDWARD
SHENK	STEVEN	FRANCIS
SHEPHERD	HARRY	
SHERMAN	CHARLES	E
SHIDEMANTLE	HAROLD	W
SHISLER	ORVILLE	
SIDEWAND	EDNA	CLEM
SIDEWAND	HARRY	H
SIDEWAND JR	HARRY	
SILVEROLI	JOHN	JAMES
SIMMONDS	JOSEPH	CARROLL
SINCLAIR	HELEN	TROY
SINDLEDECKER	CLARENCE	F
SLUSSER	WILLIAM	RAYMOND
SMART	ROWELL	VICTOR
SMITH	ARTHUR	BICKERTON
SMITH	GILBERT	A
SMITH	RICHARD	CLAYTON
SMITH	RICHMOND	EDWARD
SMITH	ROBERT	HARRY
SMITH	ROLAND	WOODROW
SMITH	WILLIAM	G
SMITH JR	RICHARD	C
SOUCHAK	NICK	JULIUS
SPRAGUE	JESSE	HARRISON
STAINBROOK	DALE	FRANKLIN
STAMBERGER	ROBERT	FREDRICK
STAMP	WILLIAM	STEVE
STANA	LEONARD	
STANDISH	SHIRLEY	ROSEMARY
STASEK	CHARLES	EUGENE
STEFANCIC	EMIL	JOSEPH
STEIGERWALD	EDWARD	FRANCIS
STEPHENS	GEORGE	HERBERT
STETLER	HARRY	
STOLLMAYER	JOSEPH	CONRAD
STONE	ALBERT	JAMES
STONEBACK	HOWARD	DETWEILER
STRICKLAND	ROBERT	ROYAL
STRICKLAND	WILLIAM	THOMAS
STUART	EUGENE	MILLER
STUBER	OLGA	
SUNAGEL	EDWARD	R
SUNAGEL	ROBERT	D
SWANK	DAVID	
SZEKELY	JOSEPH	
TAGGART	CHARLES	EDWIN
TAKACS	CHARLES	LEWIS
TAURMAN	CHARLES	LEE
TAYLOR	ALBERT	
TAYLOR	ARTHUR	
TAYLOR	CALLOWAY	

Moonlight Bathing Scene, Euclid Beach Park, Cleveland, Ohio.

20

LAST	FIRST	MIDDLE
TAYLOR	MATILDA	MARGARET
TIMPERIO	NICK	JACK
TOLAN	BERNARD	E
TUCKERMAN	WILLIAM	DAVID
TUSHAR	MARY	
TWEED	MILTON	IVAN
UHL	NORMAN	AUGUST
ULMER	HERBERT	
URANKAR	ALBIN	FRANCIS
VAN COVE	ALEX	
VAN TASSEL	JAMES	MYRON
VANCE	HOBART	CLYDE
VARGO	J	JOHN
VICKERY	JESS	WILLIAM
VOGEL	ROBERT	
WADE	WILLIAM	
WAGNER	FLORENCE	CALLAHAN
WAHL	RALPH	JOHN
WALDECK	FRANK	J
WALL	ROBERT G	L
WALTERS JR	CHARLES	MARION
WANDA	RICHARD	GEORGE
WARGOFCHIK	M	C
WARWICK	HALLIE	EMALINE
WASILEWSKI	THADDEUS	ANDREW
WATERBURY	KENNETH	R
WATKINS	EARL	THOMAS
WATSON	ERNEST	
WATSON	ROBERT	CAMPBELL
WEBB	OSLER	ALFRED
WEBER	ARCHIBALD	AUGUST
WEBER	ARCHIE	WILLIAM
WEBER	HARRIET	ETHEL
WEBER	TOM	RICHARD
WEBER	WILLIAM	
WEISS	HENRY	JOHN
WHEATCROFT	ROBERT	WILLIAM
WHITE	GEORGE	
WHITE	PHILIP	B
WHITEHOUSE	FLORENCE	CONKLE
WHITNEY	VERNON	G
WIELAND	GEORGE	SHERMAN
WIGGINS	EDWIN	RICHARD
WILCOX	ERNEST	HARLEY
WILL	GEORGE	
WILLIAMS	EDISON	GATES
WILLIAMS	WALTER	DEWEY
WILLIAMS	WILLIAM	
WILLIS	DUANE	
WILSON	BRUCE	
WILSON	LLOYD	PAUL
WINDER	KEITH	R
WINKLE	DONALD	EUGENE
WISWELL	EUGENE	HOWARD
WOODBURN	EARL	HARRISON
WOODS	FRANK	HENRY
YEAGER	ROBERT	P
YELITZ	ANDREW	
YESSO	ELIZABETH	
YOUNG	LLOYD	WESLEY
YOUNG	MOSES	
YOUNG	ROBERTA	
ZAK	EUGENE	EDWARD
ZANDER	KATHERINE	REINERT
ZELL	ROBERT	L
ZUST	FRANK	

VII.
German Village
Beer Garden.
ca.1899.

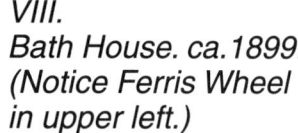

VIII.
Bath House. ca.1899.
(Notice Ferris Wheel
in upper left.)

IX.
Switchback Railway.
ca.1899.

1941
EMPLOYEE ROSTER

LAST	FIRST	MIDDLE
ADAMS	RAYMOND	C
ADKINSON	ORASTUS	H
ANDERSON	DONALD	A
ANDERSON	EMMA	KING
ANDERSON	JOHN	ANDREW
ANJESKEY	RICHARD	A
ARTELJ	ANTON	
ARTELJ	VALERIA	
ASHLEY	JIM	P
AUER	ROBERT	J
AUGUST	ROBERT	OLIN
AUGUSTINE	RAY	P
BABER	ALLAN	W
BACHER	JOSEPH	J
BACHER	WILLIAM	
BAESLACH	ROLLAND	J
BAILEY	HAROLD	
BAILEY	OLIN	A
BAKER	CILLIUS	MOSE
BAKER	KENNETH	DUNCAN
BAKER	LUCION	
BANCROFT	CHESTER	
BARAGA	FRANCES	K
BARBO	ROBERT	
BARKIN	LEONARD	M
BARRON	ROBERT	
BARTELL	HAROLD	W
BARTO	ALOYSIUS	FRANK
BARTO	BERNICE	E
BATCHLET	ROBERT	LOUIS
BAUS	EARL	A
BECHER	EMMA	SWANK
BELDIN	HENRY	F
BELL	ELMER	
BENCAR	RAY	S
BENEDICT	FRANCIS	
BERGACH	FRANK	
BERNHARDY	BERT	P
BERRIDGE	LAWRENCE	F
BETTCHER	JACK	I
BETZ	RALPH	KENNETH
BILLENS	WILLIAM	FREDRICK
BISBEE	FREDERICK	R
BISSONNETTE	DONALD	
BITKER	ALBERT	A
BLACK	WILLIAM	G
BLACK	WILLIAM	T
BLANGGER	HANS	H
BLANK	ROBERT	EUGENE
BLEAM	LULA	ELSIE
BLOEDE	WILLIAM	CARL
BLUMENTHAL	SIDNEY	P
BORK	CARL	RICHARD
BOULDIN	LUCY	
BOWHALL	ELMER	F
BOYLE	JOHN	
BRANTO	RICHARD	D
BRASCH	ANN	
BRAUN	WILLIAM	FRED
BREWER	SCIPIO	
BREYLEY	HAZEL	MAY
BREYLEY-MELKERSON	ELEANOR	MAY
BRIGGS	THOMAS	LEE
BRINK	DUANE	
BROKA	WILLIAM	ALBERT
BROOKS	DONALD	R
BROUGHTON	JAMES	LEWIS
BROWN	ELMER	J
BRUCKS	FRANK	J
BRUENING	RAY	JOHN
BRUNETTI	GEORGE	
BRUZINA	MIKE	
BUCHWALD	EDWARD	J
BUDNAR	JOHN	LOUIS
BUIKUS	ROBERT	
BURNS	PATRICIA	LEA
BURROWS	ROBERT	HARRISON
BURTON	JAMES	
BUTLER	JACK	
BUZZELLI	ELSIE	F
CALLAGHAN	BEATRICE	BERNICE
CALLAGHAN	GEORGE	EDWARD
CAMPBELL	ALEXANDER	ROBERT
CAMPBELL	FRED	
CAMPBELL	JAMES	C
CAMPBELL	MILDRED	B
CAMPBELL	RUTH	F
CAPPE	MICHAELA	
CAPRIATO	SARAH	C
CAPRIOTO	S	
CAPUTO	JOHN	
CARBONNEAU	LESTER	D
CARROLL	CLARA	LEGGON
CASE	HOWARD	J
CASTERLINE	HARRY	F
CASTLE	CLIFFORD	C
CENTA	FRANK	J
CENTA	JOSEPH	
CENTA	JOSEPHINE	ESTHER
CEPIRLO	ROBERT	C
CERCEK	JOHN	LOUIS
CERRITO	BERT	R
CHAMBERLAIN	PAULINE	MARIE
CHEMICK	BETTY	
CHINN	CYRIL	HARRIS
CHIZIK	ARTHUR	
CHRYN	NARCY	Z
CIRINO	MARY	CONSTANCE
CIRINO	ROCCO	
CLARK	ADELBERT	C
CLEARY	FLORENCE	SCOTT
COCHRANE	RUSSELL	C
COLEMAN	GEORGE	M
COLLINS	LAYTON	T
COMMANDER	COLIN	
CONNELL	WILLIAM	F
CONTENTO	GUISEPPE	
CONWAY	HARRY	F
CONWAY	ROBERT	LEO
COOK	LLOYD	HAYES
COOKE	EDWARD	E
COOPER	HOWARD	ALLEN
CORBLEY	JAMES	R
COURTER	EDWIN	E
COURTRIGHT	WILLIAM	R
COUSINS	RICHARD	T
COWAN	REX	
COWHARD	HATTIE	MAY
COX	JOHN	WALLACE
CRAIL	LEE	RICHARD
CRAIN	WILLIS	
CRAPNELL	ARTHUR	T
CRAPNELL	HAROLD	G
CROWLEY	DANIEL	
CUDAHY	WILLIAM	J
CULBERTSON	GRACE	C

LAST	FIRST	MIDDLE
DANIELSON	ROBERT	
DAUPHINELMER	VIRGIL	
DAUPHIN	VIRGINIA	RUTH
DAVENPORT	MARTHA	
DAVIES	THOMAS	J
DAVIO CAMPBELL	MILDRED	BERQUIN
DAVIS	CORBETT	R
DAVIS	ELEANOR	
DAVIS JR	ROSS	ESHER
DAYKIN	OTTO	
DEACON	DONALD	
DEADY	FREDERICK	W
DEGERIO	LOUIS	
DELAMBO	FRANK	FRANCIS
DEMORE	BERTHA	B
DEMPSEY	JOHN	F
DEVLIN	ROBERT	C
DICKEY	HERBERT	ROBERT
DICKEY	RICHARD	ALBERT
DICKSON	RICHARD	J
DIXON	ROBERT	W
DLUGOLESKI	STANLEY	WILLIAM
DODD	RICHARD	HENRY
DOOLEY	HERMAN	
DOUGHERTY	PAUL	V
DOWDELL	EMMETT	JOSEPH
DOWNER	THOMAS	SCOTT
DOWNING	WALTER	C
DOYLE	RICHARD	E
DRESSLER	JACK	H
DREYER	GEORGE	C
DUNHAM	PAUL	H
DURRETT	WARREN	
DUTCHCOT	CELIA	
DUTCHCOT	EDWARD	OSCAR
DWYER	JAMES	R
EASTER	ELBERT	H
EGAN	RICHARD	J
ELWOOD	JOHN	LEO
ENGER	CARL	C
EPAVES	FLORENCE	IRENE
EPAVES	JACK	ROBERT
EPAVES	WILLIAM	C
ERNST	JERRY	H
ESTERBROOK	LESLIE	R
EVANS	NEAL	LLOYD
EVANS	RICHARD	L
EVANS	ROBERT	F
EVERETT	CLIFFORD	E
FAINS	CHICK	
FAIRFAX	WILLIS	L
FEDELE	ANTHONY	
FEDELE	HARRY	
FEDELE	JOHN	P
FEDELE	MARY	
FERRERI	GEORGE	

FIDDES	EDWARD	JAMES
FIDDES JR	NORMAN	GERALD
FIFOLT	MILDRED	ALICE
FINDLAY	ALEX	
FINERAN	MARY	BALL
FIORTA	JOSEPH	
FISHER	ALICE	BEACH
FISHER	ELDA	
FISHER	HARLEY	DEE
FITZGERALD	PETER	MICHAEL
FIX JR	OLIVER	L
FLEMING	BRADY	K
FLEMING	JAMES	R
FLENNIKEN	HUGH	WILEY
FLETCHER	ELLIS	
FLETCHER	ELMER	
FOGARTY	GENEVIEVE	MALADY
FORD	HAROLD	
FOSTER	RALPH	N
FOWLE	ROBERT	A
FOX	OTTO	
FOX	ROBERT	LYMAN
FOYE	WILLIAM	HENRY
FRANEY	JOSEPH	M
FRANEY	MARTIN	FRANCIS
FRANLEY	HAROLD	CHARLES
FRANZEN	WILLIAM	E
FREDERICKS	FRANK	JOSEPH
FREEMAN	BETTY	
FRISCHKORN	PAUL	H
FROELICH JR	HENRY	M
FRY	FRED	
FUERST	RICHARD	
GAMBRILL JR	MERLE	C
GARBIN	ALBERT	JOHN
GARRETT	HUGH	WARREN
GARRETT	WARREN	
GASKINS	THERESA	KINDLER
GAST	FREDERICK	CARL
GAST	ROBERT	
GAST	WILLIAM	HOEPP
GEORGE	RICHARD	W
GIBBONS	GIL	
GILOY	ARTHUR	LAMBERT
GLOVER	RICHARD	
GLYNN	JOHN	
GOETTLING	JAMES	
GOINS	EDGAR	M
GORNICK	LEONARD	J
GOULDER	ALBERT	J
GOWER JR	WILBERT	N
GRAF	ROBERT	FREDERICK
GRAHAM	HAROLD	CLIFFORD
GRAHAM	RAYMOND	ROBERT
GREEN	CLAUDE	ELLSWORTH
GREENWAY	FRED	STUDER
GREGORY JR	ROBERT	
GRIEBEL	RUSSELL	JOHN
GRIFFITH	JAMES	C
GRIFFITH	WILLIAM	JOHN
GROOME	CLYDE	R
GUBANC	ROBERT	D
GUCCION	NORMAN	
GUNTON	WINIFRED	RISTOW
GUSTAFSON	RALPH	ARTHUR
GUTTER	LESTER	
HAAS	JOE	N
HACKEDORN	HORACE	
HAGERMAN	JACK	WILLIAM
HAKSOCK	PAUL	M
HALL	HELEN	MIXER
HANLEY	DANIEL	F
HANNA	LORETTA	HANNA
HANSON	JOHN	A
HARMON	THOMAS	E
HARPER	HOWARD	EARL

The Figure 8 at Euclid Beach Park, Cleveland, Ohio.

24

LAST	FIRST	MIDDLE
HARRIS	RUSSELL	T
HARROLD	THADDEUS	JAMES
HARTMAN	CARL	
HARTMAN	PETER	CARL
HARVEY	PAUL	W
HATTON	EDWARD	LOUIS
HAWKINS	JAMES	FRANKLIN
HAWKINS	THELMA	
HAYES	WILLIS	JAMES
HAZELTON	LEO	H
HEATON	MAY	
HEAVER	KATHRYN	C
HEFFNER	BEATRICE	
HENRY	CHARLES	EDWARD
HERMAN	TERRY	
HERRICK	OREN	J
HERTZEL	FRANK	
HESSLER	ROLF	
HILBRINK	WILL	RAY
HILE	WILLIAM	HENRY
HILL	WILLIAM	E
HIMEBAUGH	NORMAN	STEWART
HINSKE	JOHN	STEPHEN
HOEGLER	ALBERT	J
HOEGLER	WALTER	JOHN
HOGAN	LARRY	
HOOD	EMILY	AUGUSTA
HOOK	HENRIETTA	
HORA	SARAH	
HOUSER	FRANK	S
HOWARD	ROBERT	
HOWE	RICHARD	HEYD
HUBBARD JR	TOM	A
HUBNER	LILLIAN	BECKER
HUBNER	RICHARD	J
HUMPHREY	DORIS	ELIZABETH
HUMPHREY III	DUDLEY	SHERMAN
HUNSBERGER	EDYTHE	MAY
HUNT	BLANCHE	LEAH
HUNTER	ROY	CHARLES
IANNICELLI	ANDREW	
INGALLS	LAIRD	
JACKLITZ	ALBERT	
JACKSON	NORMAN	KARL
JAMES	DON	WILLIAM
JAMIESON	CLARENCE	BARCLEY
JAMIESON	DAVID	
JAMIESON	MARGARET	EILEEN
JANCA	GEORGE	
JANOS	JOHN	SPEROS
JANZ	JOHN	
JAROSZ	RAYMOND	R
JENKINS	CLIFF	
JENNER	ROBERT	C
JOHNSEN	GEORGE	E
JOHNSON	CHARLES	H
JOHNSON	JAMES	CARLYLE
JOHNSTON	CLIFFORD	BASIL
JOHNSTON	DONALD	HUMPHREY
JOHNSTON	WILLIAM	CLIFFORD
JOHNSTONE	JAMES	EDWARD
JONES	DAVE	O
JONES JR	JAMES	CHESTER
JORDENS	HERBERT	
JUDD	CLARENCE	HADEN
JUDD	ESTIL	LEE
KANALLY	MARGARET	
KANTRALIS	GEORGE	
KASTELIC	FRANK	MICHAEL
KECK	ROBERT	W
KELLAM	ALVIN	GUY
KELLER	FLORENCE	A
KELLY	DAISY	CURTIS
KENEALY	JAMES	EDWARD
KENEALY	RICHARD	S
KENNEDY	ARTHUR	RICHARD
KENNEDY	CLARK	
KENT	NORRIS	JOHN
KEOUGH	EDWARD	W
KERNS	LAWRENCE	
KERSHAW	RICHARD	ALLEN
KERSTEN JR	WALTER	E
KESSLER	AGNES	CECELIA
KESSLER	HERMAN	ALOIS
KESSLER JR	HERMAN	ALOIS
KESSLER JR	LEO	
KIKOLI	FRANK	
KILBY	FRANK	EVERETT
KILNER	JOSEPH	C
KING	AUD	
KINSELLA	JOHN	M
KIRBY	JOHN	THOMAS
KIRBY	RICHARD	N
KIRKPATRICK	JAMES	
KNIGHT	KARL	E
KNIGHT	PARKS	JAMES
KOVACS	JULIUS	A
KRALL	RUDOLPH	LOUIS
KRANTZ	STANLEY	J
KRAUSE	DONALD	J
KRAUSE	HARRY	
KREBS	DONALD	E
KREEGER	FRED	CARL
KRETZER JR	RALPH	V
KRIVENKI	JOHN	CARL
KRONBERGER	BEN	J
KRUEGER	ROBERT	P
KRUSELL	BEN	FRED
KRUSELL	ED	FRANK
KUBELAVIC	PETER	PAUL
KUEPPER	RAYMOND	
KULOW	DAVID	J
KUNZ	HOWARD	
KUNZ	ROBERT	
KVATERNIK	RUDY	
LAIS	DONALD	R
LAKAN	MATT	
LAMBIE JR	JOHN	EDWARD
LANG	HENRY	
LANGE	DOLOREA	J
LANGE	JOHN	FREDINAND
LASKY	TED	
LASKY	THEODORE	
LAWRENCE	DALE	H
LAWS	WALTER	HARRY
LAWYER	JAMES	D
LEACH	DONALD	E
LECHOWICZ	BENEDICT	
LECHOWICZ	RALPH	ANTHONY
LEONE	GEORGE	
LEWIS	CHARLES	
LEWIS	JOHN	ROLAND

25

LAST	FIRST	MIDDLE
LEWIS	LENORE	J
LEWIS	WILLIAM	BENJAMIN
LINDQUIST	MYRTLE	A
LINGIS	PETER	WALTER
LISTER	GEORGE	HAMPTON
LISTER	GEORGE	W
LISTER	RUTH	
LOGAN	LUCILLE	MATHILDA
LONGFIELD	TESS	EVANS
LOVETT	ARTHUR	JOHN
LOWRY	MARTIN	P
LUKAS	BEN	
LUSIN	ANNE	M
LYNAM	BILL	CHARLES
MAC WHERTER	G	W
MAC WHERTER JR	GEORGE	
MACDONALD	JOHN	
MACK	CHARLES	L
MACK	ROBERT	ANTHONY
MADER	RAY	
MAIN	FRED	L
MALAVASIC	MAX	ADOLPHE
MALONEY	THOMAS	J
MANDICH	GEORGE	
MANION	ROBERT	J
MANLEY	BILLIE	
MANLEY	JOSEPH	WILLIAM
MARKENS	EDNA	BROWNING
MARTIN	CARL	ADAM
MARTIN	JOHN	R
MASON	JACK	
MAURER	HARRY	
MAXIM	JOSEPH	JOHN
MAY	DICK	OTTO
MAY	ELIZABETH	JANE
MAY	LYDIA	
MAY	OSWALD CARL	FREDERICK
MAY	WALTER	EMIL
MCALLISTER	RICHARD	J
MCAUSLAND	DORIS	L
MCCARTHY	JOHN	P
MCCARTHY	MONTE	T
MCCAULEY	HENRY	ORMSLY
MCCAULEY	LAMBERT	
MCCLEARY	JOHN	ELWOOD
MCCLURG	JOHN	
MCCOLLUM	ROBERT	
MCCORMICK	KATHRYN	RITA
MCDONALD	WILLIAM	
MCFARLAND	PATRICIA	
MCGRATH	EDWARD	J
MCGRAW	THOMAS	M
MCGUINNESS	BERNARD	
MCKNIGHT	DAVID	K
MCMASTER	WILLIAM	GORDON
MCNIECE	GRACE	LILLIAN

LAST	FIRST	MIDDLE
MEAKER	ROBERT	THOMAS
MEANS	JOHN	WILLIAM
MEDHURST	WILLIAM	HOWARD
MEE	JEROME	T
MEGLICH	MATTHEW	
MERNAGH	THOMAS	E
MERVAR	EDWARD	J
MERVYN	WILLIAM	
METELKO	ROBERT	
MEYER	LOUIS	
MICLEA	TED	
MIELDAZIS	CHARLES	J
MIHALINEC	FABIAN	JOHN
MIHELICK	JOHN	
MILLER	CARL	JOHN
MILLER	CLARENCE	DARCY
MILLER	LOCKE	CRAIG
MILLER	LOUIS	C
MILLS	WALLACE	A
MILNER	JOSEPH	LOUIS
MIZNER	THEODORE	B
MODIC	GUS	
MODIC	PAUL	A
MODIC	ROBERT	WILLIAM
MOLLE	EUGENE	VICTOR
MOORE	FENTON	LEO
MOREL JR	MICHAEL	
MORGAN	DONALD	LEE
MORGAN	WILLIAM	F
MORLEY	EVERETT	E
MORRIS SR	WILLIAM	GEORGE
MORROW	THOMAS	NEIL
MOSES	HOMER	W
MOWER	MARK	
MUELLER	RICHARD	F
MULLALLY	ROBERT	J
MULVIHILL	THOMAS	
MURPHY	ALFRED	JOHN
MURRAY	MELVIN	A
MUSGRAVE	THEODORE	
MUSSER DON	LEWIS	
MYTROSEVICH	JOHN	
NACHTIGAL	FRED	P
NACHTIGAL	JIM	JOHN
NAGEL	RALPH	ROBERT
NEFF	RICHARD	G
NELSON	HOWARD	A
NELSON	JAMES	M
NEMETH	JOHN	
NEUBAUER	JOSEPH	MELVIN
NEUBAUER	KATHRYN	MELVIN
NEWCOMER	ROBERT	E
NEWKIRK	NORMA	FINERAN
NIELSEN	HANS	R
NOLAN	JOSEPH	MICKEY
OAKLEY	ADDIE	ELLEN
OFFENHAUSER	PHIL	L
OHARA	BERNARD	P
OHLSSON	NILS	B
OLESKI	JOHN	THOMAS
OLIVER	EDWARD	J
OLSEN	RAYMOND	GEORGE
OMERSA	JOHN	
ONISI	RICHARD	EARL
OPALICH	DANIEL	
OPALK JR	MARTIN	
ORAM	WARREN C	
OSOLIN	ROBERT	
OXER	ORLANDO	MONROE
PADDEN	EUGENE	JOSEPH
PALMER	JEAN	LAURA
PARKER	WILLIAM	
PARSONS	HERBERT	
PATALINO	ANTHONY	PETER
PELL	KELSO	
PENOVICH	ROBERT	

Derby Racer, Euclid Beach, Cleveland

LAST	FIRST	MIDDLE
PERKIN	ELLA	L
PERKO	ROBERT	CHARLES
PERME	ELMER	
PETERFY	PAUL	F
PETERS	ARNOLD	L
PETERS	ROSE	
PETERSON	JOHN	R
PETRI	EDWIN	A
PETTI	ROBERT	DONALD
PHILLIPS	DOUGLAS	JAMES
PHILLIPS	JACK	H
PHILPOTT	GEORGE	
PIRAINO	ANTHONY	FRANK
PITTMAN	KENNETH	
PITTOCK	WALTER	CARL
PIVK	HENRY	BAUER
PIVK	JOE	A
PLANTSCH	ANNA	
PLANTSCH	MICHAEL	
PORTER	ROBERT	G
POSHE	JACK	WILLIAM
POWELL	JAMES	W
PRAZNIK	ANTON	J
PRICE	ETHEL	PHILLIPS
PRICE	ETTA	O
PRICE	FLOYD	L
PRICE	FRANK	J
PRICE	MARK	THOMAS
PRICE	OTTO	ORVILL
PRICE JR	EDWARD	OWEN
PROVIDENTI	MICHAEL	A
PUCEL	FRANK	
PUGH	WILLARD	JEROME
PURPURA	MATTHEW	
QUINLIVAN	CORMACK	
RADY	STEVE	
RAMSAY	DANIEL	K
RASH	WILLIAM	F
RAY	TINA	
RAY	WANETA	E
RAYBUCK	DON	E
REA	GILBERT	WILLIAM
REED	ROBERT	CHARLES
REESE	HARRY	CARBIS
REESE	NORMAN	OLIVER
REINER	RALPH	VINCENT
REINER	ROY	PHILLIPS
REINHARD	GEORGE	ANDREW
REINHARD	GEORGE	MARTIN
RESETIC	STANLEY	
RESSLER	DONALD	
RICHMAN	ROBERT	
RITTER	ALLEN	JAMES
ROBBINS	KATHLYN	TEEPLE
ROBINSON	DEAN	G
ROBINSON	RALPH	
ROCHFORD	THOMAS	L
ROE	SUSIE	PARNELL
ROSE	GEORGE	JOSEPH
ROSE	WILLIAM	J
ROYCE	JACK	WILLIAM
RUSSELL	GEORGE	D
RUSSO	MIKE	
RUTTER	EDWARD	SAMUEL
RYAN	MILES	FRANCIS
RYAVEC	ALBERT	PETER
SAGE	EARL	E
SAMPLE	WILLIAM	J
SANDFORD	WALTER	M
SANTRY	EDWARD	T
SAUERBRUN	RICHARD	ADAM
SAUNDERS	HERBERT	E
SAXTON	CLINTON	KAWOOD
SAXTON	FOREST	J
SAXTON	JOHN	NELSON
SAXTON	PAUL	HUGO
SBROCCO	SEBASTIANO	
SCANLON	HOWARD	C
SCARLATELLA	MICHELO	
SCARLATELLI	FRANK	
SCARLATELLI	PHYLLIS	MARY
SCHAEFER	WILLIAM	M
SCHILTZ	GEORGE	B
SCHLUND	ROBERT	H
SCHMIDT	TARAS	NICHOLAS
SCHNEERER	JOHN	EDWARD
SCHNEIDER	ELDEN	EVERETT
SCHOENEN	PHILIP	
SCHUMACHER	FRED	G
SCHUMACHER	RALPH	M
SCOTT	BETTY	IDA
SCOTT	EVERETT	PALMER
SCOTT	JOHN	T
SCOTT	WILBUR	A
SEABORN	THOMAS	J
SELTNER	EDWIN	H
SESSLER	GEORGIA	L
SHAFER	JOHN	F
SHANNON	HARRIS	COOPER
SHANNON	HARVEY	PAGE
SHAVE	A	R
SHEA	ARNOLD	EDWARD
SHELTON	WAYNE	D
SHERIDAN	NEAL	M
SHERIDAN	PAUL	H
SHERLOCK	RUTH	E
SHIDEMANTLE	HAROLD	W
SHIERA	JACK	
SHIRLEY	DEWITT	
SHOEMAKER	WILLIAM	D
SIDEWAND	EDNA	CLEM
SIDEWAND	HARRY	H
SIDEWAND	RALPH	G
SINCLAIR	HELEN	TROY
SINDLEDECKER	CLARENCE	F
SINTIC	JOE	
SKELLEY	WALTER	
SKOWRONSKI	HENRY	
SLUSSER	WILLIAM	RAYMOND
SMEKEL	RICHARD	M
SMITH	GEORGE	
SMITH	GILBERT	A
SMITH	IRENE	MARIE
SMITH	PHILLIP	BURKE
SMITH	RICHMOND	EDWARD
SMITH	ROBERT	JOYCE
SMITH	STANLEY	
SMITH	WILLIAM	G
SMITH JR	RICHARD	C
SNEDDON	KENNETH	
SNEVOLD	WILLIAM	J
SNIPPERT	FRANK	B
SNOW	WEBB	CARL

27

LAST	FIRST	MIDDLE
SNOWDEN	JAMES	A
SOMRACK	FRANK	D
SOUCHAK	NICK	JULIUS
SPEARS	JERRY	G
SPIENGER	WILLIAM	A
SPRAGUE	JESSE	HARRISON
ST CLAIR	JULIUS	J
STANA	LEONARD	
STANDISH	SHIRLEY	ROSEMARY
STATES	RICHARD	CLOYD
STEFANCIC	EMIL	JOSEPH
STEIGERWALD	EDWARD	FRANCIS
STELTER	HARRY	
STEPHENS	GEORGE	HERBERT
STEPP	GENUS	W
STEVENS JR	JESSE	R
STOELTZING	WILLIAM	
STONE	ALBERT	JAMES
STONEBACK	HOWARD	DETWEILER
STROH JR	PETER	J
STROK	JACK	MACK
STUART	EUGENE	MILLER
STUBER	FLORA	ROSE
STUBER	OLGA	
SUNAGEL	EDWARD	R
SWANK	DAVID	
SWANN	LILLIAN	M
SWISHER	OTIS	
TAGGART	CHARLES	EDWIN
TALCOTT JR	M	GARDNER
TANKO	JOHN	
TANSKI	STANLEY	J
TAURMAN	CHARLES	LEE
TAYLOR	ALBERT	
TAYLOR	CALLOWAY	
TAYLOR	JOHN	
TAYLOR	MATILDA	MARGARET
TAYLOR	RICHARD	T
THOMAS	PRYSE	
THOMPSON	WILLIAM	C
TIMPERIO	ARTHUR	M
TIMPERIO	CARMEN	M
TIMPERIO	NICK	JACK
TIRABASSO	CARL	T
TUCKERMAN	WILLIAM	DAVID
TUNSTALL	ALBERT	D
TURK	JEROME	F
TUSH	REGIS	MICHAEL
TUSHAR	MARY	
TUTTLE	GEORGE	
TWEED	MILTON	IVAN
TWOHIG	JAMES	P
UHL	NORMAN	AUGUST
URANKAR	ALBIN	FRANCIS
VAN COVE	ALEX	
VAN TASSEL	JAMES	MYRON
VANCE	HOBART	CLYDE
VENCAR	RAY	S
VIDMAR	ANTON	JOE
VIDMAR	FRANK	J
VIVAS JR	JULIUS	
VOGEL	DONALD	
VOGEL	GEORGE	F
VOGEL	ROBERT	
VOLLMER	PAUL	R
VOSS	GEORGE	K
WACHSMAN	WALTER	R
WADE	GENEVA	MORGAN
WADE	WILLIAM	
WAGNER	FLORENCE	CALLAHAN
WAHL	RALPH	JOHN
WALL JR	KENNETH	P
WALTERS JR	CHARLES	MARION
WALTON	JOSEPH	P
WARGOFCHIK	M	C
WARNER JR	ROSS	OLDROYD
WARWICK	HALLIE	EMALINE
WATERBURY	KENNETH	R
WATSON	ROBERT	CAMPBELL
WEBB	JOSEPHINE	ANN
WEBER	ARCHIBALD	AUGUST
WEBER	ARCHIE	WILLIAM
WEBER	ROBERT	WILLIAM
WELKER	CHARLES	E
WELLS	ROBERT	CHARLES
WHALEY	TED	
WHEATCROFT	ROBERT	WILLIAM
WHEELER	WARDEN	E
WHITE	CHARLES	A
WHITE	CLARENCE	W
WHITE	PHILIP	B
WHITEHOUSE	FLORENCE	CONKLE
WHITNEY	HAZEL	MAE
WIELAND	GEORGE	SHERMAN
WIESLER JR	WILLARD	F
WILL	GEORGE	
WILL	KARL	WILLIAM
WILL	THEODORE	G
WILLIAMS	ALBERT	T
WILLIAMS	EDISON	GATES
WILLIAMS	MILDRED	AUGUSTA
WILLIAMS	WALTER	DEWEY
WILLIAMS	WILLIAM	
WILSON	ANNA	RILEY
WILSON	BRUCE	ALFRED
WILSON	GEORGE	D
WILSON	LLOYD	PAUL
WILSON	MARGARET	D
WILSON	RICHARD	G
WINKLE	DONALD	EUGENE
WINKLE	IDA	M
WINSLOW	WILLIAM	R
WISWELL	EUGENE	HOWARD
WITCRAFT	DALE	R
WOHLGEMUTH	CHARLES	F
WOLFE	DONALD	W
WOODBURN	EARL	HARRISON
WOODS	FRANK	HENRY
YEAGER	ROBERT	P
YOUNG	LLOYD	WESLEY
YOUNG	MOSES	
YOUNG	ROBERTA	
YUHAS	MIKE	
ZAMAN	ROBERT	V

1942
EMPLOYEE ROSTER

LAST	FIRST	MIDDLE
ACKERMAN	RICHARD	E
ADAMS	RAYMOND	C
AGIN	HERBERT	
AGNEW	CHARLIE	
AKERS	JAMES	FREDERICK
ALBRIGHT	JERRY	
ALEXANDER	DANIEL	
ALONGI	JOE	JOHN
ANDERSON	EMMA	KING
ANDERSON	JOHN	ANDREW
ANDERSON	PAUL	L
ANGELERO	RAFFAELE	
APPLEGATE	HERBERT	L
APRILE	JAMES	
ARENDT	FRED	KARL
ARMSTRONG	NORMAN	LEE
ARNHEIM	GUS	
ARNOLD	MARVIN	
ARTELJ	ANTON	
ARUNSKI	WALTER	J
ASHBROOK	LOUIS	D
ASTLEY	WALTER	E
ATWATER	ROBERT	E
AUGUSTINE	RAY	P
BABETS	DONALD	JOSEPH
BACHER	JOSEPH	J
BACHER	WILLIAM	
BACHTEL	CLAYTON	J
BACON	ERNEST	
BAILEY	HAROLD	
BAKER	CILLIUS	MOSE
BAKER	DOROTHY	E
BAKER	JAMES	
BAKER	JOE	
BAKER	LUCION	
BAKER	ROBERT	E
BARBO	ROBERT	J
BARNHARD	RICHARD	W
BARRY	DOROTHY	MARIE
BARRY	KENNETH	
BARTLETT	DONALD	LEE
BARTLEY	EDWARD	R
BARTO	ALOYSIUS	FRANK
BATTIN	LYLE	E
BAUMGART	GILBERT	
BAXT	JACK	O
BEASLEY	PAUL	
BECHER	EMMA	SWANK
BECK	BETTY	ANNE
BECK	EVELYN	
BELL	EDDIE	
BELL	HAMILTON	
BEMAN	BUD	
BEMAN	HERBERT	H
BENDER	EDWARD	A
BENEDICT	HOWARD	L
BENNHOFF	ARTHUR	C
BENNHOFF	JOHN	A
BERGER	DONALD	J
BERNHARDY	FRANCIS	J
BEYER	OTTO	L
BIELINSKI	STEVEN	
BILLENS	RAY	K
BILLENS	WILLIAM	FREDRICK
BINDER	WILLIAM	J
BITKER	ALBERT	A
BLACK	WILLIAM	G
BLACKBURN	HAROLD	
BLAHA	RICHARD	
BLAIR	BILL	A
BLEAM	LULA	ELSIE
BLOEDE	DORA	CHARLOTTE
BLOEDE	WILLIAM	CARL
BLUMENTHAL	LAWRENCE	
BLUMENTHAL	SIDNEY	P
BOGAN	CLARENCE	
BOLLING	HARRY	L
BOLLMEYER	JAMES	
BONZANG	BOB	
BORK	CARL	RICHARD
BOUNCE	FRANK	V
BOWEN	MARION	
BOWHALL	ELMER	F
BOYCE	KENNETH	F
BOYLE	JOHN	FRANCIS
BRACCIA	FRANK	
BRAIDICH	NICK	ALBERT
BRANTO	RICHARD	D
BRAUN	KARL	
BRAUN	WILLIAM	FRED
BRENCIC	ANN	MARY
BRENENSTUL	WEALTHY	E
BREST	MARION	
BREYLEY	HAZEL	MAY
BRICKER	BETTY	JANE
BRIGGS	THOMAS	LEE
BROCK	ALVIN	L
BROST	RICHARD	ROY
BROWN	ELMER	J
BROWN	HARRY	
BROWN	WILLIAM	C
BRUGMAN	CHARLES	W
BRUGMANN	LARRY	E
BRUMLEY	TOM	WILLIAM
BUCHOLTZ	CARL	
BUHNER	JOHN	EDWARD
BUIKUS	ROBERT	
BURDYSHAW	LYDIA	
BURGESS	CHARLES	V
BURGNER	ELMER	R
BURROWS	ROBERT	HARRISON
BURTON	JAMES	
BURTON	PAUL	
BUSKEY	ROBERT	GEORGE
BUZNAK	VICTOR	
BYRNE	CORNIE	JOSEPH
CAHILL	ROBERT	JOSEPH
CALL	CHARLES	C
CALLAGHAN	BEATRICE	BERNICE
CALLAGHAN	GEORGE	EDWARD
CAMP	WINFIELD	SCOTT
CAMPBELL	ALEXANDER	ROBERT
CAMPBELL	JAMES	C
CAMPBELL	MILDRED	B
CAPUTO	JOHN	
CAREY	WALTER	
CARLSON	CARL	H
CAROSELLI	ANTONE	
CARROLL	CLARA	LEGGON
CARROLL	GERTRUDE	M
CARROLL	HAZELLE	BELLE
CARROLL	JEANNE	
CARROLL	PAULINE	K
CARROLL	ROY	SANFORD
CARROLL	URSULA	F
CASTLE	ELAINE	

29

LAST	FIRST	MIDDLE
CAUDILL	OTIS	
CELIZIC	FRANK	R
CENDROWSKI	LILLIAN	D
CENTA	FRANK	J
CENTA	JOSEPH	
CENTA	JOSEPHINE	ESTHER
CERCEK	JOHN	LOUIS
CERICELLO	ART	
CHAPEK	WOODROW	
CHEK	JAMES	
CHEMICK	BETTY	
CHIANCONE	JOE	A
CHIANCONE	SAMUEL	
CHRISTIE	EUGENE	R
CICIOGOI	ANGELO	M
CIOLFI	JOE	ADAMS
CIRINO	ROCCO	
CLAGETT	HARRY	PAUL
CLARK	ADELBERT	C
CLARK	RICHARD	JAMES
CLATTERBUCK	FRANCIS	W
CLIFFORD	GEORGE	R
COBB	AHIRA	
COCHRANE	RUSSELL	C
COLE	GRANT	
COLEMAN	GEORGE	M
COLEMAN	JACK	H
COLEMAN	MARIE	E
COLEMAN	THOMAS	L
COLYER	LENNY	
COMMANDER	COLIN	
CONLIN	EDWARD	R
CONNELL	WILLIAM	F
CONOWAY	JUANITA	
CONRY	HUGH	F
CONTENTO	GUISEPPE	
CONWAY	HARRY	F
COOK	ELSIE	
COOK	LLOYD	HAYES
COSGRIFF	AMES	J
COWHARD	HATTIE	MAY
COX	GEORGE	K
CRAIG	ROBERT	JOHN
CRAIL	LEE	RICHARD
CRAPNELL	AROLD	G
CUDAHY	WILLIAM	J
CUDDIHY	ROBERT	H
CUNNINGHAM	LETTY	
CUNNINGHAM	THOMAS	F
CURRAN	SAMUEL	H
DALTON	CAL	
DANIELSON	ROBERT	
DARCY	FRANCIS	P
DAUPHIN	ELMER	VIRGIL
DAUPHIN	VIRGINIA	RUTH
DAVIES	JAMES	E
DAVIES	JOHN	R
DAVISON	JOHN	EDWARD
DEACON	DON	
DEGERIO	LOUIS	
DELAMBO	FRANK	FRANCIS
DEMO	MIKE	
DEPASQUALE	JOHN	A
DEVINE	RALPH	JOSEPH
DEVLIN	ROBERT	C
DEVOL	JOE	
DEWICK	HARRY	
DICKARD	LESTER	R
DICKEY	HERBERT	ROBERT
DIETERICH	ROBERT	F
DINKER	ED	
DLUGOLESKI	STANLEY	WILLIAM
DOOLEY	HERMAN	
DOOLEY	OLLIE	RAGAN
DOUGHERTY	PAUL	V
DOWD	BOB	THOMAS
DOWDELL	EMMETT	JOSEPH
DOWNS	ROBERT	W
DRAGE	MORRIE	
DRESSLER	H	C
DRESSLER	JACK	H
DUFFY	CLARENCE	
DUFFY	GEORGE	
DUNHAM	PAUL	H
DUNSTAN	THOMAS	E
DUTCHCOT	CELIA	
DUTCHCOT	EDWARD	OSCAR
DVORAK	CHARLES	
DVORAK	MARGARET	
DWYER	JAMES	R
EBERHARD	KENNETH	J
ECKERLE	CLIFFORD	F
EDELMAN	DARWIN	
EDUKAITIS	WILLIAM	V
ELINE	WILLIAM	BARRY
ELWOOD	GENE	JEROME
EMERSON	FRED	T
ENGLISH	GEORGE	M
ERSKINE JR	FOSTER	M
ERWIN	GENE	
ERWIN	GEORGE	
ESTERBROOK	LESLIE	R
EVANS	GEORGE	FERGUSON
EVANS	RICHARD	L
EVANS	ROBERT	F
EVERETT	CLIFFORD	E
EZZO	ANTHONY	JAMES
FABREGOT	TOMMIE	
FARMER	ANNA	VIOLA
FEDELE	ANTHONY	
FEDELE	HARRY	
FEDELE	JOHN	P
FEDELE	MARY	
FELD	MORREY	
FELDMAN	WILLIAM	F
FELLOWS	DONALD	PAUL
FELLOWS	THOMAS	F
FELTY JR	MONROE	
FEOIE	LENNY	
FERRERI	GEORGE	
FERRIS	PAUL	D
FIDDES JR	NORMAN	GERALD
FINDLAY	ALEX	
FINERAN	MARY	BALL
FISHER	ALICE	BEACH
FISHER	HARLEY	DEE
FITZGERALD	PETER	MICHAEL
FIX	GEORGE	CURTIS
FIX JR	OLIVER	L
FLATING	ROBERT	ALFRED
FLEMING	BRADY	K
FLEMING	JAMES	R

LAST	FIRST	MIDDLE
FLETCHER	ELLIS	
FOGARTY	GENEVIEVE	MALADY
FOSTER	BRUCE	WILLIAM
FOUKAL	DONALD	CHARLES
FOX	OTTO	ROGERS
FOYE	WILLIAM	HENRY
FRAME	ARCHIE	M
FRANEY	JIM	
FRANEY	JOSEPH	M
FRANLEY	HAROLD	CHARLES
FREDERICKS	FRANK	JOSEPH
FREEMAN	BETTY	
FREEMAN	CAROL	JANE
FREEMAN	JOHN	WILLIAM
FRENCH	JAMES	H
FRENCH	WILLIAM	E
FRISSELL	WILLIAM	P
FRY	FRED	
FRY	ROBERT	CHARLES
FUERST	NORMAN	A
FULLER	CHUCK	
FULLER	HAROLD	PAGE
GAINES	AL	
GALLOWAY	HUGH	ROBERT
GAMBATESE	DOMINIC	F
GAMBRILL JR	MERLE	C
GANNET	WILLIAM	JAMES
GAZDA	ALBERT	S
GEIGER	LOUIS	J
GIBBONS	GIL	
GIBBONS	ROBERT	DAVID
GILBRIDE	CLARENCE	L
GILSON	NATE	
GIRARD	HAROLD	S
GOEDEL	DORA	MAE
GOFF	LEONA	D
GOWER	HAROLD	D
GOWER JR	WILBERT	N
GRANATA	ANTHONY	
GREEN	CLAUDE	ELLSWORTH
GREENER	HUSTON	
GREENWALD	ERWIN	
GREENWALD	WILLARD	
GREENWAY	FRED	STUDER
GREGOR	EMIL	T
GREGORY JR	ROBERT	
GRIFFIN	JOE	THOMAS
GRIFFITH	JAMES	C
GRIFFITH	WILLIAM	JOHN
GROMOFSKY	JOHN	J
GUBANC	ROBERT	DAVID
GUCCION	NORMAN	
GUDIKUNST	ROBERT	E
GUNTON	WINIFRED	RISTOW
GUTTER	LESTER	
HAASE	WILLIAM	
HAGEDORN	ROBERT	FRED
HAGEMEISTER	FRED	
HALE	JACK	WARREN
HALL	KENNETH	C
HAMILTON	ALVIN	
HAMILTON	EDWARD	S
HAND	SILAS	T
HAND	THOMAS	
HANNA	LORETTA	HANNA
HANRATTY	JOAN	
HARMON	THOMAS	E
HARRIS	GEORGE	E
HARRIS	JOE	
HARROLD	THADDEUS	JAMES
HART	BOB	
HARTMAN	CARL	
HARTMAN	PETER	CARL
HARTZELL	BOB	
HATIE	JOSEPH	HENRY
HAYES	KENNETH	
HAYES	WILLIS	JAMES
HEALY	JOSEPH	
HEATON	MAY	
HEDTKY	ROBERT	
HEFFELFINGER	HARRY	
HENDERSHOTT	SHELDON	
HENES	JAMES	D
HENRICKSON	JACK	W
HERBECK	RAY	
HERM	EDDIE	
HERTZEL	FRANK	
HESENER	ROBERT	J
HIGHBERGER	J	HAROLD
HIGHBERGER	JAMES	H
HIGHBERGER	WILLIAM	J
HILL	HARRY	DAVID
HILL	JAMES	ARTHUR
HILL	JAMES	ROBERT
HILL	WILLIAM	E
HINTON	JOHN	SHEPARD
HINTON	ROBERT	
HLADKY	FRANK	C
HOFF	DOROTHY	IRENE
HOGAN	LARRY	
HOLBERT	JOSEPH	RUSSELL
HOLZHEIMER	LEONARD	E
HOOD	EMILY	AUGUSTA
HORA	SARAH	
HORNICK	FRANCES	C
HOWARD	N	C
HOWARD	ROBERT	
HOWE	DEAN	C
HOWE	RICHARD	HEYD
HOWES	WALTON	
HRIBAR	FREDERICK	J
HUBBARD JR	TOM	A
HUBNER	RICHARD	J
HUDSON	JOHN	O
HULL	RICHARD	
HUMER	KENNETH	
HUMPHREY	DORIS	ELIZABETH
HUMPHREY III	DUDLEY	SHERMAN
HUNSBERGER	EDYTHE	MAY
HUNT	BLANCHE	LEAH
HUNTER	MAYME	E
HUNTER	ROY	CHARLES
HUNTINGTON	JOHN	
HUTTON	HARRY	B
INCORVATI	ANTHONY	
INGRAHAM	LAWRENCE	R
ISENBACH	MAX	
JACKLITZ	ALBERT	
JACKSON	REBECCA	JANE
JAKLITSCH	RICHARD	E
JAMES	DON	WILLIAM
JAMIESON	CLARENCE	BARCLEY

LAST	FIRST	MIDDLE
JAMIESON	MARGARET	EILEEN
JANOS	JOHN	SPERO
JANZ	JOHN	
JELINEK	JOHN	
JENKINS	CLIFF	
JENKINS	HARDIE	
JOCA	GEORGE	ROBERT
JOCHUM	FRANK	J
JOHNS	CLAYTON	
JOHNSON	DOROTHY	IDA
JOHNSON	JAMES	MCCANN
JOHNSON	LEFTY	
JOHNSON	ROBERT	NEIL
JOHNSON JR	LISTON	A
JOHNSTON	CLIFFORD	BASIL
JOHNSTON	DONALD	HUMPHREY
JOHNSTON	WILLIAM	E
JOHNSTONE	JAMES	EDWARD
JONES	JAMES	CHESTER
JONES	RICHARD	PATRICK
JOYCE JR	JOHN	
JUDD	CLARENCE	HADEN
JUDD	ESTIL	LEE
KAMINSKI	KENNETH	L
KANALLY	MARGARET	
KANOWSKI	WILHELM	
KATZ	MICKEY	
KEGEL	EUGENE	
KEITH	KATHERINE	V
KELLER	FLORENCE	A
KELLY	CLYDE	W
KELLY	DAISY	CURTIS
KEMPTHORNE	WALTER	V
KENEALY	RICHARD	S
KENNEY	ARTHUR	W
KERNS	LAWRENCE	
KERSHAW	DEL	D
KERSHAW	RICHARD	ALLEN
KESSLER	AGNES	CECELIA
KESSLER	HERMAN	ALOIS
KESSLER JR	LEO	
KESTERSON	ROBERT	E
KIERSTEAD	EDWARD	S
KIKOL	JOSEPH	
KIKOLI	FRANK	
KILBOURNE	GEORGE	WILLIAM
KILBY	FRANK	EVERETT
KIME	HOWARD	H
KING	ALBERT	
KING	AUD	
KING	WILLIAM	L
KINKOPF	EDWARD	
KLAMERT	JOSEPH	
KLAUS	ALLEN	
KLEIN	BARNEY	
KLEMPAN	EDWARD	
KLOPFENSTEIN	CARL	G
KNAPP	AL	
KNIGHT	KARL	E
KOLBE	RICHARD	F
KOPP	WILLIAM	D
KRAFT	KARL	T
KRAUSE	HARRY	
KREEGER	FRED	CARL
KRESS	JACK	
KRICH	ELSIE	WILHELMINA
KRIVENKI	JOHN	CARL
KRULL	HERMAN	R
KUBAT	MARY	G
KUBEJA	EDWARD	
KUBELAVIC	PETER	PAUL
KUEHNE	DON	
KUHN	CHARLES	RICHARD
KULOW	DAVID	J
KUNZWARA	ADOLPH	J
KURTZ	GEORGE	E
KURTZ	RICHARD	P
KUSHNER	NATE	
LACKEY	ARTHUR	
LAGUARDIA	JOE	
LAKAN	MATT	
LAMBIE JR	JOHN	EDWARD
LANG	JAMES	VINCENT
LANGE	JOHN	FERDINAND
LASKY	THEODORE	
LAVRICH	JAMES	H
LAWYER	JAMES	D
LEACH	DONALD	E
LEBET	JULIUS	M
LECHOWICZ	BENEDICT	
LECHOWICZ	KATHERINE	M
LECHOWICZ	RALPH	ANTHONY
LEE	DORIS	
LEJOIR	PAUL	A
LEOVIC	WILLIAM	
LESKO	WALTER	
LEWANDOSKI	NORMAN	
LEWIS	JOHN	ROLAND
LINK	EMILY	PFEIL
LIPSTREU	ROY	EDWARD
LISTER	GEORGE	HAMPTON
LISTER	GEORGE	W
LLOYD	DAVID	
LLOYD	JOHN	
LLOYD	MARY	JANE
LOCKE	HUBERT	LEE
LOCKWOOD	ARTHUR	R
LOCKWOOD	CHARLES	B
LONGFIELD	TESS	EVANS
LOUCKS	DON	CLIFFORD
LUBICA	FRANK	
LUCIC	GEORGE	A
LUKAS	BEN	
LUMADUE	ARLE	LAMARR
LUMLEY	ROBERT	WILLIAM
LUND	PARKER	
LUPTON	ROBERT	
LUSIN	ANNE	M
LUSIN	MILDRED	C
LYNCH	BRYAN	EDWARD
MAC DONALD	JOHN	
MACK	CHARLES	L
MACK	ROBERT	ANTHONY
MADDOX	JAMES	J
MAIN	FRED	L
MALAGA	ROBERT	S
MALANOWSKI	FRANK	A
MALONEY	THOS	J
MANDAU	LESLIE	
MANDICH	GEORGE	
MANION	ROBERT	J
MANION	THOMAS	EMMETT

EUCLID BEACH, CLEVELAND, O.

LAST	FIRST	MIDDLE
MANNING	DOUGLAS	L
MARINO	BARNEY	
MARKENS	EDNA	BROWNING
MARTANS	PAUL	PETER
MARTIN	CARL	ADAM
MARTIN	GRACE	IRENE
MARTINS	WALTER	F
MASTERSON	PHILIP	J
MAVOR	GEORGE	
MAXIM	JOSEPH	JOHN
MAY	OSWALD	CARL FREDERICK
MAY	WALTER	EMIL
MAYER	LEW	
MAZE	DEAN	MELVIN
MCCARTHY	ROBERT	LEO
MCCARTY	RAYMOND	D
MCCAULEY	HENRY	ORMSLEY
MCCLURG	CLIFFORD	E
MCCOLLUM	JOHN	
MCCORMACK	EDWARD	J
MCCORMACK	WILLIAM	A
MCCORMICK	PAT	
MCDONALD	DONALD	
MCDONALD	JAMES	R
MCFARLAND	PATRICIA	
MCGUINNESS	BERNARD	
MCKNIGHT	DAVID	K
MCKOEWN	JAMES	
MCLAUGHLIN	FRANCES	
MCMILLAN	JOE	ROBERT
MCMILLIN	ROBERT	B
MCROBERTS	EMMA	JAYNES
MEAKER	ROBERT	THOMAS
MEANS	JOHN	WILLIAM
MEGLICH	MATTHEW	
MENDELSON	JERRY	
MENGER	WILLIAM	
MERHAR	ALBERT	J
MERVAR	EDWARD	J
METELKO	ROBERT	
MEYER	LOUIS	
MEYERS	TED	
MIELDAZIS	CHARLES	J
MILLBURN	EDITH	
MILLER	CARL	JOHN
MILLER	CLARENCE	DARCY
MILLER	H	RICHARD
MILLER	HELEN	M
MILLER	JOHN	LONG
MILLER	LOCKE	CRAIG
MINNICH	RICHARD	E
MODIC	ROBERT	WILLIAM
MOLLE	JOSEPH	S
MONET	CECIL	
MONROE	WILLIAM	T
MONTI JR	JOHN	J
MOONEY	FRANK	
MOONEY	JOHN	P
MOORE	DONLEY	E
MOORE	MAY	C
MOORE	WILLIAM	
MORGAN	DONALD	LEE
MORGAN	ROBERT	L
MORGAN JR	EDWARD	J
MORRIS SR	WILLIAM	GEORGE
MORROW	THOMAS	NEIL
MOSTER	HERBERT	C
MOTSINGER	BUDDY	
MOWER	MARK	
MUELLER	GEORGE	
MUELLER	RICHARD	F
MULLALLY	ROBERT	EDWARD
MULVIHILL	THOMAS	
MUNBAR	HAL	
MURPHY	ALFRED	JOHN
MURPHY	EUGENE	
MURRAY	BILL	
MURRAY	MELVIN	A
MUSSER	DON	LEWIS
MYERS	JAMES	HOWARD
MYERS	LEO	EDWARD
MYERS JR	WILL	S
MYTROSEVICH	JOHN	
NACHTIGAL	JIM	JOHN
NAFTULIN	LOUIS	
NASH	HY	
NEELY	DAN	EARL
NEFF	RICHARD	G
NELSON	JAMES	M
NELSON	PHIL	
NESBITT	CHARLES	W
NEUBAUER	KATHRYN	MELVIN
NEUMANN	NORM	
NEUWIRTH	WILLIS	
NEWKIRK	NORMA	FINERAN
NEWMAN	WILLIAM	H
NICHOLLS	CALVIN	C
NIMS	WALTER	
NOBLE	CLINT	
NOLAN	JOSEPH	MICKEY
OAKLEY	ADDIE	ELLEN
OBRYAN	GEORGE	E
OBRYAN	MARGARET	J
OFFENHAUSER	PHIL	L
OHLSSON	NILS	B
OLESKI	EDWARD	JOSEPH
OLSEN	JOHN	F
OMERSA	JOHN	
ONISI	CARL	JOHN
ONISI	RICHARD	EARL
OPALICH	DANIEL	
OPALK JR	MARTIN	
OWENS	JAMES	CALVIN
OXER	ORLANDO	MONROE
PALUMBO	JOHN	J
PARKER	WILLIAM	FORMAN
PARMELEE	EDITH	I
PARSONS	HERBERT	
PARSONS	RICHARD	J
PASQUALONE	RAYMOND	
PATALINO	ANTHONY	PETER
PATERSON	RICHARD	MATTHEWS
PATRICK	WILLIAM	
PATTON	PAT	
PATTON	PHILIP	A
PAULSON	RAY	
PELL	KELSO	
PETERFY	PAUL	F
PETERS	ARNOLD	L
PETERS	JULIUS	
PETERS	PHIL	
PETERS	ROSE	
PETERSEN	DONALD	EDWARD

Swimming Pool, Euclid Beach, Cleveland, Ohio

LAST	FIRST	MIDDLE
PETTAY	BOB	
PETTI	ROBERT	DONALD
PHILLIPS	CARL	
PILL	JOSEPH	
PITTMAN	KENNETH	
PIVK	JOE	A
PLANTSCH	MICHAEL	
PLO	EDWARD	FRANK
POJE	VINCE	JOHN
POLAUSKI	DUKE	
POLCYN	ROY	EDWIN
POLK	LOUIS	
POOLE	HARRY	
PORTER	JOHN	LEWIS
PORTER	THOMAS	E
POSTRANSKY	JOHN	
POTTS	EVA	
POTTS	WILLARD	
POWELL	HARRY	EDWIN
POWELL	JAMES	W
POWELL	VINCENT	JOSEPH
PRICE	ETHEL	PHILLIPS
PRICE	ETTA	O
PRICE	FLOYD	L
PRICE	HERMAN	
PRICE	LEO	E
PRICE	MARK	THOMAS
PRICE	OTTO	ORVILL
PRICE JR	EDWARD	OWEN
PRIGGE	ROBERT	
PRINGLE	ROBERT	
PRYOR	LARRY	
PUCHALLA	ROBERT	
PURPURA	MATTHEW	
RADOMSKI	WALTER	C
RADY	STEVE	
RANCOURT	ROBERT	H
RASH	WILLIAM	F
RAUPACH	KENNETH	M
RAY	TINA	
RAY	WANETA	E
REA	GILBERT	WILLIAM
READ	WILLIAM	BIGLER
REDFIELD	CUYLER	I
REED	JAMES	
REES	FLOYD	
REIDER	SEYMORE	
REILLY	JOHN	THOMAS
REINHARD	GEORGE	ANDREW
REINHARD	GEORGE	MARTIN
REINK	ROBERT	LEE
RESSLER	DONALD	
RICE	JOHN	HERBERT
RICHARDS	JAMES	E
RICHMAN	ROBERT	
RINALDI	GUSTI	ANTHONY

RITTER	ALLEN	JAMES
ROBBINS	KATHLYN	TEEPLE
ROBINSON	JESSIE	
ROBINSON	RALPH	
ROCK	RICHARD	
ROEGE	WILLIAM	CLARE
ROSENBERG	SANFORD	
ROSS	ROBERT	R
ROTH	NEIL	
RUFFING	JOHN	
RUHLMAN	ROBERT	GRIFFITHS
RUSSELL	GEORGE	D
RUSSO	MIKE	
RUTAR	JAMES	FRANK
RUTTER	EDWARD	SAMUEL
SADA	EDWARD	
SAGE	EARL	E
SANDO	ROBERT	D
SANIFER	CODY	
SANTORELLI	JOSEPH	F
SANZO	SONNY	CARMEN
SAUVAIN	EARL	
SAVINO	MICHAEL	T
SAVODA	RAY	
SAXTON	CLIFFORD	ELWOOD
SAXTON	CLINTON	KAWOOD
SAXTON	FOREST	J
SAXTON	JOHN	NELSON
SAXTON	PAUL	HUGO
SBROCCO	SEBASTIAN	O
SCARLATELLA	MICHELO	
SCARLATELLI	PHYLLIS	MARY
SCARLE	CHARLES	A
SCARPELLO	ANDREW	G
SCHAEFER	WILLIAM	M
SCHAFER	WILLIAM	W
SCHAFFER	HILDA	MARIE
SCHARLOW	R	
SCHARLOW	RAY	
SCHENTUR	ROY	EMIL
SCHIAPPACASSE	CHARLES	
SCHIRRA	JOHN	L
SCHLAPPAL	ROBERT	E
SCHNEIDER	CARL	WILLIAM
SCHNEIDER	HENRY	
SCHNUR	JOHN	
SCHOENFELD	MALVIN	W
SCHROETER	ALLEN	PAUL
SCHULE	HAROLD	L
SCHWED	CHARLES	WILLIAM
SCOTT	BETTY	IDA
SCOTT	DAVID	HUMPHREY
SCOTT	EVERETT	PALMER
SCOTT	JOHN	T
SEAMAN	MORRIS	
SERSAIN JR	EARL	ANDREW
SHAFER	JOHN	F
SHANNON	HARRIS	COOPER
SHANNON	HARVEY	PAGE
SHARPE	CHARLES	H
SHAVE	A	R
SHAVER	LORETTA	KRAUS
SHAY	GEORGE	
SHELTON	WAYNE	D
SHERGALIS	DONALD	J
SHERGALIS	LAWRENCE	D
SHERGALIS	WILLIAM	J
SHERIDAN	NEAL	M
SHERIDAN	PAUL	H
SHERMA	OTTO	
SIDEWAND	EDNA	CLEM
SIDEWAND	HARRY	H
SIDEWAND	RALPH	G
SIEDZIK	PETER	
SIEK	NORMAN	
SIMON	EARL	HOWARD
SIMONETTE	PAUL	

LAST	FIRST	MIDDLE
SINCLAIR	HELEN	TROY
SINDLEDECKER	CLARENCE	F
SIVIK	JACK	ROBERT
SKEDEL	HENRY	F
SKOWRONSKI	HENRY	
SMEKEL	RICHARD	M
SMITH	CONWAY	JAMES
SMITH	DON	J
SMITH	ELMER	CHARLES
SMITH	GILBERT	A
SMITH	IRENE	MARIE
SMITH	LEONARD	S
SMITH	MATILDA	
SMITH	PHILLIP	BURKE
SMITH JR	RICHARD	C
SNEVOLD	WILLIAM	J
SNODGRASS	PAUL	M
SNOW	WEBB	CARL
SNOWDEN	JAMES	A
SODERGREN	ROBERT	
SOMRACK	FRANK	D
SOUCHAK	NICK	JULIUS
SPEELMAN	PAUL	
SPENGLER	JOHN	
SPRAGUE	JESSE	HARRISON
ST CLAIR	JULIUS	J
STABLER	BENNY	
STANA	LEONARD	
STANDISH	SHIRLEY	ROSEMARY
STAPLETON	BUCK	
STEIGERWALD	EDWARD	FRANCIS
STELTER	ELIZABETH	H
STELTER	HARRY	
STEMPLE	KERMIT	J
STERLING	GLEN	
STEVENS	EUGENE	J
STEVENSON	EMORY	
STOCKER	DON	
STOELTZING	WILLIAM	
STOIKER	FRANKLIN	J
STONEBACK	HOWARD	DETWEILER
STRAKA	OWEN	HOMER
STRONG	VIOLET	G
STULMAKER	MORTON	
SULLENS	ROBERT	T
SVETINA	STANLEY	J
SWANK	DAVID	
SWANN	LILLIAN	M
SWINGLE	NORMAN	EDWARD
SYLVANUS	CAROL	MAY
TAFT	PHYLLIS	
TALKINGTON	JOHN	J
TALLEY	NED	A
TANSKI	STANLEY	J
TARBELL	LEONARD	E
TAYLOR	ALBERT	
TAYLOR	CALLOWAY	
TAYLOR	JOHN	
TAYLOR	MATILDA	MARGARET
TEAL	JOSEPH	E
TERBANC	THOMAS	L
TESAR	JOSEPH	M
THOMPSON	ADA	
THOMPSON	JACK	
THOMPSON	JOHN	JAMES
THRALL	WILLIAM	H
TIMPERIO	ARTHUR	M
TIMPERIO	NICK	JACK
TOMSIC	JOE	
TOTH	FRANK	
TOTH	MARSHALL	
TOTH JR	JOE	
TRAUB	GEORGE	L
TRESS	BILL	
TROMBA	JOSEPH	N
TROTTNOW	ROBERT	E
TURNBULL	JAMES	A
TURNER	RUSSELL	J
TUTTLE	GEORGE	
TWEED	MILTON	IVAN
UHL	CHARLES	E
ULEPIC	CHARLES	
USHER	BILLY	
VAN COVE	ALEX	
VAN TASSEL	DAVID	E
VAN TASSEL	JAMES	MYRON
VAN TILBURG	RICHARD	W
VANE	JOE	
VIDMAR	ANTON	JOE
VIVAS JR	JULIUS	
VOGEL	DONALD	
VOLLMER	GEORGE	W
WADE	WILLIAM	
WAGNER	FLORENCE	CALLAHAN
WAGNER	ROBERT	L
WAHL	ERWIN	
WAHL	RALPH	JOHN
WANEK	MARY	ANNE
WAPPERER	ROY	A
WARD	JAMES	EDWARD
WARD	ROBERT	M
WARWICK	HALLIE	EMALINE
WATSON	JESSE	SUMNER
WEAVER	ARNOLD	
WEBB	JOSEPHINE	ANN
WEBB JR	LEONARD	A
WEBER	ARCHIBALD	AUGUST
WEBER	HARRIET	ETHEL
WEBER	RICHARD	LAWRENCE
WEIR	JAMES	EDWARD
WEISBARTH	DON	J
WEISS	GEORGE	
WELL	GEORGE	
WENTZ	JACK	ROBERT
WHEELER	OSCAR	WILLIS
WHEELER	WARDEN	E
WHITE	BOB	
WHITE	CLARENCE	W
WHITE	PHILIP	B
WHITLAM	RICHARD	E
WHITTAKER	H	L
WILL	KARL	WILLIAM
WILLIAMS	ALBERT	T
WILLIAMS	ARTHUR	E
WILLIAMS	EDISON	GATES
WILLIAMS	WALTER	DEWEY
WILLIAMS	WILLIAM	
WILLIS	DUANE	
WILSON	ANNA	RILEY
WILSON	GEORGE	D
WILSON	LLOYD	PAUL
WILSON	MARGARET	D

LAST	FIRST	MIDDLE
WILSON	RICHARD	G
WINCHESTER	JOHN	W
WINFIELD	HERBERT	
WITCRAFT	DALE	R
WOLCOTT	EVAN	
WOLF	WILLIAM	G
WOODBURN	EARL	HARRISON
WOODIE	EDNA	
WOODS	FRANK	HENRY
WOOSLEY	RICHARD	F
WORTMAN	CARL	KEITH
WRETSCHKO	DAVID	T
WRIGHT	JAMES	V
WYANT	CLIFFORD	
WYATT	GEORGE	WILLIAM
WYCKOFF	JAMES	E
XAVIER	RAYMOND	ARTHUR
YAHRMARKET	RICHARD	
YANEY	MAUDE	MAE
YOUNG	MOSES	
YOUNG	ROBERTA	
YOUNG	ROSS	
YUILL	JAMES	HUNTER
ZAK	EUGENE	E
ZEIGLER	FRANKLIN	E
ZIELINSKI	LEO	
ZIVKO	DAN	CHARLES
ZORKO	LOUIS	
ZUPANCIC	JOHN	

1943 EMPLOYEE ROSTER

LAST	FIRST	MIDDLE
ALLEN	MARY	N
ANDERSON	EMMA	KING
ANDERSON	JOHN	ANDREW
ANGELERO	RAFFAELE	
ANGELORO	MARY	J
ARENDT	FRED	KARL
ARTELJ	ANTON	
ATKINS	WM	J
AVE	CHARLES	HAROLD
AVELLONE	CARL	BEN
BABETS	DONALD	JOSEPH
BACIAK	ALLEN	F
BAILEY	HAROLD	
BAILLIE	ELIZABETH	JEAN
BAIZEL	DANIEL	W
BAKER	CILLIUS	MOSE
BAKER	DON	ADELBERT
BAKER	LUCION	
BARON	JOHN	
BARRY	DOROTHY	MARIE
BARRY	JOHN	FRANCIS
BARZI	LOUIS	
BARZI	RICHARD	LOUIS
BASS	ISABEL	LOUISE
BAZSO	ANDREW	JOHN
BECHER	EMMA	SWANK
BECK	BETTY	ANNE
BEHREND	JAMES	EARL
BELL	MARGARET	M
BENEDICT	JAMES	NELSON
BENNHOFF	ARTHUR	C
BENNHOFF	JOHN	A
BENO	MICHAEL	
BERGANT	MARY	ANN
BEYER	OTTO	L
BIBBO	ERNEST	A
BISBEE	FREDERICK	R
BLACK	WILLIAM	G
BLACK	WILLIAM	T
BLAIR	GORDON	
BLAKE	EARL	
BLEAM	LULA	ELSIE
BLOEDE	WILLIAM	CARL
BLUMENTHAL	LAWRENCE	
BLUMENTHAL	SIDNEY	P
BODENSTEIN	ROBERT	S
BOEGE	RICHARD	G
BOGAN	CLARENCE	
BOLLING	HARRY	L
BORGMAN	HENRIETTA	M
BOUNCE	FRANK	VICTOR
BOUR	CHARLES	WILLIAM
BOWINS	ROBERT	J
BRACCIA	FRANK	
BRELEY	HAZEL	MAY
BRENENSTUL	WEALTHY	E
BREWTON	DAVID	E
BREYLEY	HAZEL	MAY
BROMELMEIER	HELEN	R
BROOKINS	EDNA	BELLE
BROWN	LOUIS	
BROZE	ARTHER	
BRYANT	THOMAS	EDWARD
BUCHANAN	DAVID	A
BUCHHOLZ	HERBERT	EDWARD
BUETTINGER	RAYMOND	
BUIKUS	ROBERT	
BULGER	FRANK	
BUNDY	RONALD	TODD
BURNS	JACK	GERALD
BURTON	JAMES	
BURWELL JR	HARVEY	HAVEN
BUSCH	EDITH	MAE
BUSCH	ETHEL	MARIE
BUSHEA	BEATRICE	
BUTLER	ANNA	MARIE
BYRNE	CORNIE	JOSEPH
CAHILL	ROBERT	JOSEPH
CALLAGHAN	BEATRICE	BERNICE
CALLAGHAN	GEORGE	EDWARD
CAMP	HAZEL	KIRKE
CAMPBELL	ALEXANDER	ROBERT
CAPPELLETTI	LOUIS	
CAPUISO	MARY	PHYLISS
CAREY	WALTER	
CARLSON	ARTHUR	C
CARPENTER	JOHN	M
CARROLL	CLARA	LEGGON
CARROLL	FRANK	D
CARROLL	HAZELLE	BELLE
CARROLL	LAUREN	J
CARROLL	LULU	E
CARROLL	ROY	SANFORD
CARTER	HUGH	RICHARD
CASEY	LEE	TOM
CERCEK	JOHN	LOUIS
CETINA	ANTON	
CHAMPA	RAYMOND	FRANK
CHEK	JAMES	
CHEMICK	BETTY	
CHESNICK	FRANK	JOE
CHESNICK	MARY	AGNES
CHESTER	LORNA	ALLYNE
CHICKETTI	MARY	
CHRISTIE	EUGENE	R
CHUBB	WILLIAM	JOHN
CHURNEY	JOHN	J
CIESICKI	HENRY	ALLEN
CIRINO	ROCCO	
CIRINO	THERESA	M
CLARK	JAMES	
CLATTERBUCK	FRANCIS	W
CLATTERBUCK	HARRY	L
CLIFFORD	GEORGE	R
CLIFFORD	ROSE	HAAS
COCANOWER	PAUL	
COCHRANE	RUSSELL	C
COLEMAN	GEORGE	M
COMMANDER	COLIN	
CONNAVINO	JOSEPHINE	
CONTENTO	GUISEPPE	
CONWAY	HARRY	F
COOK	LLOYD	HAYES
COOKE	ERNEST	F
COOPER	ROGER	S
CORLEY	ARTHUR	R
CORMIEA	PHILLIP	E
COSTLOW	EILEEN	RUTH
COTTRELL	W	R
COWHARD	HATTIE	MAY
CRAPNELL	HAROLD	G
CRNKOVICH	JOHN	
CROSBY	JACK	WHITNEY
CUDAHY	JAMES	P
CUDDIHY	ROBERT	H
CULL	PAUL	MILES

LAST	FIRST	MIDDLE
CUNNINGHAM	LETTY	
CURTIS	JOSEPH	A
DAILAIDA	EDWARD	
DANARD	ROBERTA	
DANIELSON	ROBERT	
DARCY	FRANCIS	P
DARLING	RICHARD	THOMAS
DAUPHIN ELMER	VIRGIL	
DAVERN	BERNARD	B
DAVIES	JOHN	R
DAVIS	FLORENCE	
DAVIS	GERTRUDE	L
DAVIS	H	C
DAVIS	HELEN	MARIE
DAVIS	PAUL	THOMAS
DEA	DONALD	ANDREW
DEBRAY	SADIE	J
DEGNOVIVO	ANGELA	
DELAMBO	FRANK	FRANCIS
DELFS	KENNETH	E
DEMANSKI	JOSEPH	A
DEMARIO	ANNA	MARIE
DEMINICO	FRED	A
DERCOLE	EVELYN	
DERCOLE	MARY	D
DERCOLE	ROSE	MARIE
DEVLIN	ROBERT	C
DICKARD	LESTER	R
DICKEY	HERBERT	ROBERT
DIETRICH	DUWAINE	CARL
DIETRICH	FRED	C
DILIBERTO	DONALD	R
DINUNZIO	ELIZABETH	C
DINUNZIO	NICHOLAS	J
DIXON	DONALD	ROBERT
DOESBURG	RALPH	HERBERT
DOLAN	RUTH	ANN
DOUGHERTY	AGNES	M
DOUGHERTY	JOE	
DOUGHERTY	PAUL	V
DOUGHERTY JR	FRANK	E
DOWD	BOB	THOMAS
DRAGONIC	JOHN	A
DRESSER	ARTHUR	ALVIN
DRESSLER	ALTA	DOLORES
DRESSLER	H	C
DROPIC	DOROTHY	RUTH
DUFFY	GEORGE	
DUNCANSON	ROBERT	J
DUTCHCOT	CELIA	
DVORAK	AUGUSTA	P
DVORAK	MARGARET	
DVORAK	R	A
EBERHARD	KENNETH	J
ECHOLS	SAMUEL	
EDWARDS	JOHN	ELBERT
ELDER	EDWARD	THOMAS
ELDRIDGE	MAUDE	
ELINE	WILLIAM	BERRY
ESPER	CAROL	CATHERINE
ESTERBROOK	LESLIE	R
ESTERLINE	H	B
EVANS	RICHARD	L
EZZO	ANTHONY	JAMES
FAATH	JULIA	MAE
FEDELE	ANTHONY	
FEDERICO	JOSEPHINE	MARY
FEHER	GEORGE	JOHN
FELLOWS	THOMAS	F
FELTY JR	MONROE	
FENDALL	LESLIE	A
FERRERI	GEORGE	
FETTEL	ROBERT	GENE
FETTERMAN	RONALD	L
FINERAN	MARY	BALL
FISHER	ALICE	BEACH
FISHER	HARLEY	DEE
FITZGERALD	PETER	MICHAEL
FITZ	MAURICE	THOMAS
FLEMING	BRADY	K
FLETCHER	ELLIS	
FLOOD	JOHN	
FOGARTY	GENEVIEVE	MALADY
FOLEY	MAURICE	
FORET	NORMAN	ANDREW
FOX	ALLAN	M
FOX	IRENE	BENZ
FRANEY	JOSEPH	M
FRANTZ	RICHARD	JOHN
FREDERICKS	FRANK	JOSEPH
FREEMAN	CAROL	JANE
FROMME	KENNETH	F
FRY	FRED	
GALLAGHER	WAYNE	R
GAMBRILL JR	MERLE	C
GARLAND	FRANK	C
GASKINS	RONALD	V
GENNERT	WILLIAM	
GERM	ANNE	D
GFELL	KATHERINE	M
GIBBONS	GEORGE	EARL
GIBBONS	GIL	
GIBBONS	RAY	J
GIBLIN	VINCENT	JOHN
GIERING	WAYNE	
GILBERT	LORETTA	THELMA
GILBERT	ROBERT	L
GLAWE	ANTHONY	A
GLIEBE	DONALD	R
GOLD	HARRY	
GOODWIN	GLORIA	JANE
GORCHESTER	RITA	MAE
GORHAM	JAY	M
GRANATA	ANTHONY	
GRAY	IVY	CAROLINE
GREEN	CLAUDE	ELLSWORTH
GREEN	GERALD	C
GREENWAY	FRED	STUDER
GRIFFIN	JOE	THOMAS
GRIFFIN	PATRICK	J
GRIFFITH	WILLIAM	JOHN
GROVE	CLAIRE	WILLIAM
GUENTHER	ROBERT	E
GUIDO	TRESSIE	
GUILD	GEORGE	A
GUTTER	LESTER	
HAASE	RICHARD	
HACKETT	KENNETH	C
HAGEMEISTER	FRED	C
HALL	JAMES	B
HAMILTON	ALVIN	
HAMILTON	BERLIN	

LAST	FIRST	MIDDLE
HAMPE	GENEVIEVE	E
HAMPE	GORDON	
HANN	HERMAN	H
HANNA	LORETTA	HANNA
HANRATTY	JOAN	
HANSEN	WILLIAM	G
HARDMAN	HOWARD	
HARDY	RICHARD	
HARRISON	RICHARD	
HARROLD	FLORENCE	M
HARROLD	THADDEUS	JAMES
HARTMAN	CARL	
HASSETT	ROBERT	M
HASTINGS	TOM	
HAWK	MARY	LOU
HAYES	WILLIAM	JAMES
HAZELTON	AGNES	M
HAZELTON	LOUIS	R
HEAPHEY	JIM	
HEATON	BERT	G
HEATON	MAY	
HECKMAN	FRANK	E
HEISLER	EMERSON	WAYNE
HENDERSHOTT	SHELDON	
HENN	ULYSSES	A
HENRY	BERT	
HERRICK	JOHN	E
HESENER	ROBERT	J
HIGHBERGER	J	HAROLD
HIGHBERGER	WILLIAM	J
HILL	ELSIE	LOUISE
HILL	HARRY	DAVID
HINKO	MATTHEW	
HINTON	JAMES	ROBERT
HINTON	JOHN	SHEPARD
HINTON	WILLIAM	HARRY
HIRSCH	GOTTLIEB	
HIRSCHAUER	ROGER	DON
HLADKY	FRANK	C
HODGES	JANE	
HOFF	DOROTHY	IRENE
HOGAN	ELIZABETH	
HOLLINGSWORTH	JACK	LEE
HOOD	EMILY	AUGUSTA
HOPPERT	ROBERT	JOHN
HORA	SARAH	
HOUGH	HAROLD	C
HOUSE	JOHN	FLORUS
HOWARD	TOM	JOHN
HRIBAR	FRANK	
HUBBARD JR	TOM	A
HUBBELL	HELEN	G
HUDSON	JOHN	O
HUMPHREY	DORIS	ELIZABETH
HUNSBERGER	EDYTHE	MAY
HUNT	BLANCHE	LEAH
HUNTER	MAYME	E
HUNTER	ROY	CHARLES
ILOVAR JR	JOHN	
IMPERIO	CARL	JOHN
JACKLITZ	ALBERT	
JACKSON	HAROLD	
JACKSON	REBECCA	JANE
JAKLITSCH	RICHARD	E
JAMIESON	CLARENCE	BARCLEY
JANKOWSKI	J	
JANZ	JOHN	
JERNEJCIC	FRANK	E
JETT	ZEKE	
JOCA	GEORGE	ROBERT
JOHNSON	DOROTHY	IDA
JOHNSTON	CLIFFORD	BASIL
JOHNSTON	DONALD	HUMPHREY
JOHNSTON	WILLIAM	E
JONES	GEORGE	
JONES	JAMES	
JOSEPH	RALPH	WILBUR
JUDD	CLARENCE	HADEN
JUDD	ESTIL	LEE
JUDD	JULIANA	ROTH
KAMIN	JOHN	
KAMINSKI	KENNETH	L
KANALLY	MARGARET	
KANEKI	ELMER	EDWARD
KANOWSKI	WILHELM	
KAPPELE	JOSEPHINE	BENHAM
KASUNIC	LOUIS	F
KAZMIR	WALTER	M
KEILY	FRANK	ROBERT
KELLEHER	JAMES	M
KELLEM	HARRY	JETT
KELLER	FLORENCE	A
KELLY	DAISY	CURTIS
KELLY	TOM	FRANCIS
KERES	GEORGE	
KERNS	LAWRENCE	
KESSLER	AGNES	CECELIA
KICKHAM	CHARLES	MAURICE
KIESELBACH	ALFRED	
KIKOLI	FRANK	
KILBY	FRANK	EVERETT
KIME	HOWARD	H
KING	AUD	
KING	PHILIP	
KINKOPF	EDWARD	
KINKOPH	ART	HERMAN
KIRSCH	EDWIN	S
KIRSCHNER	EDWARD	A
KLEBER	WILLIAM	KARL
KLEIN	P	
KLEMPAN	EDWARD	
KLIMA	MARGUERITE	A
KMETT	FRANCIS	P
KNAAKE	HELENE	
KOCH	CLYDE	
KOCH	HENRY	HERMAN
KOMICK	RUDOLPH	A
KOPP	WILLIAM	D
KORDICH	TOM	
KORMOS	LEONARD	R
KRAPENC	RICHARD	W
KREBS	FRED	W
KREEGER	FRED	CARL
KRESS	MITCHELL	
KRULL	HERMAN	R
KUEHNE	DONALD	L
KUNKEL	FRANK	WILLIAM
KUNZ	HERBERT	J
KUNZ	ROBERT	
LANCASTER	ROBERT	J
LANGE	JOHN	FERDINAND
LANGHIRT	MELVINA	C

LAST	FIRST	MIDDLE
LAVAN	ROBERT	EDWARD
LAVRICH	CHARLES	
LAVRICH	JAMES	H
LEASURE	KENNETH	E
LEBET	JULIUS	M
LECHOWICZ	BENEDICT	
LECHOWICZ	KATHERINE	M
LECHOWICZ	RALPH	ANTHONY
LEDENICAN	LOUIS	W
LEE	ANNA	MAY
LEES	KENNETH	A
LEFFLER	GILBERT	
LEMMO	MARY	
LENTZ	STERLING	
LEVAR	LOUIS	A
LEWANDOSKI	NORMAN	
LEWIS	HANDEL	T
LEWIS	JACK	K
LINK	EMILY	PFEIL
LION	HONEST	
LISTER	GEORGE	W
LLOYD	JOHN	
LOCKE	HUBERT	LEE
LOFTUS	THOMAS	P
LONGFIELD	TESS	EVANS
LONGTIN	EDWARD	F
LONGWINTER	ROBERT	
LOOMIS	HELEN	MARIE
LORENZO	JOSEPHINE	
LORENZO	MARIA	
LOUCKS	DON	CLIFFORD
LUKAS	BEN	
LUNAR	FRANK	A
LUSIN	ANNE	M
LUSIN	MILDRED	C
LUTSCH	JOSEPHINE	WIDMAR
LYNCH	FRED	FRANCIS
MACDONALD	JOHN	
MAHER	GEORGE	
MAHER	PHILIP	PATRICK
MALAGA	ROBERT	S
MALKIN	HARVEY	
MALONE	SAM	
MANDAU	LESLIE	
MANDAU	RUDOLPH	
MANGAN	ROBERT	LEO
MANGINO	FRANCES	
MANIERI	FRANK	
MANION	ROBERT	J
MANNING	CHARLES	T
MANNING	DOUGLAS	L
MARCOGLIESE	MADDALENA	
MARKENS	EDNA	BROWNING
MARTAUS	PAUL	PETER
MARTINS	WALTER	F
MARTINSEN	RONALD	H

MASON	WILLIAM	EDWARD
MAST	JOHN	EDWART
MAUER	ERNEST	JOHN
MAXIM	JOSEPH	JOHN
MAXIM	JULIUS	JOSEPH
MCCARTHY	MARGARET	
MCCARTHY	ROBERT	LEO
MCCAULEY	HENRY	ORMSLY
MCCAULEY	LAMBERT	
MCCORMACK	EDWARD	J
MCCORMACK	NOREEN	F
MCDERMOTT	JOHN	M
MCDONALD	JAMES	R
MCDONOUGH	DANIEL	
MCELROY	LEWIS	
MCELWEE	JOHN	T
MCGEEVER	CHAS	
MCGEEVER	PATRICK	JOHN
MCGOVERN	JOE	WILLIAM
MCGRANER	HAZEL	ANN
MCGURK	JACK	L
MCILRATH	IDA	E
MCKINLEY	WILLIAM	J
MCKNIGHT	DAVID	K
MCLAUGHLIN	FRANCIS	
MCQUAIGE	JAMES	W
MEDLING	ETHEL	
MERHAR	ALBERT	JOSEPH
METZ	ELEANOR	M
MEYER	LOUIS	
MILANICH	RUSSELL	
MILINOVICH	TOM	
MILLER	CLARENCE	DARCY
MILLER	EARL	LOUIS
MILLER	H	RICHARD
MILLER	JOHN	LONG
MILLER	QUENTIN	MILES
MILLS	WILLIAM	H
MINNICH	RICHARD	E
MODIC	ROBERT	WILLIAM
MOLINOVICH	TOM	
MOLLE	JOSEPH	S
MONROE	WILLIAM	T
MOORE	FENTON	LEO
MOORE	MAY	C
MOORSCRAFT	JOHN	
MORAN	GERALD	J
MORGAN	KATHLEEN	J
MORGAN	ROBERT	L
MORGAN	WILLARD	F
MORGAN	WILLIAM	
MORRIS	FRED	
MOSIER	FREDA	V
MOWER	MARK	
MUELLER	OSWALD	
MUELLER	RICHARD	F
MULLALLY	ROBERT	EDWARD
MULLIKIN	HARRY	F
MULVIHILL	THOMAS	
MURPHY	ALFRED	JOHN
MYERS	LEO	EDWARD
MYERS	WILLIAM	ALLEN
NACHTIGAL	EDDY	A
NACHTIGAL	WILLIAM	
NADEL	DALE	
NEALON	GEORGE	MARTIN
NEELY	DAN	EARL
NEIMAN	RICHARD	C
NEMECEK	JOHN	
NEUBAUER	KATHRYN	MELVIN
NEUWIRTH	WILLIS	
NEWTON	FRANCIS	
NEWWIRTH	WILLIS	
NICKLAS	EMMA	A
OAKLEY	ADDIE	ELLEN
OBRYAN	GEORGE	E

THE ELYSIUM, CLEVELAND, O. 10669

LAST	FIRST	MIDDLE
OCONNELL	THOMAS	
OHLSSON	NILS	B
OLEKSIUS	ROLAND	R
OLIVER	NANCYE	RUTH
OLIVER	VERDA	MARIE
OLMSTEAD	RICHARD	D
OMERSA	JOHN	
ONISI	CARL	JOHN
ONISI	RICHARD	EARL
OPALK JR	MARTIN	
OSGOOD	F	GEORGE
OSGOOD	GEORGE	
OTASEK	ROBERT	ADAM
OTOOLE	JACK	M
OXER	ORLANDO	MONROE
PADDEN	WILLIAM	
PAJEK	FRANK	VICTOR
PALLANTE	ROMEO	D
PARKER	WILLIAM	
PARKER	WILLIAM	FORMAN
PARMELEE	EDITH	I
PARR	LAWRENCE	J
PARSONS	RICHARD	J
PASQUALINE	RAYMOND	
PAUS	FRANK	
PAVLENCH	THERESA	
PELL	KELSO	
PERKINS	FLORENCE	ARLENE
PERKINS	FRIEDA	W
PERROTTI	CAROLINE	
PERRY	FLORENCE	JUNE
PERUSEK	EDWARD	F
PETERS	ARNOLD	L
PETERS	BETTY	JANE
PETERS	ROSE	
PETROVIC	RICHARD	
PHELPS JR	ELLIOTT	
PHELPS JR	MILLARD	
PHILLIPS	PAUL	J
PICKETT	RAYMOND	T
PIERCE	LOUIS	
PILL	JOSEPH	
PILL	WILLIAM	
PITTMAN	KENNETH	
PIVK	JOE	A
PLANTSCH	MICHAEL	
PODBOY	MICHAEL	J
POKRENT	THOMAS	
POLOMSKY	JOHN	V
POLZNER	DOROTHY	M
POLZNER	ROBERT	JOSEPH
POMIDORE	THERESA	MARIE
PORTER	JOHN	LEWIS
PORTER	THOMAS	E
POTTER	BARBARA	
POTTS	EVA	
POWELL	VINCENT	JOSEPH
POWERS	CHARLES	EDWARD
PRECHTL	MARY	JANE
PRENDERGAST	MICHAEL	R
PRICE	ETTA	O
PRICE	EUGENE	
PRICE	FLOYD	L
PRICE	JIM	BANFORD
PRICE	MARK	THOMAS
PRICE JR	ED	OWEN
PRIGGE	ROBERT	
PRIMEAU	JOHN	GENE
PUCHALLA	ROBERT	LEE
PULLAR	ALBERT	ANDREW
RAGONE	RAYMOND	
RAKES	PARTHENIA	LEGGON
RANALLO	MICHAEL	
RANALLO	SAM	
RAY	EDDIE	LEE
RAY	TINA	
RAY	WANETA	E
RAYER	JOSEPH	FRANK
RAYMOND	JOSEPH	JAMES
RAYMOND	PHILLIP	JAMES
READ	WILLIAM	BIGLER
REARDON	ROGER	SAMUEL
REBEK	JOE	
REDFIELD	CUYLER	I
REINHARD	GEORGE	MARTIN
RICHMAN	ROBERT	
RILEY	MARGARET	ADELE
RINALDI	GUSTI	ANTHONY
ROBBINS	KATHLYN	TEEPLE
ROBERTS	JAMES	VERN
ROBINSON	JACK	
ROTH	JULIANA	
ROYSTER	GEORGE	E
RUHLMAN	JON	RANDALL
RUHLMAN	ROBERT	GRIFFITHS
RUSCITTO	ALBERT	A
RUSCITTO	LOUIS	A
RUSS	RAYMOND	L
RUSSO	MIKE	
RUTAR	JAMES	FRANK
SAEFKOW	HOWARD	
SAGE	EARL	E
SAMSEL	JOHN	A
SANDERS	RAY	S
SANTORELLI	JOSEPH	F
SANZO	SONNY	CARMEN
SAWREY	WILLIAM	H
SAXTON	JOHN	NELSON
SAXTON	PAUL	HUGO
SBROCCO	LEONA	
SBROCCO	SEBASTIANO	
SBROCCO	SUSIE	
SCALA	JOSEPH	
SCARLATELLA	MICHELO	
SCARLE	CHARLES	
SCHAEFER	JAMES	J
SCHAEFER	TOM	BERNARD
SCHAEFER	WILLIAM	M
SCHENTUR	ROY	EMIL
SCHIRRA	JOHN	L
SCHLAGEL	RICHARD	L
SCHMIDT	BETTY	JANE
SCHMIDT JR	PAUL	
SCHOENFELD	MALVIN	M
SCHROEDER	CARL	A
SCHROEDER JR	HARVEY	
SCHUTZ	JAMES	
SCHWED	CHARLES	WILLIAM
SCHWENNER	ROBERT	
SCOTT	BETTY	IDA
SCOTT	DAVID	HUMPHREY
SCOTT	EVERETT	PALMER

A part of Euclid Beach Camping Grounds

LAST	FIRST	MIDDLE
SEACE	MARLIN	B
SEAMON	DONALD	WILLIAM
SEDLOCK	JOHN	JOSEPH
SELIA	JOHN	DAVID
SELIO	WILLIAM	D
SEVER	MARY	PAULINE
SHAFER	WILLIAM	EDWARD
SHAMPAY	ROBERT	
SHANNON	HARRIS	COOPER
SHANNON	HARVEY	PAGE
SHANTZ	ALVIN	WILSON
SHELKO	ROBERT	J
SHERGALIS	DONALD	J
SHERGALIS	WILLIAM	J
SHERIDAN	PAUL	H
SHIMA	OTTO	
SHONTZ	FRANK	CURTIS
SHONTZ	MARJORIE	MAY
SIEGEL	STANLEY	ALLEN
SIGNORELLI	JOHN	
SIVIK	DONALD	
SIVIK	JACK	ROBERT
SKENDER	JOHN	HENRY
SKOWRONSKI	HENRY	
SKRZYPKOWSKI	JOHN	C
SMEKEL	RICHARD	M
SMITH	CONWAY	JAMES
SMITH	FLORENCE	ALLEN
SMITH	GILBERT	A
SMITH	IRENE	MARIE
SMITH	JANIS	KATHRYN
SMITH	MATILDA	
SMITH	PAUL	W
SMITH	PHILLIP	BURKE
SMITH	RICHARD	CHARLES
SMITH	WILLIAM	
SMITH	WILLIAM	ARTHUR
SNOW	WEBB	CARL
SNYDER	CLARENCE	HAROLD
SOUCHAK	NICK	JULIUS
SPANGLER	BLANCHE	E
SPEELMEN	PAUL	
SPRINGBORN	GEORGE	J
SPRINGBORN	RUTH	ANN
STAHLNECKER	GEORGE	F
STAMM	JOHN	GEORGE
STANDISH	STANLEY	ROSEMARY
STEFANAC	JOHN	
STELTER	ELIZABETH	H
STEMPLE	KERMIT	J
STINE	IRVING	ELI
STITT	MARVIN	T
STOELTZING	WILLIAM	
STOIKER	FRANKLIN	J
STONEBACK	HOWARD	DETWEILER
STRONG	VIOLET	G
SULLENS	ROBERT	T
SULLENS	THOMAS	J
SULLIVAN	JOANNE	
SUMMERS	ROBERT	
SVARPA	STAN	L
SWANK	DAVID	
SWANN	LILLIAN	M
TALKINGTON	JOHN	J
TAYLOR	ALBERT	
TAYLOR	CALLOWAY	
TAYLOR	CHARLES	
TAYLOR	DAVE	GEORGE
TAYLOR	EVANS	
TAYLOR	JESSE	T
TAYLOR	JOHN	
TAYLOR	LARRY	
TERBANC	THOMAS	L
TERWILLIGER	GEORGE	E
THOMAS	EARL	M
THOMAS	PAUL	EDWARD
THOMAS	PRYSE	
THOMPSON	DOROTHY	G
THOMPSON	JOSEPH	D
THOMPSON	RICHARD	JOHN
THORNTON	GEORGE	
THRASHER	NELLIE	PEARL
TIMPERIO	CARL	JOHN
TIPPIN	WALLACE	CHAS
TITUS	GERTRUDE	
TOLL	EDWARD	WESLEY
TOMARIC	MARIE	CECILE
TOMARIC	MOLLY	JOANNE
TOMITZ	GILBERT	JOHN
TONEY	DESSIE	B
TOTH	JOHN	JOSEPH
TOW	RALPH	HERBERT
TRIVISON	CARRIE	
TRIVISON	CHARLES	A
TRIVISON	GRACE	
TRYON	JEROME	THOMAS
TUCCI	CARMELLA	
TURKOVICH	STEVE	
VAN NUIS	ALFONSO	COVAS
VERLOHN	CHARLES	J
VINSON	TROY	EDWARD
VRABEC	STANLEY	EDWARD
WAGNER	FERDINAND	
WAGNER	FLORENCE	CALLAHAN
WAHL	ERWIN	
WAHL	RALPH	JOHN
WAHL	WILLIAM	JOHN
WALKER	CHARLES	KENNETH
WALLON JR	CLIFFORD	A
WALSH	JACK	FRANCIS
WANEK	MARY	ANNE
WARD	CHARLES	STERLING
WARD	FRANK	
WARD	ROBERT	M
WARNER	NORMAN	
WARWICK	HALLIE	EMALINE
WASHBURN	WILLIAM	
WASSUM	RAYMOND	EDGAR
WATSON	ROSE	LEE
WEBER	ARCHIBALD	AUGUST
WEBER	HARRIET	ETHEL
WEBER	JAMES	HOWARD
WEBER	RICHARD	LAWRENCE
WEIL	MARSHALL	JAMES
WELCH	DOLORES	MAY
WELLS	RAYMOND	W
WENTZ	JACK	ROBERT
WHITE	CLARENCE	W
WHITE	PAULINE	
WHITEHOUSE	FLORENCE	CONKLE
WHITLAM	RICHARD	E
WHITLEY	SUSIE	WALSTON
WIDMAR	BERNICE	JOSEPHINE

LAST	FIRST	MIDDLE
WIDMAR	RUDY	P
WIESLER	HARRY	
WILES	JUNIOR	CURTIS
WILL	FRANCIS	CASE
WILL	MARY	NELL
WILLIAMS	ALBERT	T
WILLIAMS	ARTHUR	E
WILLIAMS	CHRIS	HARRY
WILLIAMS	EDISON	GATES
WILLIAMS	GRACE	WYATT
WILLIAMS	WALTER	DEWEY
WILLIAMS	WILLIAM	
WILSON	ANNA	RILEY
WILSON	FRANK	ALLEN
WOODBURN	EARL	HARRISON
WOODS	FRANK	HENRY
WOOSLEY	RICHARD	F
WORGULL	NORMAN	
WRAYNO	ROBERT	
WULBECK	TESSIE	IRENE
WYATT	WILLIAM	H
XAVIER	RAYMOND	ARTHUR
YEAGER	DON	EARL
YONKE	VICTOR	WILLIAM
YONTZ	ALBERT	F
YOUNG	MOSES	
YOUNG	ROBERTA	
ZEIGLER	JOHN	C
ZICARELLI	PETER	J
ZUPANCIC	JOHN	

43

As the Humphrey business grew, so did their Public Square stand.

The interior of the Public Square stand.

44

1944
EMPLOYEE ROSTER

LAST	FIRST	MIDDLE
AGEE	CARL	
ALBERTONE	DONALD	
ALLEN	JAMES	
AMBICKI	JOHN	
AMBLER	DAVID	L
ANDERSON	ANN	PRISCILLA
ANDERSON	EMMA	KING
ANDERSON	JOHN	ANDREW
ANGELERO	RAFFAELE	
ATKINS	JOHN	
AVELLONE	BEN	C
BABETS	DONALD	JOSEPH
BACIAK	ALLEN	F
BAILEY	GEORGE	
BAKER	CILLIUS	MOSE
BAKER	LUCION	
BALDNER	WILLIAM	
BALTERSHAT	ESTHER	W
BANDY	WILLIAM	L
BARBEE	HACKETT	
BARBEE	HARRIET	I
BARON	HYMAN	
BARRY	DOROTHY	MARIE
BARRY	LULU	MARIE
BARTLETT	ROBERT	A
BARZI	LOUIS	
BARZI	RICHARD	LOUIS
BAZSO	ANDREW	JOHN
BEECHER	GENE	
BEHREND	JAMES	EARL
BENEDICT	JAMES	NELSON
BERTRAM	THOMAS	EDWARD
BIBBO	ERNEST	A
BIDDLE	GEORGE	DWIGHT
BITKER	JOHN	E
BLACK	WILLIAM	G
BLACK	WILLIAM	T
BLAIR	GORDON	
BLOEDE	WILLIAM	CARL
BODENSTEIN	ROBERT	S
BOLDIN	FRANCES	
BOLDIN	FRANK	JOHN
BONNER	FANNIE	HARRIS
BORGA	FRED	
BOUFFARD	BART	
BOVA	PHILLIP	J
BOWINS	ROBERT	J
BOWYER	LOIS	
BOYCE	ROBERT	
BOYKIN	DENNIS	
BOYLE	CLIFFORD	
BOZICH	JOSEPH	
BRACCIA	FRANK	
BRADFORD	JOHN	
BRATOVICH	WILLIAM	
BRENNAN	HELEN	T
BREWTON	KILBY	
BRITTON	E	
BROCONE	EUGENE	C
BRODIE	TAYLOR	S
BROOKINS	EDNA	BELLE
BRUENING	LEO	
BRYANT	THOMAS	EDWARD
BUCHANAN	DAVID	A
BUCHHOLZ	HERBERT	EDWARD
BUDENZ	CARL	F
BUNDY	RONALD	TODD
BURR	CLAYTON	
BURTON	JAMES	
BUSCH	ARTHUR	EDWARD
BUTLER	ANNA	MARIE
BYRNE	BERNARD	
BYRNE	PATRICK	
CABOT	ANTHONY	V
CALLAGHAN	BEATRICE	BERNICE
CALLAGHAN	GEORGE	EDWARD
CALLALY	THOMAS	
CAMPBELL	ALEXANDER	ROBERT
CAMPBELL	DOROTHY	H
CAMPBELL	FRED	
CAPUTO JR	ERNEST	
CARLSON	GUSTAV	ERIC
CARNEY	JOHN	P
CARPENTER	JOHN	M
CARRAN	ROBERT	L
CARROLL	CLARA	LEGGON
CARROLL	HAZELLE	BELLE
CARROLL	LAUREN	J
CARROLL	ROY	SANFORD
CASEY	LLOYD	F
CAVALLARI	GUSTY	
CECIL	THERMAN	
CELESTE	PAT	
CERCEK	JOHN	LOUIS
CHANT	ROBERT	JAMES
CHEMICK	BETTY	
CHESNICK	FRANK	JOE
CHUBB	WILLIAM	JOHN
CHURNEY	JOHN	J
CIANI	ANDREW	
CICIGOI	GENE	
CICIGOI	RAY	
CIRINO	JOHN	D
CIRINO	ROCCO	
CIRINO	THERESA	M
CLARK	JAMES	
CLIFFORD	GEORGE	R
CLIFFORD	MARY	
CLINE	ROBERT	
COCANOWER	PAUL	
COCHRANE	RUSSELL	C
COLE	GRANT	R
COLLINS	BERT	
CONGIN	CARL	
CONNAVINO	JOSEPHINE	
CONTENTO	GUISEPPE	
CONTI	ANTHONY	
CONWAY	HARRY	F
CONWAY	JERRY	
COOK	EUGENE	B
COOK	LLOYD	HAYES
COOKE	ERNEST	F
COOPER	ROGER	S
CORMIEA	PHILLIP	E
COSKI JR	BERNARD	J
COSTLOW	EILEEN	RUTH
COVERT	THOMAS	
COWHARD	HATTIE	MAY
CRAPNELL	HAROLD	G
CREDICO	JOHN	
CRNKOVICH	JOHN	
CROSBY	JACK	WHITNEY
CROW	JACK	
CROW	MARGARET	E
CUDAHY	JAMES	P
CUDDIHY	ROBERT	H

LAST	FIRST	MIDDLE
CUMMINGS	SUE	JANE
CUNNINGHAM	LETTY	
CURLEY	JOSEPH	P
CUTSHALL	B	MAXINE
DAILAIDA	EDWARD	
DANIELS	ALBERT	L
DARCY	FRANCIS	P
DAY	DON	
DEA	DONALD	ANDREW
DEAN	VERNON	ALLEN
DEANER	JAMES	
DECICIO	FRANCES	
DEGNOVIVO	ANGELA	
DELAMBO	FRANK	FRANCIS
DEMARIO	ANNA	MARIE
DEMICHELE	ANTHONY	R
DENNISON	JAMES	E
DENZLER	RALPH	L
DEPETRIS	EUGENE	
DERCOLE	ANTONETTE	
DERCOLE	MARY	D
DEVLIN	ROBERT	C
DICICCO	CONCETTA	
DICKARD	JOHN	
DICKEY	HERBERT	ROBERT
DILIBERTO	DONALD	R
DINUNZIO	ELIZABETH	C
DINUNZIO	NICHOLAS	J
DISANZA	GENE	R
DIXON	JANET	H
DOESBURG	RALPH	HERBERT
DOLAN	RUTH	ANN
DONAHUE	LAURA	G
DONAHUE	MARY	ETHEL
DONAHUE	ROBERT	D
DOUGHERTY	AGNES	M
DOUGHERTY	JOSEPH	E
DOUGHERTY JR	FRANK	E
DRESSLER	ALTA	DOLORES
DRESSLER	H	C
DUFFY	GEORGE	
DUPREE	HENRY	
DUTCHCOT	CELIA	
DVORAK	MARGARET	
EAGLEN	JUANITA	ANN
EBERHARD	EILEEN T	A
EGAN	THOMAS	E
EIDNER	HARRY	
EIDNIER	REGGIE	
ELINE	WILLIAM	BARRY
EWIS	JACK	K
EZZO	ANTHONY	JAMES
FADDEN	MARGARET	M
FAZIO	TONEY	
FEDELE	ANTHONY	
FEDERICA	JOSEPHINE	
FEDERICO	JOSEPHINE	MARY
FERRERI	GEORGE	
FIER	ROBERT	
FIFOLT	MILDRED	
FINELL	JOHN	
FINERAN	MARY	BALL
FISHER	ALICE	BEACH
FISHER	HARLEY	DEE
FITZGERALD	PETER	MICHAEL
FITZMAURICE	THOMAS	
FLAHERTY	IRA	L
FLANIK	ROBERT	
FOGARTY	GENEVIEVE	MALADY
FOGEL	WILLIAM	CARL
FOLEY	DONALD	J
FOLEY	MAURICE	
FOLLETT	JOHN	RICHARD
FORET	NORMAN	ANDREW
FORONE	ANGELINA	
FOX JR	EDWARD	
FRANK	THOMAS	L
FRANKLIN	HARRY	
FRANTZ	RICHARD	JOHN
FRAZIER	ROBERT	A
FREDERICKS	FRANK	JOSEPH
FRY	MARY	ANN
GAMELLIA	ROSE	MARY
GAU	WILLIAM	HOWARD
GERM	ANNE	D
GFELL	KATHERINE	M
GIBBONS	GEORGE	EARL
GIRARDI	ROSE	MARIE
GLASS	HENRY	D
GLAWE	ANTHONY	A
GODEC	JULIA	
GOLENBERKE	KENNETH	
GORHAM	JAY	M
GOSMACK	LEON	GENE
GRANDSTAFF	JULIA	M
GREEN	CLAUDE	ELLSWORTH
GREEN	GERALD	C
GREENSHIELDS	JANE	
GREENWAY	FRED	STUDER
GRIFFITHS	MABEL	
GROSS	RICHARD	ARTHUR
GUENTHER	CHARLES	W
GUENTHER	ROBERT	E
HAAS	REGINA	
HAASE	MARGARET	
HAASE	RICHARD	
HAIGHT	FRED	
HANN	HERMAN	H
HANSEN	ROBERT	
HAPCHUK	WILLIAM	
HAPP JR	JOHN	W
HARNEY	THOMAS	ROY
HARRINGTON	JEROME	
HARROLD	FLORENCE	M
HARROLD	THADDEUS	JAMES
HARTMAN	CARL	
HARTMAN	JAMES	J
HAWKINS	MARY	KATHLEEN
HAWKINS	MAY	
HAYES	WILLIS	JAMES
HEALY	ANDREW	
HEAPHEY	JAMES	
HEATON	MAY	
HECKMAN	FRANK	E
HENN	ULYSSES	A
HERRICK	JOHN	E
HIGHBERGER	WILLIAM	J
HILL	HARRY	DAVID
HILL	JAMES	ARTHUR
HILT	WILBUR	
HINTON	WILLIAM	HARRY

LAST	FIRST	MIDDLE
HIRSCH	GOTTLIEB	
HOFF	DOROTHY	IRENE
HOGAN	ELIZABETH	
HOLZMAN	JERALD	B
HOOD	EMILY	AUGUSTA
HOOKER	SIDNEY	
HORA	SARAH	
HORN	JULIA	
HOTCHKISS	THEODORE	
HOUSE	JOHN	FLORUS
HOWE	ROSEMARY	
HUBBARD	THOMAS	
HUBBELL	HELEN	G
HUDSON	EDWIN	P
HUGHES	ELLA	M
HULING	FRANCIS	SHEPARD
HUNT	BLANCHE	LEAH
HUNT	GEORGE	FRANCIS
HUNTER	MAYME	E
HUNTER	ROY	CHRALES
IVANCIC	MARGARET	
IVANCIC	ROSE	
IVEY	WALTER	
JAMIESON	CLARENCE	BARCLEY
JANZ	JOHN	
JOHNSON	DOROTHY	IDA
JOHNSTON	CLIFFORD	BASIL
JOHNSTON	DONALD	HUMPHREY
JONES	GEORGE	
JUDD	CLARENCE	HADEN
JUDD	ESTIL	LEE
JUDD	JULIANA	ROTH
KANALLY	MARGARET	
KANOWSKI	WILHELM	
KARLOVICH	MARY	
KASUNIC	GEORGE	
KATZ	SAMUEL	
KAZMIR	WALTER	M
KELLEHER	JAMES	M
KELLER	FLORENCE	A
KELLEY	BETTY	JO
KELLY	DAISY	CURTIS
KELLY	JAMES	THOMAS
KELLY	JOHN	A
KELLY	TOM	FRANCIS
KENT	EDWARD	C
KESSLER	AGNES	CECELIA
KEST	ALAN	RICHARD
KIESELBACH	ALFRED	
KIKOLI	FRANK	
KILBY	FRANK	EVERETT
KIME	HOWARD	H
KING	ROY	
KLANCHER	ROBERT	J
KLEBER	WILLIAM	KARL
KLEIN	PAUL	
KLEINSHROT	ELMO	
KLEMPAN	EDWARD	
KLESS	ROBERT	
KLIMA	MARGUERITE	K
KLOOS	ROGER	
KNISS	PATSY	ARLENE
KOCEVAR	ANTHONY	F
KOCH	CLYDE	
KOCH	HENRY	HERMAN
KOLLAR	FRANCES	MARY
KOLLAR	RAY	
KOMICK	RUDOLPH	A
KONARSH	RAYMOND	D
KOPAITICH	STEVE	
KOSS	ROBERT	
KOTELES	CHARLES	
KRAMER	JOHN	P
KRAPENC	RICHARD	W
KREEGER	FRED	CARL
KRULL	HERMAN	R
KUBAT	MARY	G
KUBEJA	EDDIE	
KUSHAN	ALICE	
LAFFERTY	CLIFFORD	
LAMBERT	PHILIP	
LAMPE	CHAS	E
LANGE	DOLORES	J
LANGE	JOHN	FERDINAND
LARICCIA	PETER	
LATESSA	ANTOINETTE	
LATESSA	RICHARD	ROBERT
LAVRICH	CHARLES	
LEASURE	KENNETH	E
LECHOWICZ	KATHERINE	M
LECHOWICZ	RALPH	ANTHONY
LENARCIC	JOHN	
LESEFKY	GEORGE	M
LEVINE	HARRY	
LEWANDOSKI	NORMAN	
LEWIS	HANDEL	T
LEWIS	JACK	
LIGGETT	LUTHER	
LISTER	GEORGE	W
LITTLE	HELEN	
LONGFIELD	TESS	EVANS
LOOMIS	HELEN	MARIE
LOOMIS	RICHARD	
LORENZO	MARIA	
LUCCIO	RUTH	
LUKAS	BENEDICT	D
LUNDBLAD	ROBERT	A
LUPIS	FRANK	F
LUTZWEILER	ROBERT	F
LYNCH	FRED	FRANCIS
MAC DONALD	JOHN	
MADDEN	GEORGE	E
MAGNANI	EDWARD	
MAGRUDER	ELMER	D
MAHER	GEORGE	
MAHER	PHILIP	PATRICK
MALAGA	ROBERT	S
MANCINE	LOUIS	
MANCINI	LOUISE	
MANDAU	RUDOLPH	
MANGINO	FRANCES	
MANNING	CHARLES	T
MARCOGLIESE	MADDALENA	
MARKS	LULU	MAE
MARRONE	LOUIS	
MARSHALL	CADDIE	
MAST	JOHN	EDWART
MATTIX	FRANCES	
MAXIM	EMILIA	
MAXIM	JULIUS	JOSEPH
MAY	OSWALD	CARL FREDERICK
MAYER	LOUIS	

LAST	FIRST	MIDDLE
MCALLISTER	ROBERT	
MCCARTHY	ROBERT	LEO
MCCORMACK	NOREEN	F
MCDONOUGH	DANIEL	
MCELROY	FLORENCE	
MCELWEE	JAMES	FRANCIS
MCELWEE	JOHN	T
MCGOVERN	JOE	WILLIAM
MCGREGOR	JOHN	
MCGURK	JACK	L
MCILRATH	IDA	E
MCKEE	SAM	
MCLAUGHLIN	PATRICK	L
MCMILLIN	ROBERT	B
MCNABB	EARL	WAYNE
MEGLICH	ELEANOR	
MELTZER	MARGARET	
MERHAR	ALBERT	JOSEPH
METALONIS	MARIE	
METZ	ELEANOR	M
METZGER	WESLEY	C
MEYER	LOUIS	
MICHAELSON	DAN	WESLEY
MIHALINEC	ELEANOR	
MILES	LILLIAN	A
MILLER	HELEN	THERESA
MILLER	JOHN	LONG
MINNICH	RICHARD	E
MOLNAR	JAMES	AUGUST
MONROE	WILLIAM	T
MOORE	MAY	C
MOORE JR	JOHN	HENRY
MORAN	GERALD	J
MOREL	SYLVESTER	
MORGAN	KATHLEEN	J
MORGAN	ROBERT	L
MORRIS	FRED	
MORRIS SR	WILLIAM	GEORGE
MOSIER	FREDA	V
MOTISKA	RAYMOND	
MUELLER	RICHARD	F
MULIOLIS	DONALD	P
MULVEY	JOHN	W
MULVIHILL	THOMAS	
MURDOCK	WILLIAM	L
MURPHY	ALFRED	JOHN
MYERS	D	H
MYERS	LILLIAN	
NAUNCIK	FRANK	
NEFF	ARTHUR	C
NEUBAUER	KATHRYN	MELVIN
NEWNESS	SHIRLEY	MAY
NOGA	JOHN	
NOLAN	ALICE	MARY
OAKLEY	ADDIE	ELLEN
ODONNELL	THOMAS	
OECHSLE	RICHARD	
OERGEL	LAURA	JEAN
OHL	MAXINE	
OHL	VICTOR	V
OMERSA	JOHN	
ONEIL	JOHN	
OPALK JR	MARTIN	
OPRITZA JR	N	J
OROZCO JR	RAYMOND	
OST	GEORGE	CARL
OSTROWSKI	TED	
OTASEK	ROBERT	ADAM
OTT	SYLVIA	
OXER	ORLANDO	MONROE
PARKER	WILLIAM	
PARKER	WILLIAM	FORMAN
PARMELEE	EDITH	I
PARR	LAWRENCE	J
PARSONS	RICHARD	J
PARSONS	STANFORD	
PECK	DONALD	
PELL	KELSO	
PELL	KENNETH	
PERKINS	CHARLES	F
PERKINS	FLORENCE	ARLENE
PERKINS	FRIEDA	W
PERROTTI	CAROLINE	
PESSELL	DONALD	H
PETERS	ROSE	
PETERSON	ROBERT	
PILL	JOSEPH	
PISTILLO	GRACE	
PLANTSCH	MICHAEL	
PODBOY	MICHAEL	J
POLCYN	ROY	E
POLOMSKY	JOHN	V
POWELL	VINCENT	JOSEPH
PRECHTL	MARY	JANE
PRENDERGAST	THOMAS	
PRICE	ETTA	O
PRICE	FLOYD	L
PRICE	HARRY	R
PRICE	JIM	BANFORD
PRICE	JOHN	LEO
PRICE	JUNE	
PRICE	MARK	THOMAS
PRICE	RICHARD	
PROCTOR	LEONA	MARY
PRY	EMMETT	F
PUGEL	CLARENCE	
PUGEL	KATHERINE	
PUGEL	LOUIS	FRED
PULLAR	ALBERT	ANDREW
RAIMONDO	ALFRED	
RAKES	PARTHENIA	LEGGON
RANALLO	MICHAEL	
RANCOURT	RICHARD	
RANDALL	CONSTANCE	ANN
RANFT	ROBERT	
RAYER	JOSEPH	FRANK
RAYMOND	JOSEPH	JAMES
RAYMOND	PHILLIP	JAMES
REDFIELD	CUYLER	I
REILLY	FRANK	
REINHARD	GEORGE	MARTIN
RENNER	MERLIN	
RICHARDSON	BRADFORD	P
RICHMAN	ROBERT	
RILEY	MARGARET	ADELE
RITTER	ROBERT	
ROBBINS	KATHLYN TEEPLE	
ROBERTS	JAMES	VERN
ROBERTS	NICHOLA	MARIO
ROBINSON	CHARLES	H
RODERICK	ALICE	M
ROEPNACK	PAUL	ALVIN

INTERIOR OF THE LARGEST AND FINEST DANCING PAVILION IN THE WORLD, EUCLID BEACH PARK, CLEVELAND, OHIO.

LAST	FIRST	MIDDLE
RONIGER	H	E
ROSA	CARMELLA	
RUHLMAN	JON	RANDALL
RUSCITTO	ALBERT	A
RUTAR	JAMES	FRANK
RYAN	PHILIP	A
RYDMAN	ELIZABETH	
SADOWSKI	CASEY	
SADOWSKI	MARY	
SANDERS	RAY	S
SANTORELLI	JOSEPH	F
SANZO	MICHAEL	
SBROCCO	LEANA	
SBROCCO	SEBASTIANO	
SCACCIA	ALBERT	
SCARLATELLA	MICHELO	
SCHAFFER	GEORGE	
SCHELLENTRAGER	WILLIAM	
SCHIRRA	JOHN	L
SCHULTZ	RAYMOND	
SCHUMAKER	ROBERT	
SCHUTZ	JAMES	
SCHWED	CHARLES	
SCOTT	BETTY	IDA
SCOTT	EVERETT	PALMER
SEAMAN	BENJAMIN	W
SEAMON	DONALD	WILLIAM
SEIBERT	HENRI	
SHAFER	WILLIAM	EDWARD
SHAMPAY	JEAN	
SHANNON	HARRIS	COOPER
SHANNON	HARVEY	PAGE
SHARON	VIRGINIA ROSE	
SHEAFFER	ALBERT	
SHERGALIS	DONALD	J
SHERICK	RICHARD	
SHERIDAN	BETTY	JEAN
SHEWELL JR	ROBERT	
SIDEWAND	HARRY	H
SIECKER	ELIZABETH	
SIEGFRIED	GEORGE	W
SILVEROLI	JOHN	JAMES
SINCLAIR	ROBERT	
SINGER	SANFORD	M
SIVIK	DONALD	
SIVIK	JACK	ROBERT
SKEBE	WILLIAM	JOHN
SMEKEL	RICHARD	M
SMITH	ALBERT	S
SMITH	CONWAY	JAMES
SMITH	EDDY	JACK
SMITH	FLORENCE	ALLEN
SMITH	GEORGE	
SMITH	GEORGE	
SMITH	IRENE	MARIE
SMITH	MATILDA	
SMITH	MAY	F
SMITH	RICHARD	
SMITH	ROBERT	WILSON
SPORCIC	ROBERT	
SPRINGBORN	GEORGE	J
SPUZZILLO	MARY	
ST JAMES	CLARENCE	J
STAMM	JOHN	GEORGE
STARINA	FLORENCE	
STELTER	ELIZABETH	H
STELTER	ROBERT	C
STINE	IRVING	ELI
STONEBACK	HOWARD	DETWEILER
STRAUSS	DAVID	
STROETER	RICHARD	
SULLIVAN	ELEANOR	
SULLIVAN	JOANNE	
SUMSKIS	CARL	P
SWANK	DAVID	
SWANN	LILLIAN	M
SWEIGART	WILLIAM	R
TALKINGTON	CHARLES	
TAULBEE JR	MORT	
TAYLOR	CALLOWAY	
TAYLOR	DAVE	GEORGE
TAYLOR	GEORGE	ELMER
TAYLOR	HELEN	
TAYLOR	IRENE	
TAYLOR	ROBERT	MCCLELLAND
TERLEP	JOHN	
TERWILLIGER	GEORGE	E
THIEL	ROBERT	
THOMAS	PAUL	EDWARD
THOMPSON	ART	AUSTIN
THOMPSON	GEORGE	
THOMPSON	RICHARD	JOHN
THOREN	RALPH	
TIMPERIO	CARL	JOHN
TIZZANO	MARIA	
TOMITZ	GILBERT	JOHN
TOMLE	JOHN	JOHN
TOMPKINS	WILLIAM	
TOTARELLA	JOSEPH	
TRINCO	MARCO	ANTHONY
TRIVISON	CARRIE	
TRIVISON	CHARLES	A
TRIVISON	GRACE	
TRIVISONNO	MARY	
TRYON	JEROME	THOMAS
TURNER	RAYMOND	
TYSON	MARION	JEAN
UNBEHAUN	GEORGE	
VACCARIELLO	ANTONIO	
VACCARIELLO	MARY	I
VALE	WILBUR	E
VAN NUIS	ALFONSO	COVAS
VOLLMER	JACK	MONROE
VOVK	HERMAN	J
WACKER	WILLIAM	
WAGNER	FLORENCE	CALLAHAN
WAGNER	RALPH	J
WAHL	ERWIN	R
WAHL	WILLIAM	JOHN
WALLACE	WILLIAM	R
WALLETTE	GERALD	
WALSH	JACK	FRANCIS
WAPPERER	ROY	
WARD	JAMES	M
WARD	ROBERT	M
WARD	WILLIAM	CALVIN
WARNER	NORMAN	
WARWICK	WILLIAM	BERT
WAYNE	HARMON	T
WEBB	RACHEL	
WEBER	ARCHIBALD	AUGUST

LAST	FIRST	MIDDLE
WEBER	HARRIET	ETHEL
WEBER	JAMES	HOWARD
WEED	CHARLES	EARL
WEIL	MARSHALL	JAMES
WELCH	DOLORES	MAY
WELLS	RAYMOND	W
WHEATLEY	AUGUSTA	MAINE
WHITE	CLARENCE	W
WHITE	GEORGE	
WHITE	WILLIAM	
WHITEHOUSE	FLORENCE	CONKLE
WHITEMAN	CHARLES	
WIESE	MARIAN	L
WILES	JUNIOR	CURTIS
WILLIAMS	ALBERT	T
WILLIAMS	ARTHUR	E
WILLIAMS	EARL	ROGER
WILLIAMS	EDISON	GATES
WILLIAMS	GRACE	WYATT
WILLIAMS	WALTER	DEWEY
WILLIAMS JR	JOSEPH	
WILLIAMSON	LORNE	G
WILSON	ANNA	RILEY
WILSON	FRANK	ALLEN
WOHLGEMUTH	OTTO	M
WOOD	ALBERT	
WOODS	FRANK	HENRY
WRAYNO	ROBERT	LEO
WRIGHT	ROY	
XAVIER	RAYMOND	ARTHUR
YEAGER	DON	EARL
YONKE	VICTOR	WILLIAM
YOUNG	MOSES	
YOUNG	ROBERTA	
ZALLER	JOHN	
ZALOKAR	THERESA	DORIS
ZEBROWSKI	JAMES	
ZIOLKOWSKI	LEO	
ZUCHELLI	ALBERT	
ZUPANCIC	JOHN	

Bud Wilson boards the Great American Racing Derby.

Crew of the Racing Derby. ca. 1943.

50

1945
EMPLOYEE ROSTER

LAST	FIRST	MIDDLE
ADAMS	JOSEPH	
ADLER	HOWARD	
AGEE	CARL	E
AJDUKOVICH	GEORGE	
AMBLER	DAVID	L
ANDERSON	ALBERT	H
ANDERSON	ANN	PRISCILLA
ANDERSON	EMMA	KING
ANDERSON	JOHN	ANDREW
ANDERSON	NORA	H
ANGELERO	RAFFAELE	
ARCARO	MICHAEL	A
ARTHUR	IDA	MAE
ASPINWALL	HAROLD	J
ATKINS	JOHN	
AXE	PAUL	E
BACORN	DEWEY	
BACULIK	MARIE	J
BAILEY	JOHN	FINUS
BAKER	CILLIUS	MOSE
BAKER	HOWARD	H
BALDNER	WILLIAM	
BALDWIN	DEAN	DEWITT
BALLIETT	DAVID	L
BALTERSHAT	ESTHER	W
BARAGA	FRANCES	K
BARBEE	HACKETT	
BARBEE	HARRIET	I
BARBO	JOHN	
BARNES	EQUILLA	S
BARRY	DOROTHY	MARIE
BARRY	KENNETH	
BARZI	LOUIS	
BATICH	MARY	L
BAUS	HARLAN	LOUIS
BAUS	JAMES	W
BAXTER	CARL	
BAZSO	ANDREW	JOHN
BEANE	JOHN	F
BECKER	WILLIAM	P
BEECHER	EUGENE	E
BENNETT	EVELYN	
BENSCH	TOM	
BETHUY	WILLIAM	R
BICKER	HARRY	A
BITKER	JOHN	E
BLACK	WILLIAM	T
BLAIR	GORDON	
BLANCHFIELD	GEORGE	L
BLANKENSHIP	WILLIAM	E
BLEAM	LULA	ELSIE
BLISS	NETTIE	
BLOEDE	WILLIAM	CARL
BLUHM	JOHN	G
BOLDEN	ROBERT	
BORGA	FRED	
BORICK	MATTHEW	L
BOUFFARD	BART	
BOWHALL	HOWARD	
BOWINS	ROBERT	J
BOWYER	LOIS	
BOYCE	ROBERT	
BOYLE	CLIFFORD	
BRADENBURG	HAROLD	L
BRADFORD	JOHN	
BRANCHE	JAMES	R
BRANDT	JESSIE	
BRATOVICH	WILLIAM	
BRENENSTUL	WEALTHY	E
BRODNIK	RICHARD	
BROMLEY	JOSEPH	
BROOKINS	EDNA	BELLE
BROWN	PHILIP	NEAL
BROWN	RALPH	R
BROWN	WILLIAM	ORR
BRUENING	LEO	
BRYANT	FRANK	G
BRYANT JR	THOMAS	L
BRYANT SR	THOMAS	EDWARD
BUCHANAN	DAVID	A
BUCHHOLZ	HERBERT	EDWARD
BUCKLEY	OLIVER	L
BUERGER	JOSEPH	F
BURR	CLAYTON	
BURTON	JAMES	
BUTLER	JOSEPH	HACKNEY
BYRNE	CORNIE	JOSEPH
BYRNE	PATRICK	
CAFFEY	JEANNETTE	A
CALLAGHAN	BEATRICE	BERNICE
CALLAGHAN	GEORGE	EDWARD
CALLALY	THOMAS	
CALLIGHEN	FRANCIS	
CAMPBELL	ALEXANDER	ROBERT
CAMPBELL	FRED	
CAMPBELL	THOMAS	R
CAPUTO JR	ERNEST	
CARDOZA	ROBERT	R
CARETTI	FREDERICK	C
CARNEY	JOHN	P
CARPENTER	JOSEPH	
CARROLL	CLARA	LEGGON
CARROLL	HAZELLE	BELLE
CARROLL	LAUREN	J
CARROLL	PROVIA	CARTER
CARROLL	ROY	SANFORD
CARSON	HAMILTON	
CARTER JR	WILLIE	
CERCEK	JOHN	LOUIS
CHAFER	WILLIAM	
CHEMICK	BETTY	
CHRISTOPHER	MARY	
CHUBB	WILLIAM	JOHN
CHURCH JR	DON	L
CHURNEY	JOHN	J
CIANI	ANDREW	
CINUNZIO	NICHOLAS	J
CIRINO	FRANK	
CIRINO	ROCCO	
CIRINO	THERESA	M
CLARK	CHARLES	MILTON
CLATTERBUCK	FRANCIS	W
CLATTERBUCK	HARRY	L
CLIFFORD	DONALD	
CLIFFORD	GEORGE	R
CLIFFORD	MARY	
CLIFFORD	ROSE	
CLIFFORD	TERESA	
COBB	AHIRA	
COCANOWER	PAUL	
COCHRANE	RUSSELL	C
COLLINS	EDWARD	P
COLLINS	JAMES	F
COLTON	JOSEPH	
COMYNS	ARNOLD	T
CONNAVINO	JOSEPHINE	

LAST	FIRST	MIDDLE
CONNELL	TOM	BENTLEY
CONNOLLY	BETTY	JANE
CONTENTO	GUISEPPE	
CONTORNO	BENNY	
CONWAY	HARRY	F
COOK	LLOYD	HAYES
COOK	WILLIAM	T
COOKE	ERNEST	F
COOPER	PHYLLISTINE	
COOPER	ROGER	S
COPIC	JOHN	
CORMIEA	PHILLIP	E
CORSO	LUCILLE	R
COUSINEAU	THOMAS	
COWHARD	HATTIE	MAY
COWHER	HAROLD	LEROY
CRAPNELL	HAROLD	G
CRAWFORD	ALLYN	S
CROSBY	JACK	WHITNEY
CROW	CLYDE	E
CROW	JACK	
CUDDIHY	ROBERT	H
CUNNINGHAM	FRANCIS	
CURLEY	FRANK	
CUTSHALL	B	MAXINE
CZERNICKI	LEONARD	
DANCULOVIC	MARY	ANN
DANIELS	ALBERT	L
DAY	DON	
DAY	WILLIAM	RICHARD
DEA	DONALD	ANDREW
DEANER	JAMES	
DEAR	WILLIAM	
DELAMBO	FRANK	FRANCIS
DELLINGER	TOM	
DEMARIO	ANNA	MARIE
DENHAM	WILLIAM	
DENSON	CARRIE	
DENZLER	RALPH	L
DEPETRIS	EUGENE	
DERCOLE	ANTONETTE	
DERCOLE	MARY	D
DEVLIN	ROBERT	C
DICICCO	CONCETTA	
DICKEY	HERBERT	ROBERT
DILIBERTO	DONALD	R
DINEEN	JOHN	V
DINUNZIO	NICHOLAS	J
DIRRMAN	RICHARD	C
DOBAY	JAMES	
DOBRIN	RUSSELL	L
DODICK	JOE	
DOESBURG	RALPH	HERBERT
DOHERTY	THOMAS	J
DOLAN	RUTH	ANN
DOLENEC	STANLEY	
DONAHUE	LAURA	G
DONAHUE	LOIS	ANN
DONAHUE	MARY	ETHEL
DONAHUE	ROBERT	D
DONAHUE	WILLIAM	J
DOUGHERTY	JOSEPH	E
DOUGLAS	JAMES	
DOWDELL	EMMETT	JOSEPH
DRAXLER	WALTER	
DRESSLER	ALTA	DOLORES
DRESSLER	H	C
DRISCOTT	THOMAS	P
DUCHARME	FORREST	
DUNN JR	EDWIN	D
DUTCHCOT	CELIA	
DUTCHCOT	EDWARD	OSCAR
DVORAK	MARGARET	
DVORAK	RICHARD	
DWYER	WELEY	J
EADIE	JOHN	
EASTON	HOWARD	E
EASTON	LEONARD	L
EATON	HOWARD	E
EBERHARD	EILEEN	T A
EDDY	BYRON	W
EGLEY	WILLIAM	G
EIDENIER	HARRY	O
EIDNIER	REGGIE	
ELINE	WILLIAM	BARRY
EMERSON	FRED	T
ENTSMINGER	JACK	R
EVANS	HARRY	WILLIAM
FARINACCI	ROBERT	
FATICA	LORETA	
FEDERICO	CARMELA	
FEDERICO	JOSEPHINE	
FEDERICO	JOSEPHINE	MARY
FEDERICO	MARIA	NICOLA
FELDKAMP JR	EDW	J
FENDRICH	TEDDY	
FERGUSON	PATRICIA C	
FERRERI	GEORGE	
FIDDES	RAYMOND	JOHN
FIELD	RUBY	
FINERAN	MARY	BALL
FISHER	HARLEY	DEE
FITZGERALD	FRANK	T
FITZGERALD	PETER	MICHAEL
FITZMAURICE	THOMAS	
FITZPATRICK	WINIFRED	A
FOGARTY	GENEVIEVE	MALADY
FOGEL	WILLIAM	CARL
FORMICA	MARY	
FOX	JEROME	
FRANKLIN	HARRY	
FREDERICKS	FRANK	JOSEPH
FREEDER	MORTON	
FROHMBURG	ROSE	
GAINES	ALFRED	W
GALLAGHER	MICHAEL	
GANG	JAY	ORVEL
GAU	WILLIAM	HOWARD
GEITHER	WILLIAM	P
GERBER	ELLSWORTH	S
GERBER	WAYNE	O
GIBBONS	JOHN	CLARKE
GIBBONS	RAYMOND	
GILL	ELSIE	OLIVIA
GILMER	AUDREY	M
GLASS	HENRY	D
GLEASON	DAVID	M
GOLENBERKE	KENNETH	
GORHAM	JAY	M
GOWER	WILBERT N	
GRAHAM	LEASER	
GRAHAM	LOUISA	

"THE SEA SWING," EUCLID BEACH PARK, CLEVELAND, OHIO.

LAST	FIRST	MIDDLE
GREEN	CLAUDE	ELLSWORTH
GREEN	JAMES	S
GREENE	STUART	
GREENWAY	FRED	STUDER
GRIESSE	FREDERICK	W
GROSEL	ED	
GROSS	RICHARD	ARTHUR
GROVER	KATHERINA	A
GUTHRIE	JAMES	
HAASE	MARGARET	
HABA	JOHN	G
HAGEN	PAUL	GERALD
HALBURDA	DOROTHY	
HAMILTON	ALVIN	
HANN	HERMAN	H
HANNAN	THOS	J
HANSEN	ROBERT	
HARFORD	JESS	A
HARNEY	THOMAS	ROY
HARRISON	JOHN	
HARROLD	THADDEUS	JAMES
HARTMAN	CARL	
HASTINGS	ROBERT	
HASTINGS	TOM	
HASTINGS	WILLIAM	MARTIN
HAWKINS	MARY	KATHLEEN
HAWKINS	MAY	
HAYES	WILLIS	JAMES
HAYTHER	LUCILLE	
HEARN	WILLIAM	J
HEATON	MAY	
HECKMAN	FRANK	E
HEISS	EDWARD	C
HENCK	HARRIET	
HENRY	RICHARD	R
HENSEL	JOHN	T
HERRICK	ELWIN	G
HERRICK	JOHN	E
HILL	HARRY	DAVID
HILLIER	ROBERT	
HILT	WILBUR	
HOCHEVAR	RICHARD	J
HOFF	DOROTHY	IRENE
HOFFMEYER	PAUL	
HOGAN	ELIZABETH	
HOLLOWELL	ELIZABETH	
HOOD	EMILY	AUGUSTA
HORA	SARAH	
HORNYAK	EDWARD	
HOUSE	JOHN	FLORUS
HOUSE	SAMUEL	R
HUBBARD	THOMAS	
HUBBELL	HELEN	G
HUDSON	EDWIN	P
HUGHES	ELLA	M
HUMMEL	FREDERICK	H
HUNT	BLANCHE	LEAH
HUNT	GEORGE	F
HUNTER	ROY	CHARLES
HUTH	EUGENE	
HYDE	HARTER	L
HYDE	HOWARD	PHILLIP
INDA	THEODORE	
ISABELLA	DOROTHY	
IVANCIC	MARGARET	
JACKSON	HAROLD	
JACKSON	JAMES	D
JAFFEE	JOSEPH	G
JAMIESON	CLARENCE	BARCLEY
JANKE	GEORGE	A
JANSKY	EUGENIA	
JANZ	JOHN	
JARO	MARTIN	
JARUS	VIRGINIA D	
JENKINS	GEORGE	HARDIE
JESCHENIG	FRANCIS	
JOCA	GEORGE	
JOHNSON	DOROTHY	IDA
JOHNSTON	CLIFFORD	BASIL
JOHNSTON	DONALD	HUMPHREY
JONES	GEORGE	
JUDD	CLARENCE	HADEN
JUDD	ESTIL	LEE
JUMP	ROBERT	LEE
JURKO	ROBERT	C
KAMINSKI	JOHN	J
KANALLY	MARGARET	
KANOWSKI	WILHELM	
KASPER	JOHN	CHARLES
KEIFER	HARRIET	L
KELLER	FLORENCE	A
KELLEY	BETTY	JO
KELLY	DAISY	CURTIS
KENT	EDWARD	C
KESSLER	AGNES	CECELIA
KIKOLI	FRANK	
KLAAS	WILLIAM	PAUL
KLANCHER	ROBERT	J
KLEBER	WILLIAM	KARL
KLIMA	MARGUERITE	K
KLINGMAN	DON	
KOBAL	MAX	E
KOCH	HENRY	HERMAN
KOCK	CLYDE	G
KOLLAR	FRANCES	MARY
KONCILJA	VICTOR	H
KOONS	JANE	HALL
KOPINA	JOSEPH	E
KORDIC	THOMAS	
KOSHKO	WILLIAM	JOHN
KOSHOCK	EUGENE	
KOSMETOS	MICHAEL	
KOSTECKI	JOSEPH	
KOSTER	WILLIAM	J
KOZIATEK	JEROME	
KRAMER	MARTY	
KRANTZ	GEORGE	
KRAPENC	RICHARD	W
KRASOVEC	ROBERT	T
KREVES	EDWARD	
KRIVDO	JEROME	M
KRIVDO	LUKE	J
KRULL	HERMAN	R
KUBAT	MARY	G
KUMEL	JOSEPH	W
LAGASSE	ROBERT	
LAMBERT	PHILIP	F
LAMPE	CHAS	E
LANGE	HARRY	W
LANGE	JOHN	FERDINAND
LANGHAM	JOHN	DALE
LARICCIA	ARMIDA	E

45:—SCENE ON LAKE ERIE, FROM THE PIER, EUCLID BEACH PARK, CLEVELAND, OHIO.

Photo by Van Fisher

LAST	FIRST	MIDDLE
LARICCIA	MARIE	P
LARICCIA	PETER	
LASCALA	EMILIO	
LATORRE	ROBERT	N
LAUREL	RICHARD	
LEASURE	KENNETH	E
LECHOWICZ	KATHERINE	M
LECHOWICZ	MARGARET	
LECHOWICZ	ROBERT	F
LEE	ALPHEUS	
LEIFER	ROBERT	A
LENARCIC	JOHN	
LEWIS	HANDEL	T
LEWIS	JACK	K
LEWIS	PATRICK	
LISTER	GEORGE	W
LJUBI	ANTHONY	J
LONG	WILLIAM	J
LONGFIELD	TESS	EVANS
LORENZO	MARIA	
LUNDBLAD	ROBERT	A
LUPIS	FRANK	F
LYNCH	R	J
MAC DONALD	JOHN	
MACKENZIE	ELLEN	
MAGLICH JR	FRANK	
MAGRUDER	ELMER	D
MAHONY	JOHN	
MAMULA	ANNE	
MANCINE	BENJAMIN	J
MANCINE	LOUIS	
MANCINI	JOSEPHINE	
MANDAU	RUDOLPH	
MANGAN	BETTY	MAY
MANGINO	FRANCES	
MANNING	DOUGLAS	L
MANNING	RICHARD	A
MARCOGLIESE	MADDALENA	
MARINELLI	MILDRED	ALICE
MARSHALL	CADDIE	
MATHIAS	CLIFF	
MATTIX	FRANCES	
MAXIM	EMILIA	
MCARTHUR	CHARLES	W
MCBANE	TOM	
MCBRIDE	MICHAEL	E
MCCAHAN	KITTIE	
MCCLURG	ROBERT	
MCCOURT	EUGENE	J
MCDONOUGH	DANIEL	
MCELWEE	JAMES F	P
MCELWEE	JAMES	FRANCIS
MCELWEE	JOHN	T
MCGOVERN	JOE	WILLIAM
MCGREAL	JOHN	F
MCGURK	JACK	L
MCILRATH	IDA	E
MCKEE	SAM	
MCLAUGHLIN	FRANCES	
MCLAUGHLIN	JOHN	R
MCNIECE	RAYMOND	E
MEAD II	JOHN	M
MEADE	TOM	
MEDING	ROBERT	D
MEDVES	EDWARD	
MERCHANT JR	ELROY	H
MERGL	JOE	
MERKLEY	LEONARD	L
METCALF	HARRY	C
METZ	JOHN	RICHARD
MEYER	LOUIS	
MEYER	WILLIAM	ARTHUR
MIKOLICH	LOUIS	
MILINOVICH	TOM	
MILLER	DONALD	RICHARD
MILLER	GERARD	J
MILLER	JOSEPHINE	
MILLER	RONALD	GEORGE
MITCHEL	ALBERT	J
MOLE	ROBERT	A
MOLNAR	JAMES	AUGUST
MONING	HENRY	A
MONTI	CLIFF	
MOORCROFT	JOHN	W
MOORE	ERNESTINE	
MOORE	MAY	C
MORAN	GERALD	J
MORDAUNT	JACK	D
MOREL	SYLVESTER	
MORGAN	HUGH	
MORGAN	KATHLEEN	J
MORIARTY	JAMES	J
MORRIS SR	WILLIAM	GEORGE
MOSALL	ERVIN	H
MOSIER	FREDA	V
MOTSCH	GERALD	A
MOYNIHAN	GERALD	
MULLEN	JAMES	
MULVIHILL	THOMAS	
MURPHY	ALFRED	JOHN
MURPHY	EDWARD	H
MURPHY	ROBERT	IRVIN
MYERS	D	H
MYERS	LILLIAN	
NEFF	ARTHUR	C
NEGRELLI	JENNIE	M
NELSON	GRACE	
NELSON	JAMES	S
NELSON	JOHN	G
NESTIC	JAMES	
NEUBAUER	JOSEPH	MELVIN
NEUBAUER	KATHRYN	MELVIN
NEWTON	FRANCIS	
NEWTON	GEORGE	F
NOLES	ROBERT	
OAKLEY	ADDIE	ELLEN
ODONNELL	JOHN	
OLIVER	ANDREW	
OLMSTEAD	ROBERT	G
OLSON	HERB	E
OMERSA	JOHN	
ONEIL	JOHN	
OPRITZA JR	N	J
OSTRUNN	ROBERT	E
OTASEK	CHARLES	K
OTASEK	ROBERT	ADAM
OXER	ORLANDO	MONROE
PADDEN	WILLIAM	
PAJEK	STANLEY A	
PALECHEK	ROBERT	
PARKER	WILLIAM	
PARR	LAWRENCE	J

14:—THE BEACH, EUCLID BEACH PARK, CLEVELAND, OHIO.

54

LAST	FIRST	MIDDLE
PARR JR	LARRY	J
PASSALACQUA	JULIUS	
PAULIN	ELLEN	
PAYTON	HOWARD	C
PECK	DONALD	
PECK	GEORGE	L
PELL	KELSO	
PELL	KENNETH	
PERKINS	FRIEDA	W
PERME	JOHN	E
PERME	JOSEPHINE	C
PERROTT	RONALD	C
PERROTTI	CAROLINE	
PERRY	RICHARD	G
PETERS	ROSE	
PEZDIRTZ	JOHN	A
PHELPS	MILLARD	HUGH
PIKE	JERRY	
PITTENGER	FRANCES	
PLANTSCH	MICHAEL	
POLCYN	ROY	E
POWELL	ALFRED	L
PRECHTL	MARY	JANE
PRICE	ETTA	O
PRICE	FLOYD	L
PRICE	HARRY	R
PRICE	MARK	THOMAS
PRICE	OTTO	
PROCTOR	LEONA	MAY
PUGEL	KATHERINE	
PUGEL	LOUIS	FRED
PURVIS	JAMES	
QUINN	MICHAEL	
RACKAITIS	NORMA	
RADATZ	REINHARDT	
RAIMONDO	ALFRED	
RAKES	PARTHENIA	LEGGON
RANALLO	RAY	
RANCOURT	RICHARD	
RAYER	JOSEPH	FRANK
RAYER	WILLIAM	
RAYMOND	JOSEPH	JAMES
REARDON	MICHAEL	
REDFIELD	CUYLER	I
REED	MAXINE	
REED	RICHARD	H
REILLY	FRANK	
REILLY	HELEN	
REILLY	MICHAEL	
REIN	BETTY	JANE
REINHARD	GEORGE	MARTIN
RIBINSKAS	EDWARD	
RICHARDSON	BEATRICE	CUNNINGHAM
RILEY	DANIEL	
RILEY	MARGARET	ADELE
RINALDI	JOSEPH	FRANK
RINALDI	NICK	
RITTER	ROBERT	
ROBBINS	KATHLYN	TEEPLE
RODERICK	ALICE	M
ROHLOFF	ALBERT	
ROLFES	WILLIAM	R
RONIGER	H	E
ROSA	ANTHONY	C
ROWLES	DAVID	WILLIAM
ROWLEY	GRACE	MOORE
RUHLMAN	JON	RANDALL
RUTTOR	ANTHONY	
RYDMAN	ELIZABETH	
SAMPLES	CARY	
SANDS	JAMES	J
SAUNDERS	SPENCER	
SBROCCO	SEBASTIANO	
SCACCIA	ALBERT	
SCARLATELLA	MICHELO	
SCHAEFER	JAMES	J
SCHAEFER	TOM	BERNARD
SCHAEN	MARY	P
SCHELLENTRAGER	WILLIAM	
SCHILL	LLOYD	EDGAR
SCHILL JR	LLOYD	E
SCHIRRA	JOHN	L
SCHMIDT	DENNY	
SCHUTT	DONALD	ALLEN
SCHUTTE	RICHARD	
SCOTT	EVERETT	PALMER
SEAMAN	BENJAMIN	W
SEBUSCH	DON	
SEFERIAN	ROBERT	R
SEVELLO	ANNA	
SHAMPAY	JEAN	
SHANNON	HARRIS	COOPER
SHANNON	HARVEY	PAGE
SHANNON	MATILDA	JOHNSON
SHEA	PATRICK	J
SHEAFFER	ALBERT	
SHEARER	JOHN	W
SHEPPARD	CATHERINE	M
SHILLIDAY	EVERETT	P
SHIPMAN	ROBERT	
SHOEMAKER	KENNETH	W
SIECKER	ELIZABETH	
SILVEROLI	JOHN	JAMES
SILVEROLI	PETER	
SILVESTRO	ALBERT	
SILVOLA	RICHARD	
SIVIK	DONALD	
SKEBE	WILLIAM	JOHN
SKRANCE	ROSEMARY	
SKUFCA	RALPH	M
SMEKEL	RICHARD	M
SMERDEL	CHARLES	A
SMICIKLAS	JOSEPH	
SMICIKLAS	NICHOLAS	W
SMITH	ALBERT	S
SMITH	ALBERT	W
SMITH	CLEMENT	H
SMITH	HARVEY	E
SMITH	IRENE	MARIE
SMITH	MARVIN	JEROME
SMITH	MAY	F
SMITH	RICHARD	
SMITH	ROBERT	WILSON
SOMMERS	BEVERLY	J
SOMRAK	DONALD	L
SORGE	LEONA	MAY
SPARKS	WALTON	E
SPRINGBORN	GEORGE	J
SPUZZILLO	MARY	
SPUZZILLO	MARY	
SPUZZILLO	MICHAEL	A
STAKICH	DANNY	
STAMM	JOHN	GEORGE

SWIMMING POOL, EUCLID BEACH, CLEVELAND, OHIO.

LAST	FIRST	MIDDLE
STANICIC	PAUL	
STANTON	ROBERT	B
STARINA	FLORENCE	
STAVASH	JOHN	C
STEGKEMPER	ADA	E
STELTER	HARRY	E
STEPP	CLARICE	P
STEPP	GENUS	W
STERBENZ	VIOLA	A
STEWART	JAMES	
STOHLMANN	CHARLES	R
STOKEL	WILLIAM	R
STONEBACK	HOWARD	DETWEILER
STRAUSS	DAVID	J
STRMOLE	ANGELA	
STROETER	RICHARD	
STUART	EUGENE	MILLER
STUART	JOHNNY	
STUART	VIC	
STUHLER	RAY	
STUMP	DONALD	
SULLIVAN	WILLIAM	
SUMMERS	ROBERT	B
SUMSKIS	CARL	P
SUPANICK	THOMAS	
SWEIGART	WILLIAM	R
SWOPE	BRUCE	
SWOPE	BURTICE	
SZABO	LOUIS	
TAYLOR	ARTHUR	LEE
TAYLOR	CALLOWAY	
TAYLOR	DAVE	GEORGE
TAYLOR	HATTIE	HEDSON
TAYLOR	IRENE	
TAYLOR	ROBERT	MCCLELLAND
TAYLOR	THOMAS	
TAYLOR	WAYNE	
TEAL	JOSEPH	E
TERLEP	JOHN	
THOMAS	BERTHA	L
THOMAS	JAMES	ROBERT
THOMAS	PAUL	EDWARD
THOMAS	STEVEN	RAY
THOMPSON	ROBERT	
TIZZANO	MARIA	
TOMASEK	AD	T
TOMITZ	GILBERT	JOHN
TOPPIN	NOLA	
TRANCHITO	LEO	CARL
TREGENNA	WILLIAM	A
TRESS	WILBERT	D
TRIVISON	CARRIE	
TRIVISON	CHARLES	A
TRIVISON	GRACE	
TRIVISONNO	MARY	
TROMBO	MICHAEL	
TROMBO	RONALD	
TRYON	JEROME	THOMAS
TUCKERMAN	JOHN	L
TUNQUIST	CHARLES	R
TUNQUIST	ROBERT	W
TUSHAR	RAY	
ULLOM	CLEVERAL	
VACCARIELLO	MARY	I
VALENTINO	MARY	
VAN NUIS	ALFONSO	COVAS
VERBY	FLORENCE	
VERBY	VIRGINIA	
VOELKER	MELVIN	D
VOGEL	ROBERT	
VOLLMER	JACK	MONROE
VOVK	HERMAN	J
WACKER	ANTOINETTE	
WACKER	RAYMOND	G
WACKER	WILLIAM	
WAGNER	FLORENCE	CALLAHAN
WAGNER JR	PETER	
WALLACE	JAMES	E
WALTHER	PAUL	
WARD	JAMES	M
WARNER	NORMAN	
WATERS	GEORGE	
WEBB	RACHEL	
WEBB	RICHARD	C
WEBBER	EUGENE	MADISON
WEBER	ARCHIBALD	AUGUST
WEBER	HARRIET	ETHEL
WEBER	JAMES	HOWARD
WEBER	JEAN	ANN
WEED	CHARLES	EARL
WEEKS	JOHN	
WHEATLEY	AUGUSTA	MAINE
WHITE	CLARENCE	W
WHITEHOUSE	FLORENCE	CONKLE
WHITEMAN	CHARLES	A
WHITEMAN	EVELYN	S
WILLIAMS	ALBERT	T
WILLIAMS	ARTHUR	E
WILLIAMS	EARL	ROGER
WILLIAMS	EDISON	GATES
WILLIAMS	GRACE	WYATT
WILLIAMS	JACK	
WILLIAMS	WALTER	DEWEY
WILSON	ANNA	RILEY
WILSON	FRANK	ALLEN
WILSON	HAROLD	
WILSON	JOHN	W
WILSON	MABEL	
WOLCOTT	HORACE	DAVID
WOOD	RUSSELL	LLOYD
WOOD	WILLIAM	F
WOODS	FRANK	HENRY
WOODWORTH	ROBERT	
WORTHINGTON	DORSEY	
WRAYNO	ROBERT	LEO
WUERTS	GEORGE	
YATES	JEANNETTE	L
YONKE	VICTOR	WILLIAM
YOUNG	FRANK	
YOUNG	MOSES	
YOUNG	ROBERTA	
ZALOKAR	THERESA	DORIS
ZALOZNIK	JAMES	E
ZAMBONI	MARCELLO	J
ZILLES	JAMES	MACK
ZIMMERMAN	NORMAN	L
ZIOLKOWSKI	LEO	
ZOLDAK	GEORGE	J

13:—THE POOL, EUCLID BEACH PARK, CLEVELAND, OHIO.

1946
EMPLOYEE ROSTER

LAST	FIRST	MIDDLE
ABBEY	JOSEPHINE	
ABBEY	MAE	L
ABE	DONALD	F
ADAMO	CASPER	PHILLIP
ADAMS	JOSEPH	
ADAMS	RAYMOND	C
ALBAUGH	DONALD	LEE
ALLEN	CLARENCE	
ALLEN	WALTER	HUGO
AMBROSE	JOSEPH	W
AMON	EUGENE	V
AMOR	JOHN	
ANCELL	RICHARD	
ANDERSON	EMMA	KING
ANDERSON	JOHN	ANDREW
ANGELERO	RAFFAELE	
ANGELORO	CARMELA	M
ARBOGAST	WILLIAM	
ATKINS	JOHN	
AXE	PAUL	E
AYALA	JOSEPH	
BACON JR	ERNEST	S
BAILEY	JOHN	FINUS
BAKER	CILLIUS	MOSE
BAKER	RONALD	
BALDNER	WILLIAM	
BALDWIN	DEAN	DEWITT
BALTERSHAT	ESTHER	W
BANDO	EMANUEL	A
BANSHAK	GERALD	W
BARBO	JOHN	
BARR	JAMES	H
BARRETT	CHARLES	DENNIS
BARRY	DOROTHY	MARIE
BARRY	HENRY	JOHN
BARRY	JOHN	CHARLES
BARRY	KENNETH	
BARTEL	EARLE	HOWARD
BARTLETT	SYLVESTER	
BARTO	ALOYSIUS	FRANK
BARZI	RICHARD	LOUIS
BATICH	MARTHA	
BAUER	ROSE	ANN
BAUMGARTNER	ARTHUR	
BAUS	HARLAN	LOUIS
BAUS	JAMES	W
BAZNIK	CHARLES	
BECHTOLD	LLOYD	R
BECK	EDITH	ADELLE
BEEKMAN	RICHARD	P
BELANGER	JOSEPH	
BELLET	ROBERT	L
BEMAN	HERBERT	H
BENEDICT	HOWARD	L
BENEDICT	JAMES	NELSON
BENEDICT	RALPH	ODELL
BENISEK	BERNARD	J
BENJAMIN	JACK	E
BENSCH	TOM	
BENZ	FRANKLIN	L
BERGOC	JOSEPH	J
BIBBS	RAYMOND	T
BICE	CORINNE	E
BIDELMAN	RICHARD	LEE
BILLENS	RAY	K
BILLENS	WILLIAM	FREDRICK
BITSKO	LOUIS	P
BLACK	EDWARD	
BLAIN	WILLIAM	H
BLAIR	JAMES	
BLANKENSHIP	WILLIAM	E
BLASE JR	GILBERT	E
BLEVINS	WILLIAM	B
BLISS	NETTIE	
BLOEDE	WILLIAM	CARL
BLUHM	JOHN	E
BOETTCHER	ROBERT	
BOGAN	CLARENE	
BOGAN	JOHN	E
BOJEC	JOHN	
BOLDIN	FRANCES	
BOOR	ROBERT	C
BORGA	FRED	
BORK	CARL	R
BOUFFARD	BART	
BOWEN	EARL	
BOWER	THOMAS	LEWIS
BOWHALL	ELMER	F
BOWHALL	HOWARD	
BOWINS	ROBERT	J
BOWLING	RONALD	L
BOWYER	LOIS	
BOYCE	ROBERT	
BOYD	ANNA	J
BOYD	PAUL	T
BRADFORD	JOHN	
BRAHMS	WILLIAM	A
BRAHMS	WILLIAM	H
BRANCHE	JAMES	R
BRANSON	JAMES	C
BRASCHWITZ	HAROLD	J
BRATOVICH	WILLIAM	
BRENNEMAN	DONALD	M
BREWTON	KILBY	
BRIGGS	THOMAS	LEE
BRODIE	ALBERT	T
BROOKINS	EDNA	BELLE
BROW	CHARLES	F
BROWN	RALPH	R
BROWN	RICHARD	A
BROZINA	FRANCES	
BRYANT	MARY	ALICE
BRYANT SR	THOMAS	EDWARD
BUCCILLI	ANDREW	
BUCKNER	CLYDE	E
BUDAN	FRANK	J
BURDEN	BERTHA	
BURNS	BYRON	C
BURTON	JAMES	
BURY	FRANK	
BUSCH	LAVERNE	M
BUTLER	JOSEPH	HACKNEY
BUTTNER	WILLIAM	R
CALLAGHAN	BEATRICE	BERNICE
CALLAGHAN	GEORGE	EDWARD
CALLALY	THOMAS	
CALLIGHEN	FRANCIS	
CALLIGHEN	MARION	A
CAMPBELL	ALEXANDER	ROBERT
CAMPBELL	FRED	
CAMPBELL	STANLEY	T
CAPUTO JR	ERNEST	
CARDOZA	ROBERT	R
CARLSON	ETHEL	
CARNES	ELNORA	
CARNEY	JOHN	P

LAST	FIRST	MIDDLE
CARRAN	ROBERT	L
CARROLL	CLARA	LEGGON
CARROLL	HAZELLE	BELLE
CARROLL	JAMES	J
CARROLL	ROY	SANFORD
CARSON	HAMILTON	
CARTER	RILEY	
CARTRIGHT	NORMAN	C
CENTA	FRANK	J
CERCEK	JOHN	LOUIS
CHAMBERLAIN	VERN	
CHAPEK	WOODY	
CHELSETH	H	K
CHEMICK	BETTY	
CHERRY	GEORGE	F
CHRISTIAN	HARRY	
CICIGOI	RAY	
CIRINO	ANGELINE	
CIRINO	ROCCO	
CIRINO	THERESA	M
CLARK	THOMAS	ARTHUR
CLATTERBUCK	FRANCIS	W
CLATTERBUCK	HARRY	L
CLEMENCE	AGNES	M
CLEMENCE	FRANCES	V
CLIFFORD	GEORGE	R
CLINE	ROBERT	
COCANOWER	PAUL	
COLEMAN	GEORGE	M
COLLINS	JAMES	F
COMERFORD	JOHN	RAYMOND
CONNAVINO	JOSEPHINE	
CONNELL	TOM	BENTLEY
CONNOLLY-HEUER	BETTY	JANE
CONTENTO	GUISEPPE	
CONWAY	HARRY	F
CONWAY	LENA	
COOKE	ERNEST	F
COPFER	MAE	
CORCELLI	DONALD	
CORRAT	SALVATORE	
CORRIGAN	HELEN	G
CORRIGAN	OMER	O
COWHARD	HATTIE	MAY
CRAPNELL	HAROLD	G
CRAPNELL	KATHERINE	
CRAWFORD	ALLYN	S
CRETER	ROBERT	D
CUDDIHY	ROBERT	H
CUNNINGHAM	FRANCIS	
CUNNINGHAM	JAMES	E
CUNNINGHAM	RAY	
CUNNINGHAM	THOMAS	F
CURRY	ANDREW	
CUTSHALL	OLIVE	M
CYGANSKI	RICHARD	J
CZERNICKI	LEONARD	

LAST	FIRST	MIDDLE
DANIELS	GEORGE	
DAVIES	T	EARLE
DAVIS	CHARLES	VANCE
DAVIS	GEORGE F	
DAVIS	KATIE	MAY
DAWSON	ROBERT	L
DAY	DON	
DAY	WILLIAM	RICHARD
DEAN	CECIL	
DEANER	JAMES	
DEAR	WILLIAM	
DELAMBO	FRANK	FRANCIS
DELLINGER	TOM	
DEPETRIS	EUGENE	
DERCOLE	ANTONETTE	
DHONDT	DANIEL	P
DICICCO	CONCETTA	
DICKEY	HERBERT	ROBERT
DICKSON	LILLIE	
DIRRMAN	RICHARD	C
DISANTIS	JOHN	A
DOBAY	JAMES	
DOESBURG	RALPH	HERBERT
DOLAN	RUTH	ANN
DOLENCE	STANLEY	
DONAHUE	LAURA	G
DONAHUE	LOIS	ANN
DONAHUE	MARY	ETHEL
DORMENDO	ANDY	
DOUGLAS	JAMES	
DOWDELL	EMMETT	JOSEPH
DOWNS	ROY	EDWARD
DRESSLER	H	C
DRESSLER	ROBERT	J
DRINKARD	OSCEOLA	K
DUBROSKY	PAT	C
DUCHARME	FORREST	
DUNCAN	ARCHIE	LEE
DUNS	THOMAS	E
DUNSFORD	NELLIE	C
DUTCHCOT	CELIA	
DUTCHCOT	EDWARD	OSCAR
DVORAK	AUGUSTA	P
DVORAK	MARGARET	
DWYER	MARY	FRANCES
DWYER	WESLEY	J
EARLY	JULIA	A
EATON	HOWARD	E
ECKSTROM	CARL	F
EDDY	BYRON	W
EGLEY	WILLIAM	
EMERICK	KENNETH	
EVERHARD	EILEEN T	A
FABIAN	ROBERT	T
FAIRFAX	WILLIS	L
FATICA	CHARLES	
FATICA	LORETA	
FEDELE	ANTHONY	
FEDELE	JOHN	PETER
FEDERICO	JOSEPHINE	
FELLOWS	THOMAS	F
FERRERI	GEORGE	
FIDDES	RAYMOND	JOHN
FIELD	RUBY	
FINERAN	MARY	BALL
FINK	DOROTHY	M
FIORILLI	NICK	T
FISCHER	MARGARET	ANNE
FLANIK	ROBERT	R
FLATE	NORMAN	
FLEMING	BRADY	
FLETCHER	EARL	G
FLETCHER	ELLIS	
FLORENCE	HAZEL	
FOGARTY	GENEVIEVE	MALADY
FOGEL	WILLIAM	CARL
FOLTZ	ROBERT	H

25:—BATHERS, EUCLID BEACH PARK, CLEVELAND, OHIO.

58

LAST	FIRST	MIDDLE
FORMACA	MARY	
FOX	CLARA	M
FOY	ANNA	
FRANCIS	DAVID	J
FRANCIS	WINIFRED	
FRANKOVICH	NICK	PAUL
FREDERICKS	FRANK	JOSEPH
FREDERICKS	NEIL	FRANCIS
FREHMEYER	RUSSELL	
FRIEL	RICHARD	DONALD
FROMBURG	ROSE	
FRY	ROBERT	C
GAL	MARTIN	
GALLAGHER	MICHAEL	
GAMBRILL JR	MERLE	C
GARDNER	JACK	R
GARLAND	FRANK	C
GARNER	MACK	
GARRETT JR	HUGH	W
GARRETT JR	WILLIAM	S
GASKER	HARRY	R
GEHRKE	EMIL	A
GEITHER	WILLIAM	P
GEORGE	JAMES	A
GERBER	WAYNE	O
GERNAT JR	WILLIAM	
GILL	ELSIE	OLIVIA
GILPIN	ROY	BIRL
GILSON	GLADYS	E
GIRARDI	ROSE	MARIE
GLASS	HENRY	D
GLASS	RUTH	
GLEASON	JOSEPH	A
GOBISKI	ALFRED	C
GOLDEN	JOSEPH	M
GOULET	JEANNETTE	L
GRANGER	MABLE	ETTA
GRAVES	JOHN	A
GREEN	CLAUDE	ELLSWORTH
GREENWAY	FRED	STUDER
GRIFFITH	HAROLD	F
GROSS	RICHARD	ARTHUR
GROSSMAN	NORMAN	
GROVE	KATHERINA	A
GRUENIG	HELEN	
GULICH	CAROLINE	
GUNDLING	JAMES	
HAGEN	PAUL	GERALD
HALBURDA	DOROTHY	
HAMMOND	RICHARD	E
HANNAN	THOMAS	J
HARFORD	JESS	A
HARRIS	ARTHUR	
HARRIS	JOHN	
HARROLD	FLORENCE	M
HARROLD	THADDEUS	JAMES
HART	ROBERT	LEO
HARTMAN	CARL	
HARTMANN	JOHN	
HARVEY	HENRY	
HARVEY	WILLIAM	J
HASTINGS	ROBERT	
HASTINGS	WILLIAM	MARTIN
HAWKINS	JAMES	FRANKLIN
HAWKINS	MARY	KATHLEEN
HAYES	WILLIS	JAMES
HAYTHER	LUCILLE	
HEADLEY	JOHN	ALBERT
HEATON	MAY	
HEISS	EDWARD	C
HERAGHTY	PAT	
HERRICK	JOHN	E
HILBRINK	WILLIAM	J
HILT	WILBUR	
HINTON	JOHN	SHEPARD
HOCHEVAR	RICHARD	J
HOFF	DOROTHY	IRENE
HOFFMAN	JOSEPH	E
HOFFMEYER	PAUL	
HOLLIS	WILFRED	
HOLLOWELL	ELIZABET	H
HOLMES	JOHN	M
HOOVER	ROLAND	A
HOPKINS	JOHN	J
HOPKINS JR	WILLIAM	F
HOTCHKISS	CHARLES	S
HOTCHKISS	GEORGE	C
HOTCHKISS	THEODORE	G
HOWARD	N	C
HOWARD	WILLIAM	JURTA
HUBBARD	THOMAS	
HUDSON	EDWIN	P
HUGHES	ELLA	M
HUMPHREY	THOMAS	J
HUNT	BLANCHE	LEAH
HUNTER	ROY	CHARLES
HURST	DAVID	J
HUTCHENS	ROBERT	M
JACKLITZ	ALBERT	
JANZ	JOHN	
JENKINS	DONALD	THOMAS
JENKINS	EVELYN	GABLE
JOHNSON	DOROTHY	IDA
JOHNSON	JAMES	GARY
JOHNSON	LESSIE	LEE
JOHNSON	THURA	W
JOHNSON	VIDA	F
JOHNSTON	CLIFFORD	BASIL
JOHNSTON	DONALD	HUMPHREY
JOHNSTON	THOMAS	D
JONES	DONALD	C
JONES	ELMO	ROBERT
JONES	JAY	DEE
JORDAN	CINQUES	E
JUDD	ESTIL	LEE
JUSKO	ROBERT	E
KACSMAR	ANTHONY	J
KAMINSKI	JOHN	J
KANALLY	MARGARET	
KANOWSKI	WILHELM	
KARTUNAVICH	JOHN	
KAYLOR	DOROTHY	C
KEHN	DANIEL	F
KEHN	GEORGE	
KEIFER	HARRIET	L
KELLY	DAISY	CURTIS
KENT	EDWARD	C
KERMODE	GEORGE	
KESSLER	AGNES	CECELIA
KEYES JR	JOSEPH	M
KIDD	JACK	HENRY
KIKOL	JOSEPH	

150 LAKE FRONT FROM PIER, EUCLID BEACH PARK, CLEVELAND, OHIO

LAST	FIRST	MIDDLE
KIKOLI	FRANK	
KIPLINGER	GEORGE	WILLIAM
KIRKBY	DAVID	L
KITT	BESTER	
KLANCHER	ROBERT	J
KLEBER	LEON	H
KLEMA	ANDREW	
KLEMPAN	EDWARD	
KLINE	ROBERT	SAUL
KLUGH	WILLENA	
KNAPIK	JOHN	
KOELLING	ELMER	F
KOLLAR	FRANCES	MARY
KOLLER	RONALD	H
KONCILJA	VICTOR	H
KOPACZ	LAURA	
KOPINA	JOSEPH	E
KORB	RUDOLPH	C
KORENCIC	ROBERT	
KOSMETOS	MICHAEL	
KOZIATEK	JEROME	
KOZLEVCAR	STANLEY	J
KRASOVEC	ROBERT	T
KRECH	MIKE	
KRIVDO	JEROME	M
KRIVDO	LUKE	J
KRIZ	CATHERINE	
KROME	FRED	C
KRUCHEK	RICHARD	
KRUCK	EDWARD	
KRULL	HERMAN	R
KRUSELL	BEN	F
KUBIAK	LEONARD	S
KUCHIRCHUK	FRANK	F
KUEHL	ARTHUR	WALTER
KUEHM	LEWIS	JAMES
KUHN	CHARLES	RICHARD
KUSHEN	LAVERNE	
LACONTE	FRANK	
LANDIS	NORRIS	JUDSON
LANDON	RICHARD	R
LANG	JAMES	VINCENT
LANGE	JOHN	FERDINAND
LARICCIA	FRED	
LARICCIA	PETER	
LARNER	MURRAY	S
LAWSON	OTIS	D
LECHOWICZ	BENEDICT	
LEONARD	STELLA	O
LEVER	RAYMOND	V
LEVY	HAROLD	
LEWIS	JACK	K
LINDQUIST	MYRTLE	A
LISKAY	ELMER	J
LISTER	GEORGE	W
LIVINGSTONE	BERTHA	ROSS
LJUBI	ANTHONY	J
LOCKE	HUBERT	LEE
LONGFIELD	TESS	EVANS
LONGO	ELEANORE	
LORENZO	MARIA	
LUCE	ANTONIO	
LUCE	GORDON	
LUCKAY	RICHARD	
LUND	ARTHUR	F
LUNDBLAD	ROBERT	A
LUPIS	FRANK	F
LUPTON	JAMES	ALFRED
LUPTON	ROBERT	
LUPTON	WILMA	MARIE
LUX	ROBERT	
LYNCH	FREDERICK	F
LYNCH	R	J
MAC DONALD	JOHN	
MACGRUDER	ELMER	D
MACINE	LOUIS	
MACKENZIE	ELLEN	
MACKIE	LARRIE	
MAGRUDER	ELMER	D
MALLOY	ELIZABETH	A
MANCINE	BENJAMIN	J
MANCINI	JOSEPHINE	
MANCINI	SAM	L
MANDAU	RUDOLPH	
MANNING	DOUGLAS	L
MARCOGLIESE	MADDALENA	
MARNAY	ESTHER	
MARSH	ROSE	A
MARSHALL	CADDIE	
MARTIN	LARRY	J
MARTLOCK	KEN	
MARUCCI	JACOB	
MASON	KATHERINE	T
MATHEWS	BEN	
MATYJASIK	WILLIAM	J
MAURER	EDGAR	C
MAXIM	EMILIA	
MAXIM	JOSEPH	JOHN
MAXWELL	CLAUDE	R
MAY	DICK	OTTO
MAZANEC	EMIL	E
MCARTHUR	CHARLES	W
MCBANE	THOMAS	G
MCCABE	JAMES	J
MCCAFFREY	MARY	
MCCAHAN	KITTIE	
MCCLAIN	HELEN	N
MCCLURG	ROBERT	
MCDONALD	ANGIS	M
MCDONOUGH	DANIEL	
MCELWEE	JOHN	T
MCGOVERN	JOE	WILLIAM
MCGREAL	JOHN	F
MCILRATH	IDA	E
MCKEE	JAMES	A
MCKINLEY	JAMES	H
MCLAUGHLIN	FRANCES	
MCLAUGHLIN	PATRICK	
MCMILLAN	DON	C
MCNEELY	FLORENCE	
MCNIECE	RAYMOND	E
MCWATTY	RICHARD	
MEAD II	JOHN	M
MEDVES	EDWARD	
MEDVES	HELEN	R
MEGLICH	MATTHEW	
MEHOZONEK	EDWARD	
MEHOZONEK	VICTOR	
MENG	CARL	J
MERHAR	ALBERT	J
MERHAR	JOSEPH	
MERSEK	RAYMOND	
METCALF	HARRY	C

21:—Fountain and Pier, Euclid Beach Park, Cleveland, Ohio.

LAST	FIRST	MIDDLE
METZ	JOHN	RICHARD
MEYER	LOUIS	
MICHAELSON	DAN	WESLEY
MICHEL	GORDON	F
MILES	HARVEY	
MILLER	CARL	JOHN
MILLER	DONALD	RICHARD
MILLER	JOHN	
MILLER	JOHN	LONG
MILLER	JOSEPHINE	
MILLER	RONALD	G
MILLIGAN	KERMIT	M
MILLIKIN	HARRY	F
MILLS	ARTHUR	F
MILLS	EDITH	C
MOLE	ROBERT	A
MOLNAR	JAMES	AUGUST
MOLTZAN	MICHAEL	
MOORE	FENTON	L
MOORE	MAY	C
MOREL	EDDIE	
MOREL	SYLVESTER	
MORGAN	HUGH	
MORIARTY	JAMES	J
MORRIS	ROBERT	
MOSIER	FREDA	V
MOTSCH	GERALD	A
MOUGHAN	WILLIAM	
MOYERS	CLYDE	BERT
MUELLER	LOUIS	W
MULVIHILL	THOMAS	
MURPHY	ALFRED	JOHN
MURPHY	DENNIS	
MURPHY	DONALD	J
MURPHY	THOMAS	F
MUSSELMAN	AMY	C
MYERS	LEO	E
NAGY	ELEANOR	
NEFF	GEORGE	
NEGRELLI	JENNIE	M
NEUBAUER	KATHRYN	MELVIN
NOLAN	JOHN	
NOLES	ROBERT	
NYSTROM	MAURICE	M
OAKLEY	ADDIE	ELLEN
OBRIEN	JACK	
OBRIEN	JOHN	
OBRIEN	TERESA	L
OCHS	ANN	PHYLLIS
ODLAZEK	THOMAS	
OLIVER	ANDREW	
OLSON	EARL	N
OLSON	HERB	E
OMERSA	JOHN	
OPALICH	DANIEL	
OREBAUGH	MEDFORD	D
OSTRUNN	ROBERT	E
OTASEK	CHARLES	E
OTASEK	ROBERT	ADAM
OTHBERG	KENNETH	CARL
OWENS	JAMES	CALVIN
OWENS	VAN	
OXER	ORLANDO	MONROE
PADDEN	WILLIAM	
PALEVICH	LEON	J
PALLEN	RONALD	
PARENT	EDMUND	L
PARKER	WILLIAM	
PATALINO	ANTHONY	P
PATTON	LAWRENCE	J
PAWLAK	SALLY	
PAYTON	HOWARD	
PECK	DONALD	
PECK	GEORGE	L
PELL	KELSO	
PELTON	FLOYD	
PENK	CARL	W
PERME	JOHN	E
PERME	JOSEPHINE	C
PERROTT	RONALD	C
PERROTTI	CAROLINE	
PERUSEK	EDWARD	
PETERS	ARNOLD	L
PETERS	ROSE	
PHILLIPS	ERWIN	C
PIKE	JERRY	
PILWALLIS	CLARENCE	
PINKARD	LEMMET	
PITTMAN	KENNETH	
PLACK	ETHEL	M
PLANTSCH	MICHAEL	
POLCYN	ROY	E
POLLOCK	JAMES	S
POWELL	ALFRED	L
POWELL	JOHN	HENRY
PRECHTL	MARY	JANE
PRICE	EDWARD O	
PRICE	FLOYD	L
PRICE	GRACE	B
PRICE	HARRY	R
PRICE	HOWARD	G
PRICE	LEO	E
PRICE	MARK	THOMAS
PRICE	WENDELL	D
PUDIMAT	OLGA	M
PUGEL	LOUIS	FRED
PUGH	MINNIE	F
PULLAR	ALBERT	A
PULVER	JAMES	
QUAYLE	ADA	MAE
QUIRK	LESLIE	
RAGGETS	EDWARD	J
RAIMONDO	ALFRED	
RAINBOW	ROBERT	LEE
RANALLO	RAY	
RANCOURT	RICHARD	
RANCOURT	ROBERT	H
RANKER	HENRY	G
RAYER	WILLIAM	
RAYMOND	JOSEPH	JAMES
REDFIELD	CUYLER	I
REECE JR	LOUIS	
REILLY	FRANK	
REILLY	MICHAEL	
REIN	BETTY	JANE
REINER	JOSEPH	
REINHARD	GEORGE	ANDREW
REINHARD	GEORGE	MARTIN
REPLOGLE	CATHERINE	
REPP	NORMAN	
RETZER	EVELYN	
REW	DONALD	

View of Pool and Bathing Beach at Euclid Park, Cleveland, Ohio — D-25

LAST	FIRST	MIDDLE
REYNOLDS	FLORENCE	
RICARDSON	EDWIN	
RIEDEL	LAWRENCE	W
RILEY	MARGARET	ADELE
RITTER	JAMES	PHILIP
RITTER	ROBERT	
RITTER	ROY	R
ROBBINS	KATHLYN	TEEPLE
ROBINSON	RALPH	
ROBINSON	ROBERT	LEE
RODERICK	ALICE	M
RONIGER	H	E
ROSA	ANTHONY	C
ROSE	EUGENE	
ROSSOW	MABLE	
ROTH	THOMAS	HERMAN
ROTHMANN	EDWIN	
ROWAN	WILLIAM	
ROWE	KENNETH	D
RUPPERSBERG	DONALD	P
RUSSELL	GEORGE	D
RUZICKA	MARY	ROSE
RYDMAN	ELIZABETH	
SADOWSKI	MARY	
SAEFKOW	JOHN	
SANDS	JAMES	J
SATOR	ALICE	MARGARET
SAUNDERS	SPENCER	
SAYWELL JR	ED	
SBROCCO	SEBASTIANO	
SCALERO	VINCENT	N
SCHEUY	CHARLES	B
SCHILL	LLOYD	EDGAR
SCHIRRA	JOHN	L
SCHLANGER	JACK	
SCOTT	DAVID	HUMPHREY
SCOTT	EVERETT	PALMER
SCOTT	RICHARD	
SEAMAN	BENJAMIN	W
SEAMAN	JAMES	J
SEVER	JENNIE	
SHAMPAY	JEAN	
SHANNON	HARRIS	COOPER
SHANNON	HARVEY	PAGE
SHANNON	MARGARET	A
SHEA	PATRICK	J
SHEAFFER	ALBERT	
SHELKO	ROBERT	J
SHILLIDAY	EVERETT	P
SHOEMAKER	KENNETH	W
SHRINER	WILLIAM	C
SIDLEY	RUSSELL	H
SIFLING	JOHN	
SILVEROLI	JOHN	JAMES
SIMMS	EDWARD	
SIMPSON	DEAN	N

LAST	FIRST	MIDDLE
SINCLAIR	JAMES	
SIVIK	DONALD	
SKITZKI	WALLACE	J
SKLAR	MIKE	THOMAS
SKOKALSKI	FRANCES	
SLUSSER	WILLIAM	RAYMOND
SMEKEL	RICHARD	M
SMICIKLAS	JOSEPH	
SMICIKLAS	NICHOLAS	W
SMITH	ALLEN	ROBERT
SMITH	ELMA	F
SMITH	HARVEY	E
SMITH	IRENE	MARIE
SMITH	MAY	F
SMITH	RICHARD	C
SMITH	ROBERT	L
SMITH	ROBERT	WILLSON
SMITH	WILLIAM	ARTHUR
SOMMERS	ERNESTINE	
SOMRAK	ROBERT	ALLEN
SORMUNEN	HENRY	A
SPARKS	WALTON	E
SPARLIK	FRANCES	A
SPENZER	EUGENE	G
SPRING	JOE	
SPUZZILLO	MARY	
ST CLAIR	JULIUS	J
STAGNIUNAS JR	JOSEPH	A
STAMM	JOHN	GEORGE
STANICIC	PAUL	
STARINA	FLORENCE	
STEFFEN	ALFRED	DEAN
STEFOROY	ANNE	
STEGKEMPER	ADA	E
STELTER	ELIZABETH	H
STELTER	HARRY	E
STEPP	GENUS	W
STEVENS	JOSEPH	A
STEWART	RAYMOND	R
STEWART	WILLIAM	
STEYER	ALLEN	ROBERT
STOHLMANN	CHARLES	
STONE	FRANCISO	
STONEBACK	HOWARD	DETWEILER
STRAIN	WILLIAM	H
STRONG	ELEANOR	
STRONG	HAROLD	EMIL
STRUBLE	PAUL	F
STUART	EUGENE	MILLER
STUART	JAMES	MILLER
STUART	JOHN	
STUART	VIC	
STUMP	DONALD	
STUMPF	CORA	MARIE
SULLIVAN	EVERETT	L
SULLIVAN	SOPHIE	
SULLIVAN	WILLIAM	
SUMSKIS	CARL	
SUNAGEL	EDWARD	R
SUNAGEL	ROBERT	DALE
SVARPA	STANLEY	
SVET	TONY	
SVETLIK	LLOYD	
SWOPE	BRUCE	
SWOPE	BURTRICE	
SYABO	LEWIS	
TAYLOR	EVANS	L
TAYLOR	IRENE	
TAYLOR	THOMAS	
TAYLOR	WAYNE	
TAYLOR	WILLIAM	K
TERLEP	EDWARD	
TERLEP	JOHN	
TERRELL	ROBERT	
THOMAS	BERTHA	L
THOMAS	PAUL	EDWARD
THOMAS	STEVEN	RAY

42:—SURPRISE HOUSE, EUCLID BEACH PARK, CLEVELAND, OHIO.

Photo by Van Fisher

LAST	FIRST	MIDDLE
THOMAS	WILMA	S
THOMPSON	GEORGE	
THOREN	RALPH	
THORNWELL	BESSIE	M
THUNHURST	EDYTHE	
TIMPERIO	ANTHONY	A
TIMPERIO	NICK	JACK
TOLER	ALBERT	H
TOLHURST	FOSTER	J
TOMITZ	GILBERT	JOHN
TOTH	GEORGE	T
TRANCHITO	LEO	CARL
TRIVISON	CARRIE	
TRIVISON	CHARLES	A
TRIVISON	GRACE	
TRIVISONNO	JOHN	D
TRIVISONNO	MARY	
TROMBO	FLORENCE	E
TROMBO	MICHAEL	
TROMBO	RONALD	
TUCHOLSKI	GEORGE	J
TUNQUIST	CHARLES	R
TURNER	PAUL	
TUTAY	MAE	
TWEED	MILTON	IVAN
TYSON	MARION	JEAN
ULLOM	CLEVERAL	
VACCARIELLO	MARY	I
VAGO	JULIUS	
VALE	GLORIA	
VALE	WILBUR	E
VAN COVE	ALEX	
VAN NUIS	ALFONSO	COVAS
VANDERZYDEN	ANN	
VEHOVEC	FRANK	
VENCL	JOHN	
VESCUSO	BENNY	
VIDMAR	TILLIE	GRETA
VITEZ	JOHN	
VLACH	GEORGE	
VLANDOWSKI	RAYMOND	A
VOGEL	ROBERT	
VOLLMER	JACK	MONROE
WACKER	WILLIAM	
WADE	JEAN	MARIE
WAGNER	FLORENCE	CALLAHAN
WAGNER	RALPH	J
WAGNER	ROY	H
WAGNER JR	PETER	
WALK	KEITH	AMOS
WALKER	WILLIAM	CLYDE
WALL JR	HAROLD	E
WALLACE	JAMES	E
WALLSTROM	ROLAND	
WALTERS	RICHARD	R
WARNER	NORMAN	
WASHEL	ANDREW	
WATSON	WILLIAM	R
WAYLES III	WILLIAM	
WEBB	RACHEL	
WEBBER	EUGENE	MADISON
WEBER	ARCHIBALD	AUGUST
WEED	CHARLES	EARL
WEIMER	GEORGE	W
WELKER	PAUL	ALBERT
WENBERG	HUGO	
WERTZ	DONALD	M
WHEATLEY	AUGUSTA	MAINE
WHITEHOUSE	FLORENCE	CONKLE
WHITEMAN	EVELYN	S
WHITLOW	ROBERT	CARL
WHYTE	ERMA	W
WICKS	GERALDINE	
WILL	PAUL	JOHN
WILLIAMS	ARTHUR	E
WILLIAMS	EARL	ROGER
WILLIAMS	EDISON	GATES
WILLIAMS	GRACE	WYATT
WILLIAMS	JACK	
WILLIAMS	MILDRED	AUGUSTA
WILLIAMS	WALTER	DEWEY
WILSON	HAROLD	
WILSON	LLOYD	PAUL
WILSON	RICHARD	G
WINKLE	IDA	M
WINKLE	WILLIAM	E
WODA	ELMER	J
WOOD	EARL	D
WOOD	JAMES	HENRY
WOOD	RUSSELL	LLOYD
WOODS	FRANK	HENRY
WRAYNO	ROBERT	LEO
YONKE	VICTOR	WILLIAM
YOST	MORRIS	R
YOUNG	MOSES	
YOUNG	ROBERTA	
YUSHKA	ALBERT	P
ZABUKOVEC	HENRY	
ZAK	EUGENE	EDWARD
ZAKRAJSEK	JOHN	D
ZAPPIA	STEVE	
ZIBERT	RUDOLPH	
ZIMMERMAN	F	E
ZOLDAK	GEORGE	J
ZULEWSKI	ALFONSE	
ZUPANCIC	FRANK	

WEDNESDAY, AUGUST 14, 1946

CORNELIA CURTISS WRITES OF:

Doris Humphrey of Euclid Beach

Real Enthusiasm for Popcorn Characterizes Park Founder's Granddaughter

Doris Humphrey would rather eat popcorn at Euclid Beach than a fancy supper in a night club. She prefers dancing in the casino at the Beach, to waltzing at a country club.

When people say, "Don't you ever take a vacation?" she answers, "The whole summer is a vacation for me."

In short, Doris who pinch hits at times as public relations department for her family's enterprise, is the Beach's best publicity agent every waking minute.

As she herself says, she was brought up in the tradition set by her late grandfather, founder of the amusement park. He believed in clean entertainment, with "nothing to depress or demoralize."

PICKS GOLF, SKATING FOR RECREATION

She has always lived an active life. When she isn't around the park, she is usually out playing golf. In the winter, figure skating is her pastime.

Doris devotes mornings to office work and knows all about the way the wheels revolve to keep the park pastimes moving. She took a business course at Katharine Gibbs School after finishing at Smith College so she had a good basis for a job at the Humphrey office. Business fascinates her and she would like to teach courses in the subject if she could check out of them by April.

That's when the family begins actively to plan for the next season, though it's rather their main interest all the year.

Her parents, Mr. and Mrs. Harvey J. Humphrey, have a home on the eastern fringe of the park, overlooking Lake Erie and Doris grew up there. The Beach has never lost its lure for her. Probably the best patron the popcorn stand has, she admits she eats the fluffy white kernals constantly.

EACH VISITOR A CHALLENGE

A visitor is a challenge. Rides and the new thrillers she shows off with pride and enthusiasm. I'll never know just how I avoided at least a trip on the merry-go-round, for Doris was inspired to take me on all the chutes and slides. We compromised by riding on the small train which meanders around the grounds and I listened to Doris tell of life as the daughter of Euclid Beach Park. We nibbled popcorn and ice cream cones.

A date with her usually turns out to be right there, no matter whether the masculine side of the twosome may have considered a whirl at the Terrace Room or the Bronze Room. He is apt to find himself, instead, imbibing a frozen custard at the park, rather than a frozen daiquari at a night spot.

This week she is taking a little time off to supervise registration at the Women's Western Amateur Golf Tournament at the Country Club. As for her own game, her handicap is 15 and she has decided not to worry about lowering it. When she plays it's for fun, not to aim for championship honors.

Doris loves to whip up cakes and turns out two or three each week. Usually they are white cakes, with several layers, "because my family won't eat any other kind."

The tricky business of lifting a cake cleanly from its tin has been conquered by this en erprising cook. She cuts wax paper not only large enough to cover her cake pan, but to overlap it with "ears." "Then you just take the cake out," she told me.

Once she thought she might try to sell the idea to a wax paper company but gave it up. "I decided not to bother," she concluded.

Saturday evening she played hostess to all the women taking part in this week's tournament. Everybody was "tagged" and given tickets, and, as the saying goes, a good time was had by all—and certainly it was a novel evening for most of them who would have vowed themselves beyond the carrousel age.

Doris never does things by halves so she is not only a faithful member of Fourth Church of Christ, Scientist, but secretary of the Sunday school.

One of her accomplishments is being able to keep records and sets of books and her church and golf friends recognize this. Doris casually points to the fact that it stems from her Beach work. In fact, you can't get her far from the subject at any time.

DORIS HUMPHREY

First permanent popcorn stand at Euclid Beach Park.

Early Park Managers. (Note original oval windows in the log cabin.)

1947
EMPLOYEE ROSTER

LAST	FIRST	MIDDLE
ABBEY	JOSEPHINE	
ABBEY	MAE	L
ADAMOWICZ	SALLY	
AKINS	BLAND	
ALBAUGH	DANIEL	C
ALBRIGHT	MARTIN	
ALMASHY	IDA	MAE
AMOR	JOHN	
ANCELL	RICHARD	
ANDERSON	JOHN	ANDREW
ANDREUCCI	FRED	
ANGELERO	RAFFAELE	
ANGELORO	CARMELA	M
ARNDT	PAUL	C
ARTHUR	IDA	MAE
ASHLEY	JAMES	D
ATKINS	JOHN	
AUSTIN	HUGH	B
AXE	PAUL	E
AZMAN	JOHN	
BAKER	CILLIUS	MOSE
BAKER	JOHN	R
BAKER	RONALD	
BALDWIN	DEAN	DEWITT
BALTERSHAT	ESTHER	W
BALTZ	PAUL	H
BARDO	JOHN	
BARRY	DOROTHY	MARIE
BARRY	HENRY	JOHN
BARRY	KENNETH	
BARTO	ALOYSIUS	FRANK
BATIC	MARIA	
BAZNIK	CHARLES	
BEHREND	JAMES	E
BENCIN	CHARLES	
BENEDICT	HOWARD	L
BENEDICT	JAMES	NELSON
BENEDICT	RALPH	ODELL
BENES	DALE	C
BENJAMIN	IVAN	E
BENTLER JR	JOSEPH	J
BENZ	FRANKLIN	L
BENZ	LEROY	H
BERAN	RAY	A
BERARDINELLI	ANGELINE	
BERGOC	JOSEPH	J
BETZ	JOHN	M
BETZ	LOUISE	ELLA
BIDELMAN	RICHARD	LEE
BILLENS	RAY	K
BLACKWELL	DOROTHY	E
BLACKWELL	JACK	
BLAKEMORE	WILLIAM	
BLEAM	LULA	ELSIE
BLOEDE	WILLIAM	CARL
BLUHM	JOHN	E
BOETTCHER	IRENE	A
BOLAN	JOHN	A
BOUFFARD	BART	
BOWEN	EARL	
BOWER	THOMAS	LEWIS
BOWHALL	ELMER	F
BOWHALL	HOWARD	
BOYCE	ROBERT	
BRADFORD	JOHN	
BRADLEY	LONNIE	
BRASCHWITZ	HAROLD	J
BRATTON	HARRY	H
BRAUNLICH	WILBUR	K
BRIGGS	THOMAS	LEE
BROOKINS	EDNA	BELLE
BROOKS	EARL	JOHN
BROWN	AMANDA	
BROWN	LLOYD	
BROWN	RALPH	R
BROZINA	FRANCES	
BRUNNER	CHARLES	JERRY
BRYANT JR	THOMAS	L
BRYANT SR	THOMAS	EDWARD
BUCCILLI	ANDREW	
BUICK	THEODORE	
BURDEN	BERTHA	
BURGESS	ALBERT	S
BURGESS	HARRY	J
BURTON	JAMES	
BUTLER	JOSEPH	HACKNEY
BYINGTON	HOWARD	H
CALLAGHAN	BEATRICE	BERNICE
CALLAGHAN	ROBERT	
CALLIGHEN	FRANCIS	
CALLIGHEN	MARION	A
CAMPBELL	ALEXANDER	ROBERT
CAMPBELL	FRED	
CAMPEAU	JOE	
CAPPS	ALBERT	DAVID
CAPUTO JR	ERNEST	
CARETTI	FREDERICK	C
CARICO	ANNIE	R
CARNEY	JOHN	P
CARRAHER	LEO	F
CARROLL	CLARA	LEGGON
CARROLL	GERTRUDE	M
CARROLL	HAZELLE	BELLE
CARROLL	JAY	F
CARROLL	ROY	SANFORD
CARROLL	URSULA	F
CARSON	HAMILTON	
CELESTE	MARY	
CHAMBERLAIN	VERN	
CHELSETH	H	K
CHEMICK	BETTY	
CHRISTOPHER	JOSEPH	
CIASULLO	EUGENE	J
CIPRA	ROBERT	J
CIRINO	ROCCO	
CIRINO	THERESA	M
CLARK	CASS	O
CLATTERBUCK	FRANCIS	W
CLATTERBUCK	HARRY	L
CLEMENCE	FRANCES	V
CLIFFORD	GEORGE	R
COCANOWER	PAUL	
COLLINS	JAMES	F
COMERFORD	EVELYN	
COMYNS	ARNOLD	T
CONN	PAUL	J
CONN	THOMAS	V
CONNAVINO	JOSEPHINE	
CONNICK	DONALD	J
CONNOLLY HEUER	BETTY	JANE
CONTENTO	GUISEPPE	
CONWAY	HARRY	F
CONWAY	LENA	
COOPER	GEORGE	R
CORRIGAN	HELEN	G
CORRIGAN	OMER	O

LAST	FIRST	MIDDLE
COWAN	JOHN	D
COWHARD	HATTIE	MAY
CRABTREE	JAMES	L
CRAPNELL	HAROLD	G
CRETER	ROBERT	D
CRITTENTON	WILLIE	
CROWELL	WILLIAM	R
CRUMLEY	JACK	F
CUDDIHY	ROBERT	H
CUNNINGHAM	RAY	
CURLEY	FRANK	L
CURRY	ANDREW	
CUTSHALL	B	MAXINE
CUTSHALL	OLIVE	M
DANIEL	CALVIN	P
DAUPHIN	VIRGINIA	
DAVIES	THOMAS	J
DAVIS	KATIE	MAY
DELAMBO	FRANK	FRANCIS
DELCALZO	VITO	A
DENISON	DAVID	V
DEPETRIS	EUGENE	
DERCOLE	ANTONETTE	
DESAN	ANTHONY	M
DICICCO	CONCETTA	
DICKEY	HERBERT	ROBERT
DONAHUE	LOIS	ANN
DONAHUE	WILLIAM	J
DOUGLAS	JAMES	
DOWDELL	EMMETT	JOSEPH
DRESSLER	H	C
DRESSLER	ROBERT	J
DRISCOLL	THOMAS	P
DUBROSKY	PAT	C
DUCHARME	FORREST	
DUDZIC	WALTER	
DUFFY	JEAN	
DVORAK	MARGARET	
DWYER	MARY	FRANCES
DWYER	WESLEY	J
DYANZIO	NANCY	JANE
EARLY	JULIA	A
EBERHARD	EILEEN T	A
EDDY	BYRON	W
EVANS	WILLIAM	B
FATICA	CHARLES	
FEDERICO	JOSEPHINE	
FELLOWS	THOMAS	F
FERRELL	MOSES	
FERRERI	GEORGE	
FERRIS	GERTRUDE	
FIEDLER	KENNETH	L
FINERAN	MARY	BALL
FINK	DOROTHY	M
FOGARTY	GENEVIEVE	MALADY
FOX	CLARA	M
FRANCIS	WINIFRED	
FRANKFORD	HARRY	L
FRANKOVICH	NICK	PAUL
FREDERICKS	FRANK	JOSEPH
FROHMBURG	ROSE	
FULLER	HARRY	
GALL	JOSEPH	
GALLAGHER	MICHAEL	
GALLEY	JACK	W
GANDEE	GUY	
GARDNER	JACK	R
GARDNER	JOSEPH	D
GASTER	EUGENE	
GEDULDIG	RAYMOND	
GEISWEIDT	DONALD	L
GENTILE	JAMES	A
GENTRY	HOBERT	
GEORGE	JAMES	A
GIBSON	OLLIE	J
GILL	ELSIE	OLIVIA
GILL	RUTH	S
GILMORE	RALPH	E
GLASS	HENRY	D
GLASS	RUTH	
GOLDEN	JOSEPH	M
GORHAM	JAY	M
GORMAN	JEAN	M
GOULET	JEANNETTE	L
GRANGER	MABLE	ETTA
GREEN	CLAUDE	ELLSWORTH
GREEN	GERALD	C
GREENWAY	FRED	STUDER
GREGER	ELMER	C
GREGORY JR	ROBERT	
GRIM	WENDEL	
GROVER	KATHERINA	A
GRUMBACH	NICK	
HACKERT	JEAN	
HAGEDORN	JACK	
HALBURDA	DOROTHY	
HAMMOND	ROBERT	A
HANN	HERMAN	H
HAPP JR	JOHN	W
HARFORD	JESS	A
HARRINGTON	JEROME	
HARRIS	MATTHEW	
HARRIS	WILLIAM	
HARRISON	MICHAEL	F
HARROLD	A	THOS
HARROLD	CHRISTOPHER	
HARROLD	FLORENCE	M
HARTMAN	CARL	
HAUCK	GEORGE	WILLIAM
HAUDE JR	OSCAR	
HAWKINS	HENRY	
HAYES	WILLIS	JAMES
HAYTHER	LUCILLE	
HEATON	MAY	
HEISS	EDWARD	C
HELM	GENE	
HERRICK	JOHN	E
HINDS JR	ROBERT	
HINTZ	ROBERT	J
HOCHEVAR	RICHARD	J
HOFF	ALICE	A
HOFF	DOROTHY	IRENE
HOFFMAN	RONALD	C
HOFFMAN	WILBERT	
HOGG	WILLIAM	R
HOWARD	WILLIAM	JURIA
HUDSON	EDWIN	P
HUGHES	ELLA	M
HUNT	BLANCHE	LEAH
HUNTER	ROY	CHARLES
ISABELLA	DOROTHY	

43:—UP IN THE CLOUDS, EUCLID BEACH PARK, CLEVELAND, OHIO.

Photo by Van Fisher

LAST	FIRST	MIDDLE
JACKLITZ	ALBERT	
JAMES	JESSE	
JANZ	JOHN	
JAROS	WALTER	C
JASENSKY	STANLEYR	
JEFFERSON	RICHARD	
JENKINS	EVELYN	GABLE
JENNE	CHARLES	
JOHNSON	DONALD	LEE
JOHNSON	DOROTHY	IDA
JOHNSON	HUBERT	C
JOHNSON	JAMES	GARY
JOHNSON	JOHN	
JOHNSON	ROBERT	J
JOHNSON	RONNIE	H
JOHNSON	THURA	W
JOHNSON	VIDA	F
JOHNSON	WILLIAM	
JOHNSTON	CLIFFORD	BASIL
JOHNSTON	DONALD	HUMPHREY
JOHNSTON	ROBERT	H
JONES	THOMAS	A
JORDENS	HERBERT	
JUDD	ESTIL	LEE
KACZYNSKI	EDWARD	J
KAMINSKI	JOHN	J
KARLIN	PAUL	W
KASTELIC	RUDOLPH	J
KEENEY	NORA	
KEIFFER	CHARLES	J
KELLY	DAISY	CURTIS
KELLY	JAMES	THOMAS
KENNEDY	HERMAN	O
KENNEDY	RICHARD	W
KIKOLI	FRANK	
KIRKBY	DAVID	L
KLAAS	WILLIAM	PAUL
KLEIS	ELIZABETH	
KLEIS	THOMAS	H
KLEMENTS	BERNARD	J
KLEMPAN	EDWARD	
KNUCKLES	WILLIAM	C
KOELLING	ELMER	F
KOLLER	RONALD	H
KRAUSE	HARRY	
KREVES	EDWARD	J
KRIVDO	LUKE	J
KRIZETI	GEORGE	
KRULL	HERMAN	R
KUEBLER	RICHARD	H
KULL	RONALD	R
KURPIEWSKI	GEORGE	
LACONTE	FRANK	
LANG	H	WILLIAM
LAPP	LEONARD	JAMES
LARICCIA	FRED	
LARICCIA	PETER	
LAWYER	JAMES	D
LEE	WILLIAM	EDGAR
LEMMO	RICHARD	P
LEWALLEN	JAMES	
LIPOVSKY	STEVE	G
LIVINGSTONE	BERTHA	ROSS
LIVINGTON	S	H
LOCKE	HUBERT	LEE
LONGFIELD	TESS	EVANS
LONGO	ELEANORE	
LORENZO	MARIA	
LUCIANO	DORIS	M
LUCKAY	RICHARD	
LUKAS	FRANK	J
LUKS	DONALD	J
LUNDBLAD	ROBERT	A
LUPTON	ROBERT	
MAC DONALD	JOHN	
MAC DONALD	LENA	
MACHAMER	CHAS	ROBERT
MACHAMER	JOHN	Q
MACKENZIE	ELLEN	
MACKIN	NANCY	JAYNE
MALONEY	THOMAS	J
MANCINE	LOUIS	
MANDAU	RUDOLPH	
MANDEL	IDA	
MANLEY	EDWARD	B
MANNING	DOUGLAS	L
MARCOGLIESE	MADDALENA	
MARINO	DOMINIC	
MARINSEK	ANTON	
MARSHALL	CADDIE	
MARTIN	DONALD	C
MARTIN	ROBERT	JOHN
MARTIN JR	WILLIAM	H
MARTINS	RALPH	A
MARTINS JR	WALTER	F
MASON	ALLEN	H
MATESIC	EILEEN	B
MATTESON	ROBERT	D
MCBANE	THOMAS	G
MCCABE	JAMES	J
MCCLAIN	HELEN	N
MCCLURG	ROBERT	
MCDONALD	ANGELINE	
MCDONALD	ANGIS	M
MCGREGOR	JOHN	J
MCGURER	ROY	ALLEN
MCILRATH	IDA	E
MCKINLEY	JAMES	H
MCLAUGHLIN	FRANCES	
MCLAUGHLIN	PATRICK	
MCMAHAN	ROBERT	
MCMANAMON	RUTH	
MCMASTER	CHARLES	W
MCWATTY	RICHARD	
MEAD II	JOHN	M
MEADE	ROBERT	E
MEADE	TOM	
MEDVES	EDWARD	
MEDVES	HELEN	R
MEHOZONEK	VICTOR	
MERHAR	ALBERT	J
MERHAR	JOSEPH	
METCALF	HARRY	C
METZ	JOHN	RICHARD
MEYER	LOUIS	
MICHALEK	JAMES	T
MIGHT	GERALD	W
MIGHT	SALLY	
MIKOLICH	LOUIS	
MILLER	MELVIN	L
MINNILLO	CATHERINA	
MOLE	ROBERT	A
MOORE	MAY	C

35:—The Bug, Euclid Beach Park, Cleveland, Ohio.

LAST	FIRST	MIDDLE
MOREL	SYLVESTER	
MORGAN	UGH	
MOSIER	FREDA	V
MOUGHAN	WILLIAM	
MOYNIHAN	GERALD	
MROSS	RONALD	V
MUELLER	LOUIS	W
MUHIC	ANN	
MULLALY	WILLIAM	J
MULLEN	JAMES	
MULLEN	ROBERT	E
MULVIHILL	THOMAS	
MURPHY	THOMAS	
MURPHY	THOMAS	F
MUSSELMAN	AMY	C
MYERS	STEVE	PAUL
NAGY	ELEANOR	
NASON	RICHARD	
NEGRELLI	JENNIE	M
NEUBAUER	KATHRYN	MELVIN
NEWTON	FRANCIS	T
OAKLEY	ADDIE	ELLEN
OBRIEN	JACK	
OLSON	EARL	N
OMERSA	JOHN	
ONEIL	JOHN	
ONEIL	WILLIAM	
OPALICH	DANIEL	
OREBAUGH	MEDFORD	D
OTCASEK	CHARLES	K
OXER	ORLANDO	MONROE
PAGE	JOHN	E
PALENIK	MARY	JANE
PARADISO	CANDIDA	
PARASKA	FRANK	W
PARHAM	FRED	R
PARKER	WILLIAM	
PARKER	WILLIAM	FORMAN
PARKS	ALBERT	L
PARKS	DONALD	
PATTI	JOHN	
PATTI	WALTER	F
PAYNE	CLAUDE	E
PECK	GEORGE	L
PELL	KELSO	
PELL	KENNETH	
PERROTT	RONALD	C
PERROTTI	CAROLINE	
PERZ	PAULINE	
PETERS	HERBERT	C
PHILLIPS	HUGH	JONES
PHILLIPS	JAMES	F
PITTENGER	FRANCES	
PITTMAN	KENNETH	
PLANTSCH	MICHAEL	
PLESNICAR	VICTOR	T
POLLOCK	JOHN	A
POLSON	ALLYN	S
PORTER	JOHN	LEWIS
PORTOLOS	PANAGHIS	
POWELL	ALFRED	L
PRATT	WILLIAM	
PRICE	EDWARD	O
PRICE	FLOYD	L
PRICE	GRACE	B
PRICE	HARRY	R
PRICE	MAY	B
PUDIMAT	OLGA	M
PUGH	MINNIE	F
PUGH	WINNIE	F
PURVIS	JAMES	
QUINN	WILLIAM	J
RAIMONDO	ALFRED	
RAMASKA	PETER	E
RANKER	HENRY	G
RAYER	WILLIAM	
RAYMOND	JOSEPH	JAMES
REAGAN	WILLIAM	P
REDDAWAY	JOHN	D
REDFIELD	CUYLER	I
REECE	CLARENCE	
REED	MARIE	E
REILLY	MICHAEL	
REINER	JOSEPH	
REINHARD	GEORGE	ANDREW
REINHARD	GEORGE	MARTIN
REPELLA	THOMAS	
RICHENS	JAMES	F
RINALDI	NICK	
ROBBINS	KATHLYN	TEEPLE
ROBITAILLE	ART	J
ROBITAILLE	JOHN	A
RODERICK	ALICE	M
ROEDIGER	KATHRYN	
ROEPNACK	PAUL	A
ROEPNACK	ROBERT	A
ROGERS	PEARL	
ROGERS	ROBERT	
ROOME	HOWARD	
ROSA	ANTHONY	C
ROSE	EUGENE	
ROSE	HENRY	E
ROTHMANN	EDWIN	
RUSSELL	CHARLES	
RYDMAN	ELIZABETH	
SADOWSKI	MARY	
SAFREED	ROBERT	
SALVADORE	ROGER	
SAMUELS	DAVID	J
SATOR	ALICE	MARGARET
SBROCCO	LEONARD	
SBROCCO	SEBASTIANO	
SCAFIDI	JOE	
SCALERO	VINCENT	N
SCARLATELLI	PHYLLIS	
SCHEFFEL	HENRY	
SCHIRRA	JOHN	L
SCHOLLE	GLENN	
SCHWARTZ	RICHARD	
SCOTT	DAVID	HUMPHREY
SCOTT	EVERETT	PALMER
SCOTT	JENNETTE	
SEAMAN	BENJAMIN	W
SEAMAN	JAMES	J
SEVELLO	ANNA	
SHANNON	HARRIS	COOPER
SHANNON	HARVEY	PAGE
SHARP	WILLIAM	G
SHEAFFER	ALBERT	
SHEPARD	JEROME	J
SHILLIDAY	EVERETT	P

THE GREAT AMERICAN RACING DERBY, EUCLID BEACH, CLEVELAND. FIFTH CITY.

LAST	FIRST	MIDDLE
SHUBERT	MORTON	
SIDEWAND JR	HARRY	
SIECKER	ELIZABETH	
SILVEROLI	JOHN	JAMES
SIVIK	DONALD	
SKITZKI	WALLACE	J
SLUSSER	EUGENE	G
SLUSSER	WILLIAM	RAYMOND
SMITH	CONWAY	JAMES
SMITH	DONALD	J
SMITH	GEORGE	A
SMITH	HARVEY	E
SMITH	IRENE	MARIE
SMITH	MATILDA	
SMITH	MAY	F
SMITH	ROBERT	JOYCE
SMITH	WALTER	
SMITH	WARREN	G
SOMMERS	ERNESTINE	
SOWINSKI	LEONA	
SPAUR	GLENN	DALE
SPENZER	EUGENE	G
SPRINGBORN	RUTH	ANN
SPUZZILLO	MARY	
SPUZZILLO	MICHAEL	A
ST CLAIR	JULIUS	J
STAKICH	DANNY	
STAMM	JOHN	GEORGE
STANG JR	GEORGE	H
STARINA	FLORENCE	
STEPP	GENUS	W
STEPPIC	CONRAD	S
STERNAD	LOUIS	J
STEVENS	JOSEPH	A
STEWART	WILLIAM	
STINEDURF JR	LEROY	
STONEBACK	HOWARD	DETWEILER
STRAUSS	JACK	
STRONG	ELEANOR	
STROSS	THERESA	
STUART	JAMES	MILLER
STUMP	DONALD	
STUMPF	CORA	MARIE
SULLIVAN	SOPHIE	
SUNAGEL	EDWARD	R
SUNAGEL	ROBERT	DALE
TAGGART	GLENN	
TAYLOR	THOMAS	
TAYLOR	WAYNE	
TELISMAN	RUDOLPH	
THOMAS	EARL	SMITH
THOMAS	PAUL	EDWARD
THOMPSON	CLARENCE	M
THOMPSON	GEORGE	
THORNWELL	BESSIE	M
THUNHURST	EDYTHE	
TIMPERIO	ANTHONY	A
TIMPERIO	NICK	JACK
TONNE	RICHARD	W
TREBEC	FRANK	
TRIVISON	CHARLES	A
TRIVISONNO	MARY	
TROMBO	MICHAEL	
TUREK	JOHN	R
VACCARIELLO	MARY	I
VALE	WILBUR	E
VALIQUETTE	JAMES	V
VAN HORN	STELLA	
VITEZ	JOHN	
VOGEL	ROBERT	
WACKER	WILLIAM	
WADE	CLARICE	MAE
WAGNER	FLORENCE	CALLAHAN
WAGNER JR	PETER	J
WALKER	DOUGLAS	
WALKER	LUTHER	
WALKER	ROBERT	
WALSH	JOHN	F
WAPPERER	ROY	
WARD	ROBERT	M
WARNER	NORMAN	
WEBB	RACHEL	
WEBBER	EUGENE	MADISON
WEBER	ARCHIBALD	AUGUST
WELTER	MARTIN	H
WERTZ	DONALD	M
WEST	CHARLES	W
WEST	NETTIE	M
WEST	WILLIAM	J
WHALEN	MILDRED	
WHITEHEAD	FRANK	
WHITEHOUSE	FLORENCE	CONKLE
WHITLOW	ROBERT	CARL
WHITNEY	PAUL	R
WHITNEY	WALTER	H
WICKS	GERALDINE	
WILLIAMS	ARTHUR	E
WILLIAMS	EARL	ROGER
WILLIAMS	GRACE	WYATT
WILLIAMS	JACK	
WILLIAMS	MARION	L
WILLIAMS	MILDRED	AUGUSTA
WILLIAMS	WALTER	DEWEY
WILSON	LLOYD	PAUL
WILSON	RICHARD	G
WILSON	ROBERT	S
WINKLE	IDA	M
WINKLE	WILLIAM	E
WINTERMUTE	JOHN	C
WOOD	F	C
WOODARD	LOUIS	C
WOODS	FRANK	HENRY
WRIGHT	ETHEL	MAY
WURM	CHARLOTTE	L
YOHO	FOREST	O
YOUNG	MOSES	
YOUNG	ROBERTA	
ZAPPIA	STEVE	
ZILCH	ROBERT	D
ZIMMERMAN	EDWARD	L
ZIMMERMAN	F	E
ZIMMERMAN	GEORGE	C

19:—SOME OF THE RECREATIONS, EUCLID BEACH PARK, CLEVELAND, OHIO.

X. Artists drawing, early 1900's. (Notice the water slide into the lake.)

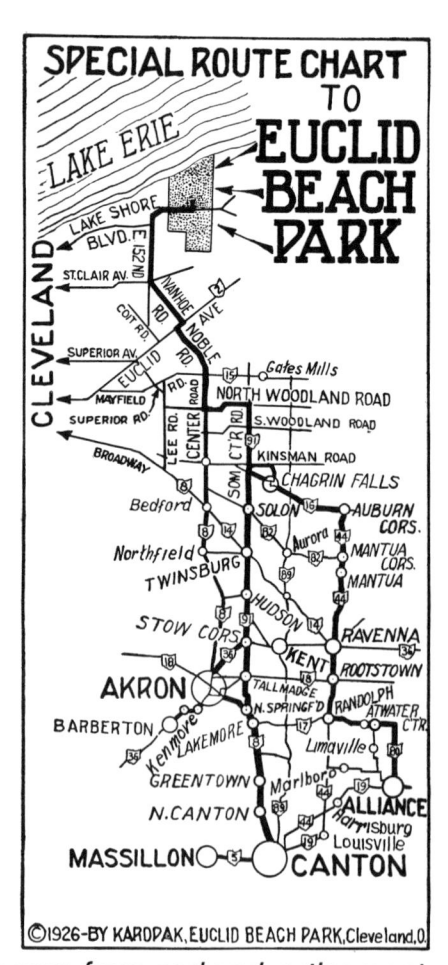

XI. Photo of early incinerator building.

XII. Route map from early advertisement.

1948
EMPLOYEE ROSTER

LAST	FIRST	MIDDLE
ABBEY	JOSEPHINE	
ABBEY	MAE	L
ABELA	EDWARD	F
ADAMS	EARL	
ADAMS	JAMES	
ADKINS	GRETA	L
ADKINS	NORMAN	B
AGRESTA	PHYLISS	
ALBRIGHT	MARTIN	
ALLEN	MAMIE	BERRY
AMOR	JOHN	
ANDERSON	JOHN	ANDREW
ANDERSON	WILLIAM	C
ANDREUCCI	FRED	
ANGELERO	RAFFAELE	
ANGELORO	CARMELA	M
ARNDT	WILLIAM	C
ARTHUR	IDA	MAE
ASHBA	JOSEPH	H
ASHLEY	JAMES	D
AXE	PAUL	E
BAILEY	GEORGE	K
BAILEY	LANE	L
BAIRD	WILSON	
BAKER	CILLIUS	MOSE
BAKER	JOHN	R
BALDWIN	DEAN	DEWITT
BALLNOW	JOHN	F
BALZONI	BESSIE	N
BARNES	JAMES	L
BARRY	DOROTHY	MARIE
BARRY	HENRY	JOHN
BARRY	KENNETH	
BARTO	ALOYSIUS	FRANK
BAZNIK	CHARLES	
BEAN	JOE	W
BEARD	JERRY	W
BECKLER	HOWARD	
BEHRA	VERNON	J
BENTLER JR	JOSEPH	J
BESHARA	GEORGE	E
BESSICK	JOHN	
BETZ	LOUISE	ELLA
BICHIMER	ALBERT	F
BILLENS	RAY	K
BLAIN	HEROLD	W
BLAKEMORE	WILLIAM	
BLAZEVIC	MAX	J
BLEADINGHEISER	J	H
BLEAM	C	M
BLEAM	LULA	ELSIE
BLEVINS	NERIMIAH	
BLISS	NETTIE	
BLOEDE	WILLIAM	CARL
BLUHM	JOHN	E
BOEHNLEIN	JOSEPH	W
BOETTCHER	IRENE	A
BOGAN	MISSOURI	
BOHAN	PATRICK	
BOOTH	MONROE	T
BOWHALL	ELMER	F
BOWHALL	HOWARD	
BOWMAN	WALTER	E
BOYCE	ROBERT	
BOYD	ANNA	J
BOYD	CHARLES	CLARK
BRADAC	JOE	A
BRADFORD	JOHN	
BRDAR	MILAN	G
BRICKNER	EDWARD	J
BRIGGS	THOMAS	LEE
BROOKINS	EDNA	BELLE
BROOKS	SAMUEL	J
BROWN	AMANDA	
BROWN	JOHN	
BROWN	RALPH	R
BROWN	RICHARD	A
BROWN	WALTER	R
BROZINA	FRANCES	
BRUNARSKI JR	PAUL	
BRYAN	JOHN	HENRY
BRYANT JR	THOMAS	L
BUCHANAN	DAVID	A
BUCKHARDT	MAYNARD	
BURDEN	BERTHA	
BURGESS	ELIZABETH	E
BURKHOLDER	GARY	
BURMEISTER	JAMES	R
BURTON	JAMES	
BUTLER	DWIGHT	H
BUTLER	JOSEPH	HACKNEY
CALLAGHAN	BEATRICE	BERNICE
CALLAGHAN	GEORGE	EDWARD
CALLIGHEN	FRANCIS	
CAMERON	RICHARD	
CAMPBELL	ALEXANDER	ROBERT
CAMPBELL	FRED	
CAPPE	JAMES	V
CAPPS	ALBERT	D
CARDER	CHARLES	
CARETTI	FREDERICK	C
CARNEY	JOHN	P
CARNICKI	JOSEPH	
CARROLL	GERTRUDE	M
CARROLL	HAZELLE	BELLE
CARROLL	LAUREN	J
CARROLL	ROY	SANFORD
CARROLL	URSULA	F
CASON	ALMA	
CENTA	HARRY	
CHAMBERLAIN	VERN	
CHAPMAN	WILLIE	A
CHARLES	WALTER	W
CHELSETH	H	K
CHEMICK	BETTY	
CHESTER	PAUL	A
CHIARELLI	JOSEPH	
CHITTENDEN	ROBERT	E
CHRISTOFORO	NICOLA	
CHRISTOPHER	JOSEPH	
CIASULLO	EUGENE	J
CICIGOI	RAY	
CIPITI	THOMAS	F
CIRINO	ROCCO	
CIRINO	THERESA	M
CISCO	JOAN	L
CLARDY	RUTH	
CLARK	ALEC	G
CLATTERBUCK	HARRY	L
CLIFFORD	GEORGE	R
CLIPSTON	HERBERT	
COCANOWER	PAUL	
COGAN	CHARLES	K
COLE JR	JOE	WILLIE
COLISTER	ROBERT	P
COLLINS	DOLORES	C

71

LAST	FIRST	MIDDLE
COLLINS	JAMES	F
COLLINS	ROSEMARY	
COMERFORD	EVELYN	
COMYNS	JAMES	A
CONNAVINO	JOSEPHINE	
CONTENTO	GUISEPPE	
CONWAY	HARRY	F
CONWAY	LENA	
COOLIDGE	GLADYS	M
CORBETT	RICHARD	M
CORRAI	SALVATORE	
CORRIGAN	HELEN	G
CORRIGAN	OMER	O
CORSI	VICTOR	R
COSTLOW	DONNA	
COWAN	JOHN	D
COWHARD	HATTIE	MAY
COX	WILLIAM	R
CRAPNELL	HAROLD	G
CROOKS	EMMA	
CROSS	JOHN	A
CROSS	VERA	L
CUNNINGHAM	RAY	
CURLEY	FRANK	L
CURRY	ANDREW	
CURRY	JOHN	MYERS
CUTSHALL	LAVON	G
DADE	ARTHUR	C
DAUGHERTY	LEROY	
DAUPHIN	VIRGINIA	
DAVIES	WILLIAM	M
DAVIS	HARRY	E
DAVIS	ROGER	R
DAY	WILLIAM	R
DEAMICO	TONY	
DECKER	LYLE	S
DEE	JAMES	E
DEL BROCCO	RAY	
DELAMBO	FRANK	FRANCIS
DELCALZO	AMERICO	J
DELLINGER	TOM	
DENISON	DAVID	V
DEPETRIS	EUGENE	
DESAN	ANTHONY	M
DESCENZO	DAN	
DETELICH	ROBERT	N
DEVEL	GENE	PIERRE
DEWILLE	ARTHUR	
DEWITT	LEROY	W
DICKEY	HERBERT	ROBERT
DIJULIUS	LEWIS	R
DILIBERTO	DONALD	R
DITTMAN	MARGERY	E
DIZIG	LOUIS	J
DOBAY	JAMES	A
DODELSON	BERT	B

DOTY	DONALD	
DOUGHTY	JOSEPH	
DOUGLAS	JAMES	
DRAKE	RICHARD	
DRESSLER	ALTA	DOLORES
DRESSLER	H	C
DUBROSKY	PAT	C
DWYER	MARY	FRANCES
EARLY	JULIA	A
EBERHARD	EILEEN T	A
ELLIS	HAROLD	B
ELLISON	STEWART	E
ENOCHIAN	ENOCH	
ETCHELL	DAVID	
EVANS	ROBERT	J
EVANS	WILLIAM	B
FAIRWEATHER	GEORGE	E
FATICA	EUGENE	E
FEENEY	BLANCHE	
FERRERI	GEORGE	
FIGLER	ETHEL	A
FINERAN	MARY	BALL
FIORETTA	THERESA	
FISHER	CHARLES	W
FLICK	ARTHUR	R
FORTUNA JR	JAMES	
FOSSELMAN	COLETTA	
FRANCIS	DAVID	J
FRANCIS	WINIFRED	
FRANK	GEORGE	W
FRANTZ	LORETTA	
FREDERICKS	ELIZABETH	
FREDERICKS	FRANK	JOSEPH
FREYER	CHARLES	E
FROHLICH	MARGARET	
FROHMBURG	ROSE	
GAFFNEY	LEROY	
GALLAGHER	ANNE	ROSE
GAMBATESE	FLORENCE	
GAMMIERO	LUKE	
GARDNER	JOSEPH	D
GASKINS	MARION	R
GAY	JAMES	W
GEISWEIDT	DONALD	L
GEORGE	JAMES	A
GERGELY	BERT	R
GEROME	LENA	
GHANN	JAMES	W
GIBSON	KATHRYN	F
GIBSON	OLLIE	J
GIERMAN	WILLIAM	F
GILL	ELSIE	OLIVIA
GILL	WILLARDR	
GILMORE	RALPH	E
GIPSON	LOWELL	
GLASS	HENRY	D
GLASS	LEAH	RUTH
GLASS	RUTH	
GOETZ	WILLIAM	R
GOLDEN	JOSEPH	M
GORMAN	JEAN	M
GRAHAM	LEASER	
GRANGER	MABLE	ETTA
GREEN	CLAUDE	ELLSWORTH
GREEN	GERALD	C
GREENWAY	FRED	STUDER
GREGORICH	FRANK	
GREGORY JR	ROBERT	
GREULICH	DAVID	P
GROVER	KATHERINA	A
GULLIFORD	CHARLES	
GUNN	EDWARD	F
GUY	ALVIN	R
HAGEDORN	JACK	T
HALBURDA	DOROTHY	

LAST	FIRST	MIDDLE
HALLFORD	RUSSELL	H
HAMLIN	VIRGIL	
HANNAN	LEO	W
HANSON	CLARENCE	H
HARFORD	JESS	A
HARRISON	CHARLES	L
HARROLD	FLORENCE	M
HAUCK	GEORGE	WILLIAM
HAYES	WILLIS	JAMES
HECKMAN	JAMES	R
HEINZ	MILTON	G
HERRICK	JOHN	E
HEYSE	WALTER	E
HICKOX	RICHARD	L
HINES	ROY	C
HINKELMAN	PAUL	
HINSKE	WILLIAM	B
HINTON	WILLIAM	H
HINTON	WILLIAM	R
HLAVSA	ALBERT	J
HOCHEVAR	RICHARD	J
HOGAN	LAWRENCE	E
HOHMAN	RICHARD	
HRADISKY	DONALD	
HRADISKY	ROBERT	
HROSAR	JOSEPH	T
HUDAK	JOHN	M
HUDSON	EDWIN	P
HUGHES	ELLA	M
HUNT	BLANCHE	LEAH
HUNTER	ROY	CHARLES
HUTTER	JOSEPH	
IANNETTA	ANNA	
IFFARTH	WILLIAM	F
INDIANO	S	S
IRVINE	ROBERT	W
JACKLITZ	ALBERT	
JACKSON	RALPH	LEE
JANKO	WILLIAM	L
JANZ	JOHN	
JASENSKY	STANLEY R	
JEROME	ROSE	
JOHNSON	DOROTHY	IDA
JOHNSON	JAMES	GARY
JOHNSON	THURA	W
JOHNSTON	CLIFFORD	BASIL
JOHNSTON	DONALD	HUMPHREY
JONES	CHARLES	G
JONES	RUBY	
JORDENS	HERBERT	
JUDD	ESTIL	LEE
KAJFEZ	MATTHEW	J
KARLIN	PAUL	W
KAUFMAN	BURTON	
KEEFE	JACK	F
KEELER	ELIZABETH	
KELLY	DAISY	CURTIS
KELLY	LOUIS	A
KEMMETT	JOHN	W
KENNEDY	HERMAN	O
KENT	EDWARD	C
KERBY JR	JOSEPH	B
KERR JR	JOHN	E
KIKOLI	FRANK	
KILLEEN	FRANK	W
KING	GLENN	D
KING JR	JACK	
KINSELLA	TOM	J
KLEMENTS	BERNARD	J
KMIT	GEORGE	
KOMRAUS	ELEANOR	R
KORDIC	ELIZABETH	C
KOZLEVCAR	STANLEY	J
KRISLOV	SAM	
KRIVOY	RAYMOND	
KROCKER	JAMES	P
KRULL	HERMAN	R
KRUPSKY	ANTHONY	
KUHNS	DAVID	P
LACONTE	FRANK	
LANDIG	RONALD	
LANG	H	WILLIAM
LANG	JAMES	V
LAPASH	JAMES	
LAPE	CLARK	E
LAVRICH	CHARLES	
LAVRICH	JAMES	H
LAWRENCE	JOHN	P
LAYMON	SAMUEL	RANDALL
LENZ	FRED	A
LESLIE	JAMES	C
LEVEREAUX	DUANE	L
LEWALLEN	JAMES	R
LEWIS	JOHN	J
LEXA	DONALD	A
LIGGETT	HARRIET	
LIGHT	MARY	E
LILLIS	STEPHEN	J
LINDER	ROBERT	J
LINDSTROM	SAMUEL	G
LINN	PETER	O
LLOYD	CHARLES	M
LOCKE	HUBERT	LEE
LOCKE	ROBERTTIEN	
LONGFIELD	TESS	EVANS
LORD	HARRY	F
LORENZO	MARIA	
LORENZO	MARY	I
LOVSIN	JOHN	
LUCIANO	DORIS	M
LUCKAY	RICHARD	
LUKAS	FRANK	J
LUNDBLAD	ROBERT	A
LUPTON	JAMES	ALFRED
LUPTON	ROBERT	
LUTZ	HOWARD	
LYNCH	DENIS	B
LYNCH	FRED	FRANCIS
MAC DONALD	JOHN	
MAC DONALD	LENA	
MACGILLIV	RAY	HAROLD
MACHAMER	CHAS	ROBERT
MACKENZIE	ELLEN	
MADISON	BENNETTE	L
MAHER	JAMES	S
MAHONEY	EARL	
MALONEY	MELVIN	P
MALONEY	THOMAS	J
MANCINE	BENJAMIN	J
MANDAU	MIKE	
MANDAU	RUDOLPH	
MANDEL	IDA	

16:—COMING DOWN THE DERBY DIPS, EUCLID BEACH PARK, CLEVELAND, OHIO

LAST	FIRST	MIDDLE
MANNING	DOUGLAS	L
MARCOGLIESE	MADDALENA	
MARKENS	EDNA	BROWNING
MARLINI	ALBERT	
MARSHALL	CADDIE	
MARTIN	ANGELA	P
MARTIN	DONALD	C
MARTIN	JOHN	M
MASTROPIETRO	ANGELA	
MATTHEWS	DAVE	
MATTHEWS	LENA	
MAYS	RICHARD	
MCCABE	JAMES	J
MCCLAIN	HELEN	N
MCCLURE	BETTY	L
MCDONALD	CARRIE	M
MCDONOUGH	DANIEL	
MCGUIRE	GLADYS	E
MCGUIRE	THEODORE	R
MCHARG	RICHARD	
MCILRATH	IDA	E
MCINTYRE	ROBERT	
MCKINLEY	JAMES	H
MCKINNEY	SADIE	
MCLAUGHLIN	FRANCES	
MCMAHAN	ANN	
MCNIECE	RAYMOND	E
MEAD II	JOHN	M
MEDVES	HELEN	R
MEINHARDT	ART	R
MENDE	JOHN	E
MERHAR	JOSEPH	
MERRIMAN	ROBERT	
MESAROS	LOUIS	
METCALF	HARRY	C
METZ	JOHN	RICHARD
MEYER	LOUIS	
MEZGET	JOHN	
MILETI	RAYMOND	A
MILLS	ARTHUR	F
MOORE	MAY	C
MOORE	ROBERT	E
MOREL	SYLVESTER	
MORGAN	CATHERINE	
MORGAN	HUGH	
MORRIS SR	WILLIAM	GEORGE
MORROW	HARRY	C
MORROW	WILLIAM	F
MORTON	DEEMS	B
MOSIER	FREDA	V
MOSNIK	RAYMOND	
MOUGHAN	TERRY	
MOUGHAN	WILLIAM	
MUELLER	LOUIS	W
MULLALY	WILLIAM	J
MULLEN	JAMES	
MULVIHILL	THOMAS	
MUMAW	SUSAN	B
MUNSON	RALPH	E
MURPHY	THOMAS	F
MYERS	STUART	S
MYSYK	EDITH	R
NAGY	ELEANOR	
NEFF	ARTHUR	C
NEMEC	JOSEPH	G
NEUBAUER	KATHRYN	MELVIN
NEWKIRK	NORMA	
NEWTON	SOPHIA	P
NICKERSON	PAGE	H
NIESS	MAY	
NOLAN	LARRY	
NOVARRO	EARL	D
NUNEMAKER	VIVIAN	
OAKLEY	ADDIE	ELLEN
OLESKI	EDWARD	J
OLIVER	IRENE	B
OMERSA	JOHN	
ONEAL	JAMES	D
ONEIL	JOHN	
ONEIL	WILLIAM	
ORR	ROBERT	S
OSHABEN	JOHN	G
OSTROWSKI	PETER	T
OTCASEK	CHARLES	K
OTHBERG	KENNETH	C
PACK	RALPH	W
PADGETT	ROBERT	L
PALUCKAS	ANTHONY	L
PARADISO	CANDIDA	
PARKER	WILLIAM	
PARKS	ALBERT	L
PARMELEE	CHARLES	W
PASQUALE	CARMELLA	F
PATTERSON	ROBERT	
PATTI	JOHN	
PATTI	WALTER	F
PATTON	PATRICIA K	
PAVLINA	EDWARD	G
PECK	DONALD	
PECK	GEORGE	L
PECKO	CARL	
PELL	KELSO	
PEPPEREL	RICHARD	N
PEPPLE	VIRGINIA E	
PERISH JR	JOSEPH	R
PERROTT	RONALD	C
PERROTTI	CAROLINE	
PERTZ	FRANK	W
PETERS	ARNOLD	L
PETRIE	JOSEPH	L
PHELPS	ROBERT	J
PISTILLO	GRACE	
PLANTSCH	MICHAEL	
PLESHINGER	HELEN	C
PLESNICAR	VICTOR	T
POJE	CHARLOTTE	
POOLE	DALLAS	
POTOCNIK	ROBERT	
POWELL	ROBERT	H
PRICE	EDWARD	O
PRICE	FLOYD	L
PRICE	GRACE	B
PRICE	HARRY	R
PUDIMAT	OLGA	M
PURVIS	JAMES	
RAE	JAMES	
RAIMONDO	ALFRED	
RAINES	OSCAR	
RAINS	HERMAN	
RAINS	OVA	
RANKER	HENRY	G
RASH	VERNER	W

49:—SUNSET ON LAKE ERIE, EUCLID BEACH PARK, CLEVELAND, OHIO.

Photo by Van Fisher

LAST	FIRST	MIDDLE
RASH	WILLIAM	F
RAYER	WILLIAM	
REARDON	MICHAEL	
REARDON	ROGER	
REDFIELD	CUYLER	I
REECE JR	LEWIS	
REED	MARIE	E
REESE	LEONARD	C
REINER	JOSEPH	
REINHARD	GEORGE	ANDREW
REINHARD	GEORGE	MARTIN
REPELLA	THOMAS	
REPINE	LUTHER	W
RICHARDS	ANDREW	P
RICHARDS	PATRICIA	
RICHARDSON	LULA	BELLE
RICHENS	JAMES	F
RIDER	RICHARD	C
RIENDEAU	RAYMOND	K
RIGA	LAWRENCE	
RILEY	JERRY	M
RILEY	MILDRED	L
RINALDI	SUE	M
RITTER	ROBERT	W
RITTER	WARREN	E
ROBBINS	KATHLYN	TEEPLE
ROBINSON	ROBERT	L
RODERICK	ALICE	M
ROEDIGER	KATHRYN	
ROEPNACK	ROBERT	A
ROGERS	ROBERT	G
ROMANCHIK	RICHARD	
RONIGER	H	E
ROSA	ANTHONY	C
ROSE	EUGENE	
ROSKIN	ALVIN	N
ROSSODIVITA	DAVID	
ROTH	THOMAS	H
ROTT	JOSEPH	L
ROYON	DAVID	H
RUSSELL	GEORGE	D
RYDMAN	ELIZABETH	
SABOT	FRANK	J
SACK	DON	J
SACKS	CHARLES	S
SAMSON	DONALD	E
SANDE	JOHN	R
SANDRICK JR	THOMAS	
SANDS	WILLIAM	T
SANTORELLI	ANNA	
SANTORELLI	JOSEPH	F
SATOR	ALICE	MARGARET
SBROCCO	LEONARD	
SCAFIDI	JOE	
SCALERO	VINCENT	N
SCHARLAU	RAY	E
SCHEFFEL	HENRY	
SCHILL	LLOYD	EDGAR
SCHILL JR	LLOYD	E
SCHIRRA	JOHN	L
SCHOENBECK	KENNETH	H
SCHRAN	JEAN	M
SCHUEREN	IODORA	B
SCHULTZ	HATTIE	E
SCHUTT	DONALD	ALLEN
SCHWEIKERT	FREDERICK	
SCHWEIKERT	RICHARD	
SCOTT	EVERETT	PALMER
SEBRASKY	RICHARD	
SEDLACEK	ELDON	J
SEDLOCK	ALFRED	
SEXTON	LARRY	
SHANKLIN	MOSES	
SHANNON	HARRIS	COOPER
SHANNON	HARVEY	PAGE
SHARP	WILLIAM	G
SHERWOOD	ANNE	D
SHERWOOD	WILLIAM	M
SHILLIDAY	EVERETT	P
SHIPLEY	WILLIAM	R
SHRIVER	JAMES	
SIFLING	JOHN	
SILVERMAN	MEYER	
SILVEROLI	JOHN	JAMES
SIMA	JAMES	J
SINCLAIR	HELEN	T
SIRL	CHARLES	F
SIVIK	DONALD	
SIVILLO	MARY	
SKEANS	MAY	
SKLAD	CHESTER	M
SKODA	JOSEPH	J
SKUR	JOSEPH	
SMITH	DONALD	K
SMITH	IRENE	MARIE
SMITH	JAMES	A
SMITH	MATILDA	
SMITH	MAY	F
SMITH	ROBERT	JOYCE
SMITH	WARREN	G
SNYDER	EUGENE	
SOLTES	DOROTHY	
SOMMERS	ERNESTINE	
SOUSA	EUGENE	
SPISCAK	BETTIE	
SPUZZILLO	MARY	
ST CLAIR	JULIUS	J
STALLA	HARVEY	
STALLARD	BETTY	ANN
STAMM	JACK	G
STAMM JR	J	G
STANA	LEONARD	J
STANG JR	GEORGE	H
STANICIC	PAUL	
STANTON	CLARENCE	M
STARINA	FLORENCE	
STECKER	ALFRED	G
STEELE	ARTHUR	W
STEPP	GENUS	W
STEUER	HERBERT	S
STEVENS	JOSEPH	A
STINEDURF JR	LEROY	
STIRLING	PAUL	M
STOHLMAN	RICHARD	E
STONEBACK	HOWARD	DETWEILER
STRAUSS	JACK	
STRONG	ELEANOR	
STRONG	HAROLD	E
STRUMBLE	JOSEPH	
STUMPF	CORA	MARIE
STUPKA	EMMA	C
SULLENS	SAMUEL	
SULLIVAN	WILLIAM	J
SUNAGEL	EDWARD	R

LAST	FIRST	MIDDLE
SUNAGEL	ROBERT	DALE
SURAD	MARTIN	E
SWAGGER	JOHN	E
SWEENEY	ROBERT	W
TANNO	THERESA	
TAYLOR	JOHN	
TAYLOR	THOMAS	
TERBANC	THOMAS	L
THOMAS	PAUL	EDWARD
THUNHURST	EDYTHE	
TIZZANO	IDA	
TIZZANO	MARIA	
TOIGO	JOSEPH	D
TRAPP	JAMES	R
TRIVISON	CHARLES	A
TRIVISONNO	MARY	
TROMBO	RONALD	
TRUMPHOUR	WILLIAM	
TUCKER	FRANK	J
UHL	PETER	
ULMER	MARTIN	
VACCARIELLO	MARY	I
VALARDO	NICK	
VALE	WILBUR	E
VAN HOUTEN	GERALD	
VANCE	JOHN	H
VEITCH	ROBERT	L
VOGEL	BENEDICT	B
VOGEL	ROBERT	
VRANEKOVIC	JOS	D
WADE	CLARICE	MAE
WAGNER	FLORENCE	CALLAHAN
WAITE	JOHN	W
WALKER	GEORGE	E
WALKER	LUTHER	
WALKER	RAY	
WALKER	ROBERT	
WALKER	WILLIAM	CLYDE
WALLACE	FRANK	
WARDELL	DONALD	Q
WARNER	NORMAN	
WARNICK JR	GLENN	W
WASCHURA	JACK	B
WATKINS	JOHN	
WATSON	DORIS	MAE
WATSON	E	D
WATSON	WILLIAM	R
WATSON JR	WILLIAM	R
WAYDA	EDWARD	E
WEBB	RACHEL	
WEBB JR	JOHN	E
WEBER	ARCHIBALD	AUGUST
WEED	CHARLES	EARL
WEILAND	JAMES	
WEILAND	JOE	J
WESTERFIELD	WADE	
WHEATLEY	AUGUSTA	MAINE
WHITE	JACK	W
WHITE	JOHN	M
WHITNEY	PAUL	R
WHITNEY	WALTER	H
WICKS	GERALDINE	
WILBERSCHEID	JOE	
WILLIAMS	EARL	ROGER
WILLIAMS	GRACE	WYATT
WILLIAMS	WALTE	DEWEY
WILLIAMS	WALTER	L
WILSON	LLOYD	PAUL
WILSON	RICHARD	G
WISE	DOROTHY	
WOLASKI	MARTHA	
WOLF	FRED	
WOLTMAN	RAYMOND	
WOODS	FRANK	HENRY
WOODS	ROBERT	H
WRIGHT	JAMES	J
WURM	CHARLOTTE	L
XAVIER	RAYMOND	ARTHUR
YAFANARO	ALBERT	
YOUNG	LLOYD	WESLEY
YOUNG	MOSES	
YOUNG	ROBERTA	
ZALLER	JOHN	E
ZDROJEWSKI	EDWARD	
ZILCH	ROBERT	D
ZIMMERMAN	F	E
ZIMMERMAN	GEORGE	C
ZIMMERMAN	NORMAN	L
ZINGALE	FREDRICK	
ZITKO	ALICE	L
ZORUMBA	PETER	L
ZUPANCIC	JOHN	

"Balloons, Euclid Beach Park, Cleveland, Ohio"

ONE OF THE MANY AMUSEMENT DEVICES FOR CHILDREN, EUCLID BEACH PARK, CLEVELAND, OHIO

Children's Playground, Euclid Beach Park, Cleveland, Ohio

1949
EMPLOYEE ROSTER

LAST	FIRST	MIDDLE
ABBEY	MAE	L
ADAMOWICZ	SALLY	
AGRESTA	PHYLISS	
ANDERSON	JOHN	ANDREW
ANGELORO	CARMELA	M
ANTHONY	ROBERT	J
ARENDT	GEORGE	L
ARTHUR	IDA	MAE
ASH	JOE	
AVERILL	EDGAR	N
AVERILL SR	EDGAR	NATHAN
AXE	PAUL	E
BADAR	LAWRENCE	
BAILEY	GEORGE	K
BAILEY	LANE	L
BAKER	CILLIUS	MOSE
BALDNER	WILLIAM	
BALDWIN	DANIEL	
BARNHART	RICHARD	
BARRY	DOROTHY	MARIE
BARTO	ALOYSIUS	FRANK
BAZNIK	CHARLES	
BEHRA	ROGER	
BENCINA	WILLIAM	
BETZ	HAROLD	C
BEVIER	JOHN	A
BLAKEMORE	WILLIAM	
BLEAM	LULA	ELSIE
BLOEDE	DOROTHEA	
BLOEDE	WILLIAM	CARL
BOGAN	CLARENCE	
BOGAN	MISSOURI	
BOWHALL	ELMER	F
BOWHALL	HOWARD	
BOWMAN	CHARLES	E
BOYD	ANNA	J
BOYD JR	PAUL	
BRADFORD	JOHN	
BRENNAN	DANIEL	J
BRENNAN	MARGARET	
BRIGGS	THOMAS	LEE
BROOKINS	EDNA	BELLE
BROWN	JOHN	
BRYAN	JOHN	HENRY
BUCKLEY	THURMAN	R
BURDEN	BERTHA	
BURDEN	EUGENE	J
BURKHARDT	MAYNARD	
BURMEISTER	JAMES	R
BURTON	JAMES	
BURTON	WILLIE	D
BUTLER	JOSEPH	HACKNEY
CALLAGHAN	BEATRICE	BERNICE
CALLIGHEN	FRANCIS	
CAMPBELL	ALEXANDER	ROBERT
CAMPBELL	FRED	
CAPRA	CHRISTINA	
CAPRA	LOUISA	
CARROLL	GERTRUDE	M
CARROLL	HAZELLE	BELLE
CARROLL	LAUREN	J
CARROLL	ROY	SANFORD
CARROLL	URSULA	F
CATALANO	FRANK	A
CELESTINI	GUY	
CENTA	HARRY	
CHAMBERLAIN	VERN	
CHARLES	WALTER	W
CHEMICK	BETTY	
CHOFFIN	ALFRED	L
CHRISTOFORO	NICOLA	
CHRISTOPHER	JOSEPH	
CIGAN	ERNEST	C
CIPITI	THOMAS	F
CIRINO	ROSE	MARIE
CIRINO	THERESA	M
CLIPSTON	HERBERT	
COCANOWER	PAUL	
COLVIN	JESSIE	
COMERFORD	EVELYN	
COMYNS	ARNOLD	T
COMYNS	JAMES	A
CONNAVINO	JOSEPHINE	
CONNELL	TOM	
CONTENTO	GUISEPPE	
CONWAY	HARRY	F
CONWAY	LENA	
COOK	NORMAN	S
COOLIDGE	GLADYS	M
COWAN	JOHN	D
CRAPNELL	HAROLD	G
CRAXTON	LEROY	L
CRIMALDI	ELEANORA	
CROOKS	EMMA	
CUDAHY	WILLIAM	J
CURRY	ANDREW	
CURRY	JOHN	MYERS
CUTSHALL	LAVON	G
DAUPHIN	VIRGINIA	
DAY	WILLIAM	R
DELAMBO	FRANK	FRANCIS
DELLINGER	TOM	
DEMORE	IRENE	TAYLOR
DESAN	ANTHONY	M
DILIBERTO	DONALD	R
DLOUGY	JOSEPH	J
DOPSLAF	ARTHUR	
DRESSLER	H	C
DUNBAR	RICHARD	
EGAN	MARY	JO
EPAVES	RICHARD	A
EUCKER	PAUL	J
FAIRFAX	WILLIS	
FALKNER	EDWIN	H
FEDERICO	JOSEPHINE	
FERGUSON	NORMAN	N
FERRERI	GEORGE	
FLETCHER	WILLIAM	R
FLOWERS	KENNETH	
FOUTS	CLARENCE	E
FRANCIS	DAVID	J
FRANCIS	MOZART	
FRANCIS	WINIFRED	
FREDERICKS	ELIZABETH	
FREDERICKS	FRANK	JOSEPH
FROHMBURG	ROSE	
GAMBATESE	FLORENCE	
GARDNER	JOSEPH	D
GENTILE	JAMES	A
GENTILE	JOHN	R
GIBSON	OLLIE	J
GILL	ELSIE	OLIVIA
GLASS	RUTH	

LAST	FIRST	MIDDLE
GOVANG	MATIE	B
GRABOWSKI	RICHARD	A
GRAHAM	CHARLES	A
GRANGER	MABLE	ETTA
GRAY	DON	
GREEN	CLAUDE	ELLSWORTH
GREENWAY	FRED	STUDER
GREMILLION	AUGUSTE	
GREMILLION	LEE	
GRUGLE	VALERIE	C
HALL	WINDSOR	S
HALLFORD	RUSSELL	H
HANSON	CLARENCE	H
HAPP	GEORGE	W
HAPP JR	J	W
HARDING	BLANCHE	E
HARFORD	JESS	A
HARRIS	WILL	
HARROLD	FLORENCE	M
HELMS	HARRY	C
HILL JR	JAMES	S
HINES	ROY	C
HINTON	WILLIAM	R
HIRSCHAUER	ROGER	D
HOCEVAR JR	JOSEPH	R
HOFFMAN	RONALD	C
HOGAN	LAWRENCE	E
HOLLIS	JOHN	P
HRADISKY	DONALD	
HROSAR	JOSEPH	T
HUBBELL	HELEN	G
HUDSON	EDWIN	P
HUGHES	ELLA	M
HUNT	BLANCHE	LEAH
HUNTER	ROY	CHARLES
IAMMATTEO	MICHELINA	
INGENDORF	CHARLES	T
INTIHAR JR	VICTOR	
JACKLITZ	ALBERT	
JANZ	JOHN	
JENNINGS	LOUELLA	
JEROME	ROSE	
JOHNSTON	CLIFFORD	BASIL
JOHNSTON	DONALD	HUMPHREY
JONES	LOIS	EILEEN
JUDD	ESTIL	LEE
JULIAN	MARGARET	H
JUNKINS	ALAN	D
KANTOR	MICHAEL	
KEELER	ELIZABETH	
KEELER	RONALD	D
KELLER	CARL	J
KENNEDY	MAGGIE	W
KENNEDY	ROSIE	W

KENT	EDWARD	C
KERBY JR	JOSEPH	B
KIKOLI	FRANK	
KING	FREDERICK	
KING JR	JACK	
KINSELLA	TOM	J
KIRCHNER	KEITH	E
KLEMENTS	BERNARD	J
KOCH	HENRY	HERMAN
KRULL	HERMAN	R
KRUPSKY	ANTHONY	
KUCHTA	JAMES	L
KUJAT	RICHARD	R
LAIRD	EDWARD	J
LAMBERT	BESSIE	M
LANG	H	WILLIAM
LAVRICH	CHARLES	
LAVRICH	JAMES	H
LAYMON JR	SAMUEL	R
LEWIS	JACK	K
LIGGETT	HARRIET	
LONGFIELD	TESS	EVANS
LORENZO	MARY	I
LOVE	J	C
LOVE	KATIE	
LUOMA	JAMES	A
LUPTON	ROBERT	
LUTZ	HOWARD	
MAC DONALD	JOHN	
MAC DONALD	LENA	
MACHAMER	CHAS	ROBERT
MACKENZIE	ELLEN	
MACLEOD	LAVINA	J
MAHONEY	EARL	
MAHONEY	RONALD	H
MALONEY	MELVIN	P
MANCINE	BENJAMIN	J
MANCINE	LOUIS	
MANDAU	RUDOLPH	
MANDEL	HELEN	E
MANDEL	IDA	
MANNING	DOUGLAS	L
MANNION	JOHN	J
MARCOGLIESE	MADDALENA	
MARIANO	BESSIE	
MARINO	ANTHONY	M
MARINO	GLORIA	
MARKENS	EDNA	BROWNING
MARKLE	KATHLEEN	H
MAROLD	FRANK	C
MARSHALL	CADDIE	
MARTINO	ARMAND	
MAYOCK	WILLIAM	P
MCGUIRE	ARTHUR	D
MCILRATH	IDA	E
MCKINLEY	JAMES	H
MCLAUGHLIN	FRANCES	
MCMAHAN	ANN	
MEDVES	HELEN	R
MEYER	LOUIS	
MEYER SR	LOUIS	
MILLER	RONALD	
MIOZZI	RAMON	C
MOORE	MAY	C
MORGAN	CATHERINE	
MORGAN	CHARLES	
MORGAN	HUGH	
MOSIER	FREDA	V
MOUGHAN	TERRY	
MUELLER	LOUIS	W
MULVIHILL	THOMAS	
MUMAW	SUSAN	B
MUSTARD	JANET	M
MYERS JR	WILL	S
MYSYK	EDITH	R

LAST	FIRST	MIDDLE
NEKICH	ROBERT	
NESTICK	JOHN	
NEUBAUER	JOSEPH	MELVIN
NEUBAUER	KATHRYN	MELVIN
NEWTON	SOPHIA	P
NICOLL	ROBERT	G
OAKLEY	ADDIE	ELLEN
OLESKI	JOHN	T
OLIVER	IRENE	B
OMERSA	JOHN	
ONEIL	JOHN	
ORRID	CLARENCE	C
PALMER	FLORA	
PALUCKAS	ANTHONY	L
PARKER	WILLIAM	
PARKS	CARL	R
PARKS	DONALD	
PARKS	WARREN	H
PARMELEE	CHARLES	W
PAVLINA	EDWARD	G
PECK	DONALD	
PELL	KELSO	
PEPPLE	VIRGINIA E	
PERROTTI	CAROLINE	
PERROTTI	JOSEPH	
PETERS	ARNOLD	L
PHILLIPS	MARIE	
PIOVARCHY	LAVERNE	
PLANTSCH	MICHAEL	
PORTER	THOMAS	C
POTTER	CHARLOTTE	
PRESTERL	RICHARD	E
PRICE	EARL	GEORGE
PRICE	FLOYD	L
PRICE	GRACE	B
PRICE	HARRY	R
PUINNO	NICHOLAS	J
RABE	PETER	L
RAIMONDO	ALFRED	
RASH	VERNER	W
REARDON	MICHAEL	
REDFIELD	CUYLER	I
REED	MARIE	E
REINHARD	GEORGE	MARTIN
REPELLA	THOMAS	
RICHARDS	CHARLES	R
RICHARDS	FRED	H
RICHARDS	FREDDIE	D
RICHENS	JAMES	F
RIENDEAU	RAYMOND	K
RIGHTNOUR	KENNETH	
ROBERTS	RICHARD	L
ROBINSON	ALFONSO	
RODERICK	ALICE	M
ROGERS	ANNA	E
ROLLINS	ERNEST	L
RONIGER	H	E
ROSA	ANTHONY	C
ROSE	EUGENE	
ROSSODIVITA	DAVID	
ROUNDTREE	IOLA	
RYDMAN	ELIZABETH	
SACK	DON	J
SAMSON	DONALD	E
SATOR	ALICE	MARGARET
SCAFIDI	JOE	
SCHAFFER	ALLAN	A
SCHEFFEL	HENRY	
SCHILL JR	LLOYD	E
SCHMIEDING	HAROLD	W
SCHNUR	ROBERT	P
SCHRECK	CECILIA	G
SHANK JR	CHARLES	

LAST	FIRST	MIDDLE
SHANNON	HARRIS	COOPER
SHARP	WILLIAM	G
SHILLIDAY	EVERETT	P
SHILLIDAY	JOHN	B
SIMA	JAMES	J
SIMMS	CHARLES	L
SINCLAIR	HELEN	T
SKENDER	CHARLES	
SKODA	JOSEPH	J
SMITH	ALAN	H
SMITH	FRED	C
SMITH	IRENE	MARIE
SMITH	MAY	F
SMITH	ROBERT	JOYCE
SMITH	WARREN	G
SNYDER	LUCY	J
SOMMERS	ERNESTINE	
SOUSA	EUGENE	
SOUTHWICK	FRANCIS	DAVID
SPUZZILLO	MARY	
SPUZZILLO	MICHAEL	F
ST CLAIR	JULIUS	J
STANG JR	GEORGE	H
STARINA	FLORENCE	
STARRE	EUGENE	
STAVOLE	ANGELO	J
STEELE	ARTHUR	W
STEPP	GENUS	W
STEVENS	JOSEPH	A
STEVENS	VALENTINE	G
STEVERDING	VINCENT	E
STEWART	B	MAXINE
STEWART	JAY	
STIELAU	EDWARD	R
STONEBACK	HOWARD	DETWEILER
STRAUSS	JACK	
STRONG	ELEANOR	
STRUMBLE	JOSEPH	
STUART	EFFIE	MAY
STUART	JAMES	M
STUMPF	CORA	MARIE
STUMPF	NANCY	J
TAYLOR	JAMES	E
TAYLOR	WILLIAM	K
TERRELL	ROBERT	L
THOMAS	PAUL	EDWARD
THOMAS	RICHARD	
THUNHURST	EDYTHE	
TIMPERIO	JOHN	C
TOIGO	JOSEPH	D
TOTH	GEORGE	A
TOWNER	JAMES	K
TRIVISON	CHARLES	A
TRIVISONNO	MARY	
TROMBO	MICHAEL	
TROMBO	RONALD	
TRUMPHOUR	WILLIAM	
TUCCI	ANTHONY	J

LAST	FIRST	MIDDLE
TUCCI	JOE	
TURNER	CHARLES	L
UPTON	DOROTHY	O
VACCARIELLO	MARY	I
VAN HOUTEN	GERALD	
VAN SCODER	DICK	
VESEL	FRANK	L
VIETMEIER	ALFRED	A
VOGEL	ROBERT	
VOIGT	RICHARD	J
VRANEKOVIC	JOS	D
WADE	CLARICE	MAE
WAGNER	FLORENCE	CALLAHAN
WAHL	CHARLOTTE	POJE
WAITE	JOHN	W
WALKER	LUTHER	
WALKER	ROBERT	
WEBB	RACHEL	
WEBB	RAYMOND	T
WEBER	ARCHIBALD	AUGUST
WEILAND	JAMES	
WEILAND	JOE	J
WHITE	JACK	WILLIAM
WHITE	JOHN	M
WHITNEY	PAUL	R
WHITNEY	WALTER	H
WHITTAKER	BENJAMIN	H
WICKS	GERALDINE	
WIDOWSKI	JAMES	J
WILLIAMS	ARTHUR	E
WILLIAMS	EARL	ROGER
WILLIAMS	ELIZABETH	
WILLIAMS	GRACE	WYATT
WILLIAMS	VASHTI	
WILLIAMS	WALTER	DEWEY
WILLIAMSON JR	JAMES	
WILSON	LLOYD	PAUL
WILSON	RICHARD	G
WITZKE	EMIL	D
WOLTMAN	RAYMOND	
WOODS	FRANK	HENRY
YOUNG	LLOYD	WESLEY
YOUNG	MOSES	
YOUNG	ROBERTA	
ZANDER	PAUL	H
ZAY	WILLIAM	J
ZBOROWSKI	PETE	
ZIMMER	MARY	
ZIMMERMAN	GEORGE	C
ZUPANCIC	JOHN	
ZUSY	RAYMOND	E

Interior of popcorn stand at Euclid Beach.

Doris Humphrey and Mr. Wiffel of The Cleveland Press test the Bumble Bounce. ca. 1939.

1950 EMPLOYEE ROSTER

LAST	FIRST	MIDDLE
ABBEY	MAE	L
ADAMS	EARL	
ADKINS	CHARLES	V
AGRESTA	PHYLISS	
AINGWORTH	ELSIE	
ANDERSON	JOHN	ANDREW
ARNEY	ROXIE	E
ARNEY	RUTH	M
ARTHUR	IDA	MAE
AVERILL	EDGAR	N
AVERILL	NEAL	
AXE	PAUL	E
BABER	ALICE	Z
BAKER	CILLIUS	M
BALAS	RONALD	F
BALDNER	WILLIAM	
BALDWIN	DANIEL	
BARBA	SAM	
BARBATO	JERRY	
BARBATO	JOHN	F
BARNARD	ROBERT	L
BARNES	MAUDE	L
BARON	GEORGE	A
BARRY	DOROTHY	MARIE
BARRY	HENRY	JOHN
BARTO	ALOYSIUS	FRANK
BARTY	JOSEPH	
BAUMGARD	NORMAN	
BECK	MONICA	
BEHRA	ROGER	
BEHRA	VERNON	
BERKOVITZ	MEYER	
BERNAKY	THOMAS	F
BERNDSEN	CLYDE	E
BETZ	HAROLD	C
BILLINGHURST	EDWARD	E
BISSON JR	JOHN	F
BLAKEMORE	WILLIAM	
BLEAM	LULA	ELSIE
BLOEDE	WILLIAM	CARL
BOGAN	GEORGE	
BOGAN	MISSOURI	
BOLDIN	FRANK	
BORGA	FRED	
BOWHALL	ELMER	F
BOWHALL	HOWARD	
BOWMAN	CHARLES	E
BOWMAN	CHARLES	M
BOYCE	ROBERT	
BOYD	ANNA	J
BOYD JR	PAUL	
BRADFORD	JOHN	
BRENNAN	DANIEL	J
BRICKLEY	GEORGIA	L
BROOKINS	EDNA	BELLE
BROWN	HENRY	A
BROWN	JOHN	
BRYAN	JOHN	HENRY
BRYSON	LOUIE	
BUCKLEY	THURMAN	R
BURNS	THOMAS	
BURR	GERALD	E
BURR	RAYMOND	E
BURTON	JAMES	
BURTON	WILLIE	D
BUTLER	JOSEPH	HACKNEY
BYRNE	FRANK	P
CALI	ANTHONY	B
CALLAGHAN	BEATRICE	BERNICE
CALLIGHEN	FRANCIS	
CAMPBELL	ALEXANDER	ROBERT
CAMPBELL	FRED	
CAPASSO	JENNIE	
CARLSON	WALTER	
CARPENTER	GILBERT	
CARROLL	GERTRUDE	M
CARROLL	HAZELLE	BELLE
CARROLL	ROY	SANFORD
CARROLL	URSULA	F
CASALINA	FRANK	A
CASALINA	RAYMOND	V
CASALINA	RICHARD	
CASAMATTA	NORMAN	
CASCIATO	ROBERT	A
CASE	FRANK	E
CASE	HOWARD	J
CASIATO	ROBERT	A
CASSERLY	JAMES	
CASTLE	JOSEPH	C
CENTA	HARRY	
CERAR	VINCENT	J
CHAMBERLAIN	CAROL	
CHAMBERLAIN	VERN	
CHANDLER	MARIAN	
CHAPMAN	DUSTIN	C
CHARLES	WALTER	C
CHARLES	WALTER	W
CHRISTOFORO	NICOLA	
CHRISTOPHER	JOSEPH	
CIPITI	THOMAS	F
CIRINO	JENNIE	
CIRINO	ROSE	MARIE
CIRINO	THERESA	M
CLARK	HALE	W
CLINTON	VERNON	W
COCANOWER	PAUL	
COHAN	MITCHELL	
COLE	JOHN	W
COLVIN	JESSIE	
COMYNS	ARNOLD	T
COMYNS	JAMES	A
CONLON	HOWARD	E
CONNAVINO	JOSEPHINE	
CONNOLLY	GRACE	L
CONRY	JOHN	F
CONTENTO	GUISEPPE	
CONTORNO	ROBERT	
CONWAY	HARRY	F
CONWAY	LENA	
COOK	EDDIE	
CORLETT	GARRETT	R
CRAPNELL	HAROLD	G
CRNKOVICH	JOHN	J
CROMWELL	ROBERT	C
CROOKS	EMMA	
CROTTY	LAURENCE	A
CROW	BOB	
CROWE	BURLEY	N
CURRY	ANDREW	
CURRY	JOHN	MYERS
DARKIN	STEVE	
DAUPHIN	VIRGINIA	
DAY	WILLIAM	R
DEANGELIS	ROBERT	
DEANGELIS	RUSSELL	
DEHIL	ANDREW	

LAST	FIRST	MIDDLE
DELAMBO	FRANK	FRANCIS
DELLINGER	TOM	
DEPETRIS	EUGENE	
DEPTOLA	PHILIP	R
DESAN	ANTHONY	M
DETHOMAS	DOMENIC	
DILIBERTY	DONALD	R
DINERO	ANTHONY	
DIPALMA	IGINA	
DLOUHY	JOSEPH	J
DOBERDRUK	FRANK	
DOLEY	ANTHONY	
DOUGLAS	OLLIE	MAE
DOZIER	GRETHEL	E
DRESSER	ROBERT	H
DRESSLER	H	C
DUNCAN	JAMES	
DUNCAN	JOHN	
DURBIN	LEO	PAUL
EDWARDS	BERT	
EGAN	MARY	JO
FARMER	FLOYD	E
FARMER	J	C
FARNER	GEORGE	H
FARONE	ROBERT	
FEDERICO	JOSEPHINE	
FERRERI	GEORGE	
FINDLAY	ALEXANDER	
FISHER	CHARLES	W
FISHER	ROBERT	
FLETCHER	WILLIAM	R
FRANCIS	JOHN	S
FRANCIS	MOZART	
FRANK	GEORGE	W
FRASCH	LAURENCE	
FREDERICKS	ELIZABETH	
FREDERICKS	FRANK	JOSEPH
FROHMBURG	ROSE	
GAIER	LAWRENCE	G
GAMBATESE	FLORENCE	
GAMIERO	MICHAEL	
GARDNER	JOSEPH	D
GEIGER	ROBERT	
GENTILE	JAMES	A
GIBSON	OLLIE	J
GILMORE	RALPH	E
GOOD	RUTH	
GOODE	LENA	
GOVANG	MATIE	B
GRAHAM	JOHN	A
GRANGER	MABLE	ETTA
GRAY	ARTHUR	
GRAY	DON	
GREEN	ANNA	E
GREEN	CLAUDE	ELLSWORTH
GREEN	GEORGE	W
GREENWAY	FRED	STUDER
GRESSLE	KEITH	
GRIGGS	EDNA	MAE
GROVER	KATHERINA	A
GRUGLE	VALERIE	C
GULLIFORD	CHARLES	E
HADZICK	WILLIAM	
HALL	WINDSOR	S
HANSON	CLARENCE	H
HAPP	GEORGE	W
HARDING	BLANCHE	E
HARROLD	FLORENCE	M
HASKINS	TIM	M
HENRY	CHARLES	E
HENRY	RALPH	
HENWOOD	GEORGE	R
HIBLER	ALBERT	L
HILL	MARY	
HILL	ROY	M
HINES	ROY	C
HINTON	ELSIE	L
HINTON	WILLIAM	HARRY
HINTON	WILLIAM	R
HIRSCHAUER	ROGER	D
HOCHEVAR	RICHARD	
HOFFMAN	MARGARET	R
HOGAN	LAWRENCE	E
HOLLEY	NOVELLA	
HOLLIS	JOHN	P
HOLZHEIMER	LAWRENCE	
HOOPINGARNER	DON	
HOPKINS	ROBERT	R
HORVAT	VALENTINE	
HOVEY	JOHN	A
HOWSER	SAMUEL	L
HROSAR	JOSEPH	T
HUBNER	RICHARD	J
HUDSON	EDWIN	P
HUDSON	JOHN	B
HUGHES	ELLA	M
HUNT	BLANCHE	LEAH
HUNTER	ROY	CHARLES
HURLEY	MINNIE	
HYLAND	LILLIAN	
IMMKE	WILBUR	L
IRISH	CHARLES	L
JACKLITZ	ALBERT	
JACOB	JOHN	S
JAHLICKA	LOUIS	W
JOHNSON	DONALD	H
JOHNSTON	CLIFFORD	BASIL
JOHNSTON	DONALD	HUMPHREY
JONES	LOIS	EILEEN
JUDD	ESTIL	LEE
JUDD	WILLIAM	F
JULIAN	MARGARET	H
KAJFEZ	MATTHEW	J
KAMINSKI	RAYMOND	
KANDLE	WILLIAM	E
KANTER	ALBERT	
KEEFE	JACK	F
KEELER	ELIZABETH	
KEELER	RONALD	D
KEELER	RUSSELL	J
KELLER	CARL	J
KELLER	CHARLES	J
KELLY	TERRANCE	
KENNEDY	HERMAN	O
KENT	EDWARD	C
KERBY JR	JOSEPH	B
KERMAN	ROBERT	R
KESEGICH	JOHN	S
KIKOLI	FRANK	
KILBANE	BART	

LAST	FIRST	MIDDLE
KING	WILLIAM	J
KING JR	JACK	
KINSELLA	TOM	J
KIRKWOOD	ROBERT	E
KLEIN	ROBERT	
KNAUSS	VIRGINIA	
KOLARIK	ALLAN	
KORTZ	WALTER	A
KRALIC	RICHARD	
KRANTZ	GEORGE	E
KRULL	HERMAN	R
KRUPSKY	ANTHONY	
KUCHTA	JAMES	L
KURCHOCK	STELLA	
LAMANNA	ROSALENA	
LANG	H	WILLIAM
LANGAN	EDWARD	
LARSON	RICHARD	C
LAWSON	LOUIS	L
LAYMON JR	SAMUEL	R
LEA	EDITH	
LEARY	ROSE	B
LEVSTIK	JOHN	
LIVINGSTONE	BERTHA	ROSS
LONG	JACK	BILL
LONGFIELD	TESS	EVANS
LONGWELL	JOSEPH	
LORENZO	MARY	I
LOVE	J	C
LOVE	KATIE	
LUDEMAN	ANN	MARIE
LUPTON	ROBERT	
LUTZ	HOWARD	
MAC DONALD	JOHN	
MAC DONALD	LENA	
MACEK	VLADIMIR	
MACKENZIE	ELLEN	
MAHLER	LEE	
MAHONEY	EARL	
MAHONEY	RONALD	H
MANCINE	BENJAMIN	J
MANCINE	LOUIS	
MANDAU	RUDOLPH	
MANDEL	IDA	
MANLEY	JESSIE	
MANLEY	THOMAS	
MANNING	DOUGLAS	L
MANNING	JOHN	E
MAPLE	WILLIAM	
MARGIOTTI	CARMIN	
MARKENS	EDNA	BROWNING
MARKLE	KATHLEEN	H
MAROLD	FRANK	C
MARSHALL	CADDIE	
MARUNIAK	DONALD	
MATESIC	EILEEN	B
MATHENY	ALBERT	
MATYJASIK	WILLIAM	J
MAYER	ROBERT	R
MAYS	RICHARD	
MCCARTHY	FRANK	
MCCORMICK	GEORGE	
MCCOY	CATHERINE	
MCCRUDDEN	RICHARD	
MCDONOUGH	DANIEL	
MCGAR	WILLIAM	H
MCILRATH	IDA	E
MCINTOSH	LAWRENCE	S
MCKINLEY	JAMES	H
MCLAUGHLIN	FRANCES	
MCMAHAN	ANN	
MCNELLY	DONALD	N
MCVEEN	JERRY	M
MCVEEN	MILFORD	
MERRIMAN	ROBERT	
MERZDORF	EDWARD	
METZ	JOHN	RICHARD
MEYER	LOUIS	
MEYER	LOUIS	
MILDENBERGER	DOROTHY	
MILLER	ROBERT	W
MILLER	RONALD	
MILLS	WAYNE	E
MINER	EDWIN	D
MIOZZI	RAMON	C
MODIC	HAROLD	
MOKLEY	JOHN	W
MOLESKY	STANLEY	
MOLNAR	JOSEPH	M
MONROE	JAMES	
MOORE	MAY	C
MOORE	MILDRED	
MORGAN	CATHERINE	
MORGAN	CHARLES	
MOSIER	FREDA	V
MOUGHAN	TERRY	
MOZELSKI	RALPH	A
MUELLER	WILBUR	
MUICK	PAUL	C
MUMAW	SUSAN	B
NEKICH	ROBERT	
NESTICK	JOHN	
NEUBAUER	JOSEPH	MELVIN
NEUBAUER	KATHRYN	MELVIN
NEVILLE	WILLIAM	T
NEWTON	SOPHIA	P
NICOLL	ROBERT	G
NYKEL	RAY	R
OAKLEY	ADDIE	ELLEN
OLDENBURGH	BERNARD	A
OLESKI	EDWARD	J
OLSON	HERB	E
ORNE	HARRY	D
ORR	FREDIA	C
OTT	JOHN	J
OTTO	WALLACE	D
OZOG	CHESTER	V
PALUMBO	SAMUEL	
PANUSKA	JOSEPH	
PAOLONI	GINO	
PARKER	WILLIAM	
PARKER JR	JOHN	A
PARKS	ALBERT	L
PARKS	CARL	R
PARKS	DONALD	
PARMELEE	CHARLES	W
PASQUALE	CARMELLA	F
PAVLINA	EDWARD	G
PECK	ROBERT	D
PECK	WILLIAM	C
PELL	KELSO	

General View of Euclid Beach, Cleveland

LAST	FIRST	MIDDLE
PEPPLE	VIRGINIA E	
PERROTTI	JOSEPH	
PETERS	ARNOLD	L
PISCIONERI	JOSEPH	J
PLANTSCH	MICHAEL	
PLESHINGER	HELEN	C
PONTONI	MAE	
POTOCNIK	ROBERT	
POTTER	CHARLOTTE	
POWELL	VINCENT	J
PRESTERL	RICHARD	E
PRICE	EARL	G
PRICE	FLOYD	L
PRICE	GRACE	B
PRICE	HARRY	R
PRIJATEL	FRANK	L
PRY	DORIS	BETTY
PUINNO	NICHOLAS	J
PUTZBACH	CHARLES	
QUILTY	DANIEL	F
RAGBORG	ROBERT	J
RAIMONDO	ALFRED	
RAINES	OSCAR	
RANCATORE	PETER	M
RANEY	LUTHER	
RASH	VERNER	W
RASH	WILLIAM	F
RAWSON	MYRON	A
REARDON	ROGER	
REBMAN	CLIFFORD	
REBMAN JR	CLIFFORD	
REDFIELD	CUYLER	I
REED	GLENN	G
REED	MARIE	E
REINHARD	GEORGE	MARTIN
REPELLA	THOMAS	
RHODES	CELIA	
RICHARDS	JACK	D
RICHARDSON	ARNOLD	J
RICHARDSON	HAROLD	H
RICHENS	JAMES	F
RIDER	GILBERT	L
RIENDEAU	EUGENE	
RIENDEAU	RAYMOND	K
RIGHTNOUR	KENNETH	
RITCHEY	ALVIN	B
ROBERTS	SIMON	
ROCKWELL	STILLMAN	
RODEN	KENNETH	W
RODERICK	ALICE	M
ROGERS	PEARL	
ROLLINS	ERNEST	L
RONIGER	H	E
ROOSE	DAVID	
ROSA	ANTHONY	C
ROSE	ALBERT	C
ROSE	EUGENE	
ROSSODIVITA	DAVID	
ROSSODIVITA	MARY	
ROTHEY	CARL	
ROUCHE	NORMAN	
ROUNDTREE	IOLA	
RUMMEL	RUDOLPH	
RUSKAY	FRANK	
RUSS	STANLEY G	
RYDMAN	ELIZABETH	
RYON	EDWIN	S
SALOVON	JACK	
SAMSON	DONALD	E
SATOR	ALICE	MARGARET
SCACCO	CARMEN	
SCAFIDI	JOE	
SCARANO	ANGELA	MARY
SCHAFFER	ALLAN	A
SCHAFFER	FRANK	E
SCHEFFEL	HENRY	
SCHNUR	ROBERT	P
SCHRECK	CECILIA	G
SCOTT	WAKTER	W
SEFCHECK	MICHAEL	
SELERS	WILLIAM	S
SENICH	ALBERT	
SEXTON	LARRY	
SHANK JR	CHARLES	
SHANNON	HARRIS	COOPER
SHARP	WILLIAM	G
SHILLIDAY	EVERETT	P
SHILLIDAY	JOHN	B
SHIRER	DONALD	
SIMA	JAMES	J
SIMMS	CHARLES	L
SIMMS	LOU	CARSON
SISSON	GEORGE	
SKODA	JOSEPH	J
SKRANCE	JOHN	
SMITH	IRENE	MARIE
SMITH	MAUREEN	E
SMITH	ROBERT	E
SMITH	ROBERT	JOYCE
SMITH	TIMOTHY	
SMITH	WARREN	G
SNYDER	LUCY	J
SOMMERS	ERNESTINE	
SONNIE	WALTER	O
SOUSA	EUGENE	
SPUZZILLO	MARY	
SPUZZILLO	MICHAEL	F
ST CLAIR	JULIUS	J
STANG JR	GEORGE	H
STANLEY	LYLE	
STARINA	FLORENCE	
STARMAN	LOUIS	
STAVOLE	ANGELO	J
STEPHENS	HARRY	J
STEPP	GENUS	W
STEVENS	VALENTINE	G
STEVERDING	VINCENT	E
STEWARD	JAY	
STIELAU	EDWARD	R
STONEBACK	HOWARD	DETWEILER
STOSS	EDWARD	
STRONG	ELEANOR	
STUMPF	CORA	MARIE
STUMPF	NANCY	J
SUCHY	FRANK	
SULLIVAN	JOHN	H
SUPPLE	JOHN	F
SUTKAYTIS	DONALD	F
SUTTON	DALE	
SWOR	FRANCIS	D
SZABADOS	JOSEPH	
TABOR	ALBERT	P
TAGLIA	AMELIO	J

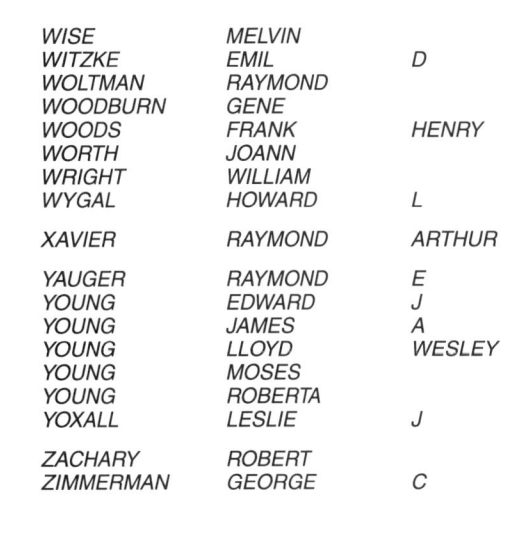

LAST	FIRST	MIDDLE
TALLETT	MARGARET	R
TAYLOR	JAMES	E
TAYLOR	THOMAS	
TAYLOR	WILLIAM	K
TEITELBAUM	DONALD	
TERRELL	ROBERT	L
TESKE	DONALD	C
THOMAS	FRANK	D
THOMAS	PAUL	EDWARD
THOMAS	PRYSE	
THOMAS	STEVEN	
THOMPSON	IRVIN	C
THUNHURST	EDYTHE	
TIMPERIO	CARL	JOHN
TIMPERIO	JOHN	C
TIZZANO	MARIA	
TOIGO	JOSEPH	D
TOWNER	JAMES	K
TRIVISON	CHARLES	A
TRIVISONNO	MARY	
TRIVISONO	LOUIS	
TRIZZINO	ROSE	L
TUCCI	JOE	
TURNER	DEAN	
TURNER	JAMES	W
TURNER	JERRY	
VACCARIELLO	URBAN	
VALENTINE	LEWIS	
VALERIO	RONALD	
VEGSO	STEVEN	J
VEIL	ROBERT	W
VERBECKY	DOLORES	
VERDERBER	ROY	E
VESEL	FRANK	L
VIETMEIER	ALFRED	A
VOGEL	ROBERT	
VRANEKOVIC	JOS	D
WADE	CLARICE	MAE
WAGNER	FLORENCE	CALLAHAN
WAHL	CHARLOTTE	POJE
WAID	JOHN	C
WAITE	JOHN	W
WALKER	ROBERT	
WALTERS	EUGENE	F
WARREN	PAUL	E
WATKINS	ROSA	BELLE
WAXMAN	SOLOMON	
WEATHERSPOON	LEATIS	
WEAVER	JOHN	H
WEBB	RACHEL	
WEBB JR	RAYMOND	T
WEBER	ARCHIBALD	AUGUST
WEBER	JAMES	H
WEBER	THOMAS	R
WELCOME	JAMES	L
WENIG	LEONARD	
WETMORE	ROBERT	
WHITNEY	WALTER	H
WHITTAKER	BENJAMIN	H
WICKES	JAY	E
WIDOWSKI	JAMES	J
WILEY	RICHARD	K
WILLIAMS	EARL	ROGER
WILLIAMS	FRANCIS	
WILLIAMS	GRAYCE	WYATT
WILLIAMS	MARGARET	
WILLIAMS	WALTER	DEWEY
WILLIAMS	WILLIAM	G
WILSON	FRED	C
WILSON	LLOYD	PAUL
WILSON	PAUL	
WILSON	RICHARD	G
WINKLE	WILLIAM	E
WINTRODE	EDWIN	

LAST	FIRST	MIDDLE
WISE	MELVIN	
WITZKE	EMIL	D
WOLTMAN	RAYMOND	
WOODBURN	GENE	
WOODS	FRANK	HENRY
WORTH	JOANN	
WRIGHT	WILLIAM	
WYGAL	HOWARD	L
XAVIER	RAYMOND	ARTHUR
YAUGER	RAYMOND	E
YOUNG	EDWARD	J
YOUNG	JAMES	A
YOUNG	LLOYD	WESLEY
YOUNG	MOSES	
YOUNG	ROBERTA	
YOXALL	LESLIE	J
ZACHARY	ROBERT	
ZIMMERMAN	GEORGE	C

1935 and nobody got out of line.

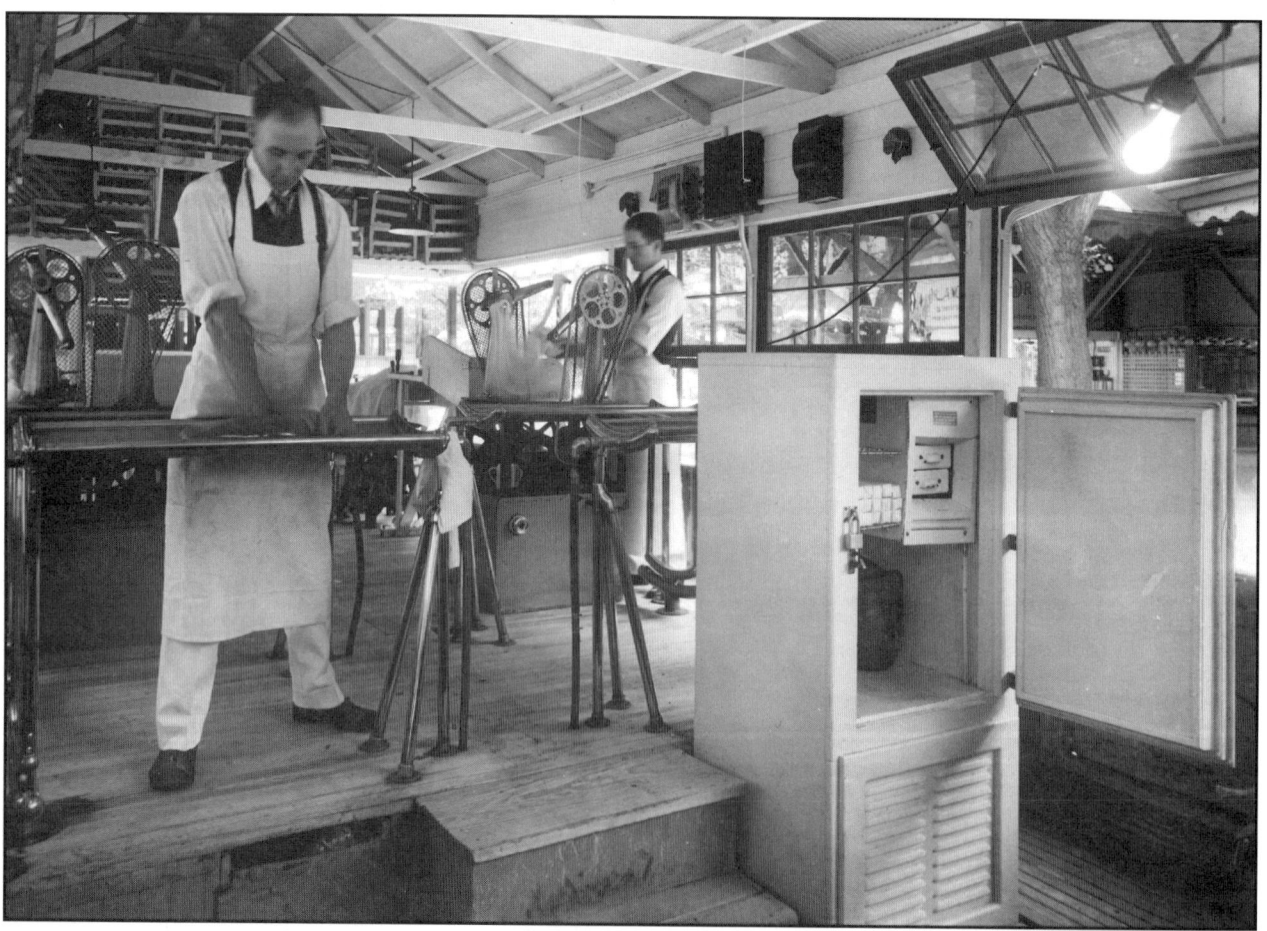

Interior of the candy kiss stand at Euclid Beach.

1951
EMPLOYEE ROSTER

LAST	FIRST	MIDDLE
ABBEY	MAE	L
ACHIN	JAMES	
AGRESTA	PHYLISS	
ALBRIGHT	GEORGE	H
ALBRIGHT	MARTIN	
ANDERSON	JAMES	R
ANDERSON	JOHN	ANDREW
ANDREWS	EUGENE	
ANDREWS	TOM	
ANGELORO	CARMELA	M
ANTOLICK	FRANCIS	J
AQUILA	JAMES	
ARNEY	WILFRED	
AUSTIN	IRENE	
AVERILL	EDGAR	N
AXE	PAUL	E
BAIRD	MARY	E
BAKER	CILLIUS	M
BAKER	EDWARD	
BAKER	RAY	H
BAKER	WILLIAM	D
BALDNER	WILLIAM	
BALDWIN	DANIEL	
BALDWIN	FRANK	D
BARBA	SAM	
BARBER	HUGH	R
BARKSDALE	KATHERINE	
BARNETT	DALLAS	
BARRY	DOROTHY	MARIE
BARRY	HENRY	JOHN
BARTHOLOMEW	CORLISS	
BARTO	ALOYSIUS	FRANK
BEAUDRY	BRUCE	H
BEHRA	ROGER	
BEHRA	VERNON	
BEHREND	EARL	J
BELASKA	EDWARD	
BENNETT	HENRY	R
BENOVICH	JOHN	J
BERKOVITZ	MEYER	
BERNDSEN	CLYDE	E
BIFANO	EMIL	J
BISHOP	ROBERT	W
BITKER	KONSTON	
BITTEL	ROGER	A
BLAKEMORE	WILLIAM	
BLANKSHINE	CARL	W
BLEAM	C	M
BLEAM	LULA	ELSIE
BLOEDE	WILLIAM	CARL
BLUMENTALIS	EDWARD	
BOGACKI	LEONARD	P
BOGAN	MISSOURI	
BOLDIN	FRANK	
BONDI	JAMES	P
BOOTH	LEE	DONALD
BORGA	FRED	
BORGSTEADT	MYRON	
BOTIN	RONALD	C
BOWHALL	ELMER	F
BOWHALL	HOWARD	
BOWMAN	CHARLES	M
BOYCE	ISAAC	P
BOYD	ANNA	J
BOYD JR	PAUL	
BRADFORD	JOHN	
BRADLEY	JAMES	H
BRANCELY	FRANK	J
BRANCH	WILFRED	
BRANT	ORA	M
BRAYKOVICH	JUANITA	
BRENNAN	JAMES	E
BRIGGS	THOMAS	LEE
BROOKS	WILBUR	D
BROWDER	LUDIE	
BROWN	GEORGE	H
BROWN	JOHN	
BROWN	WILLIAM	S
BRYAN	JOHN	HENRY
BRYANT	JACK	G
BRYLAND	VIRGINIA	
BUCHANAN	DAVID	A
BURNS	THOMAS	
BURTON	JAMES	
BUTLER	JOSEPH	HACKNEY
CALI	ANTHONY	B
CALLAGHAN	BEATRICE	BERNICE
CAMPBELL	ALEXANDER	ROBERT
CAMPBELL	FRED	
CAMPOLI	NANCY	
CANADA	JAMES	W
CANCELSLLOR	LILLIE	
CANTRELL	GARY	
CAPASSO	JENNIE	
CAPRA	CHRISTINA	
CAPRA	LOUISA	
CARDILE	FRANK	J
CARLSON	WALTER	
CARPENTER	GILBERT	
CARRIERE	MICHAEL	
CARROLL	GERTRUDE	M
CARROLL	HAZELLE	BELLE
CARROLL	ROY	SANFORD
CARROLL	URSULA	F
CARTER	CHESTER	G
CARUSO	FRED	
CARWARDINE	EDWARD	
CASAMATTA	NORMAN	
CASCIATO	OTTO	
CASCIATO	ROBERT	A
CASSERLY	JAMES	
CASTLE	JOSEPH	C
CAUSER	SHIRLEY	ANN
CERAR	VINCENT	J
CERRI	NICHOLAS	A
CHAMBERLAIN	CAROL	
CHAMBERLAIN	OMA	
CHAMBERLAIN	VERN	
CHARLES	WALTER	W
CHIARELLI	JOSEPH	
CHILDERS	ORVIL	
CHRISTIE	WILLIAM	W
CHRISTOFORO	NICOLA	
CHRISTOPHER	ANNA	G
CHRISTOPHER	JOSEPH	
CIELINSKI	DORA	
CIPITI	THOMAS	F
CIRINO	ROSE	MARIE
CIRINO	THERESA	MARIE
CLARKE	TEX	
CLATTERBUCK	FRANCIS	W
COBURN	BEUFORD	
COCHRANE	DAVE	
CONLEY	BURNARD	
CONNAVINO	JOSEPHINE	
CONTENTO	GUISEPPE	

LAST	FIRST	MIDDLE
CONWAY	LENA	
COOPER	MATILDA	
COPEN	LEWIS	
CORRIGAN	BOB	
COX	JAMES	E
COX	LAWRENCE	H
CROMWELL	ROBERT	C
CROOKS	AUDREY	J
CROOKS	EMIL	
CROOKS	EMMA	
CROSS	JOHN	W
CROTTY	LAURENCE	A
CURRY	ANDREW	
CVELBAR	THERESA	M
DAUNCH	DAVID	E
DAY	WILLIAM	R
DEANGLIS	ROBERT	
DEANGLIS	RUSSELL	
DEBENEDICTUS	JOHN	
DEBOW	BRADFORD	
DELAMBO	FRANK	FRANCIS
DELLINGER	TOM	
DERUBEIS	BENEDICT	
DIBBLE	LAWRENCE	L
DIETRICH	DONALD	E
DIFRANCO	ANGELO	
DINERO	ANTHONY	
DIPALMA	IGINA	
DODD	MATTHEW	A
DODGE	HARRY	C
DODGE	MARGOT	
DOLEY	ANTHONY	
DONALD	BESSIE	
DORRINGTON	W	P
DOTY	HAZEL	L
DOUGLAS	JAMES	W
DRESSLER	H	C
DUCCA	ALEXANDER	
DVORAK	NORMAN	R
DVORAK	WILLIAM	
EDWARDS	BERT	E
EDWARDS	DOCIA	L
EDWARDS	ELBERT	G
EDWARDS	ROBERT	D
EDWARDS	THOMAS	G
EGAN	MARY	JO
EIDAN	HARRY	
ELKINS	KENNETH	R
ETTER	MARTIN	R
EVANS	MINNIE	LEE
FANTINI	FLORENCE	
FEDERICO	CATHERINE	
FEDERICO	JOSEPHINE	
FERRELL	LAURA	
FERRERI	GEORGE	
FILLER	CALVIN	
FINK	DON	R
FISHER	CHARLES	W
FLETCHER	MILTON	L
FOX	CLARA	M
FOX	OWEN	L
FRAME	ARCHIE	M
FRANCIS	JOHN	S
FREDERICKS	ELIZABETH	
FREDERICKS	FRANK	JOSEPH
FRENCH	MILDRED	R
FROHMBURG	ROSE	
FUNK	ROBERT	N
GABLE	THELMA	
GAERTNER	JAMES	L
GALDWIN	HAROLD	E
GALIK	JOHN	
GAMBATESE	FLORENCE	
GANNET	WILLIAM	J
GARDNER	JOSEPH	D
GENTILE	JOHN	R
GERTZ	RALPH	C
GETZ	JOHN	P
GETZIEN	HARRY	
GIERMAN	WILLIAM	F
GILMORE	RALPH	E
GLIEBE	KARL	
GLOWE	LEONARD	
GOOD	RUTH	
GOULD	GEORGE	H
GRABINSKI	LEONARD	
GRANGER	MABLE	ETTA
GREEN	CLAUDE	ELLSWORTH
GREENWAY	FRED	STUDER
GREGORY JR	ROBERT	
GRIGGS	EDNA	MAE
GUNNERMAN	GEORGE	
GUNTON	WINIFRED	L
HAAS	FRED	W
HADDON	WILLIAM	
HAHN	ROBERT	
HALLGREN	HAROLD	H
HAMILTON	VIRGINIA A	
HANDLEY	OTHO	N
HANEY	EDWARD	W
HANSON	RICHARD	
HARDING	BLANCHE	E
HARMON	CHARLES	J
HARRIER	HENRY	
HARROLD	FLORENCE	
HARTE	RICHARD	E
HARWOOD JR	JOHN	W
HASSELBACH	ROGER	N
HAYES	JOHN	
HAYWARD	CLAYTON	L
HEGLAW	ROBERT	
HENDRICKS	JYTTE	J
HERZDORF	HELEN	
HINES	ROY	C
HINKO	JERRY	E
HINTON	ELSIE	L
HINTON	WILLIAM	HARRY
HINTON	WILLIAM	R
HIVELY	CHARLES	K
HIVLY	NELLIE	F
HODNIK	JOHN	
HOFFMAN	JAMES	A
HOFFMAN	WILLIAM	A
HOLLAR	ROBERT	
HOLLEY	NOVELLA	
HOLMES	ROBERT	
HOLT	GEORGE	W
HOPKINS	JOHN	J
HORRIGEN	LAWRENCE	A
HORVAT	VALENTINE	
HORWATH	PETER	P
HOUSE	EFFIE	A

THE AUTO TRAIN, EUCLID BEACH PARK, CLEVELAND, OHIO.

LAST	FIRST	MIDDLE
HOWARD	CHARLES	F
HROSAR	JOSEPH	T
HUBBARD	GERRACE	N
HUDSON	EDWIN	P
HUGHES	ELLA	M
HUGHES	KENNETH	
HUNT	BLANCHE	L
HUNTER	OLIVER	
HUNTER	ROY	CHARLES
HURD JR	CARL	B
HURLEY	ELIZABETH	E
HURLEY	ROBERT	
IAMMATTEO	MICHELINA	
INTIHAR	RICHARD	R
ISHEE	ROBERT	
JACKLITZ	ALBERT	
JACKSON	HARRY	C
JANES	PETER	J
JARAS	WILLIAM	P
JEFFRIES	WAYDE	
JEHLICKA	LOUIS	W
JENKINS	MARION	E
JENNE	CHARLES	
JOHNSON	ARLINA	
JOHNSON	MARTIN	G
JOHNSON	OSCAR	
JOHNSON	RICHARD	L
JOHNSON	WILLIAM	R
JOHNSTON	CLIFFORD	BASIL
JOHNSTON	DONALD	HUMPHREY
JOPKINS	JOHN	H
JOSLIN	A	E
JUDD	ESTIL	LEE
JULIAN	MARGARET	H
KALLENBORN	VERN	
KAMINSKI	RAYMOND	
KANIA	JOSEPH	
KARNES	HUGH	T
KARPY	STEPHEN	J
KATCHER	WILLIAM	
KEELER	ELIZABETH	
KEELER	RONALD	D
KELLER	SAM	
KELLY	JOHN	
KENT	EDWARD	C
KERN	SYLVESTER	
KIKOLI	FRANK	
KILPATRICK	JOHN	R
KILROY	JAMES	J
KING	JAMES	H
KING	LAWRENCE	
KINNEY	LEE	ROY
KINSELLA	TOM	J
KIRKWOOD	ROBERT	E
KISTHARDT	MARIE	L
KOCH	JOHN	D
KOCKA	JAMES	L
KOERNER	JOE	
KOLARIK	ALLAN	
KORTA	WALTER	A
KOSKEY	JOHN	J
KOTNIK	NORMAN	L
KOVACH	JOSEPH	
KOVAL	LEO	J
KRAFT	FRANK	
KRALIC	RICHARD	
KRAUSE	HARRY	
KREBS	JOHN	P
KREUGER	RICHARD	P
KRULL	HERMAN	R
KUCHTA	JAMES	L
LACH	RALPH	D
LACY	CHARLES	E
LAMANNA	ROSALENA	
LANG	HELMUT	M
LANGAN	NELLIE	
LAWSON	ALFRED	
LAYMON JR	SAMUEL	R
LEARY	ROSE	B
LEE	WILLIAM	E
LEFFEL	JOHN	
LEVENDOSKI	FRANK	
LEVON	ARTHUR	
LITMAN	CHARLES	L
LOGAN	BARCLAY	
LONGABERGER	DONALD	
LONGFIELD	TESS	EVANS
LONGWELL	JOSEPH	
LOPATICH	ANTON	
LOVELACE	LILLIE	B
LUCICH	WALTER	C
LUDEMAN	ANN	MARIE
LUDEMAN	JOANNE	
LUPTON	ROBERT	
LYNN	DALE	B
LYNN	H	A
LYNN	JESSE	E
LYNN	STEWART	G
LYONS	ELIZABETH	C
MAC DONALD	JOHN	
MACHAK	STEVE	F
MACKENZIE	ELLEN	
MAHON	HAROLD	
MAHONE	WILLIE	B
MAHONEY	RONALD	H
MAITLAND	DUNCAN	
MANCINE	LOUISE	
MANDEL	IDA	
MANIERI	ROCCI	F
MANNING	DOUGLAS	L
MANNING	EDWARD	
MANSELL	JAMES	
MARGHERET	GENE	
MARINE	CARLO	
MARKEL	JOSEPH	
MARKENS	EDNA	BROWNING
MARKLEY	EDWARD	C
MAROLD	FRANK	C
MARRA	ANTHONY	
MARSHALL	CADDIE	
MARTIN	PARKER	D
MASTERSON	WARREN	
MASTRANGELO	FRED	A
MAURER	RICHARD	
MCCARSKY	EDW	J
MCCARTHY	FRANK	
MCCARTHY	JAMES	D
MCCARTHY	MARGHERITA	
MCCLELLAND	FRANK	
MCCLEMENS	JOHN	D
MCCREADY	KATHLEEN	

123. THE DODGEM, EUCLID BEACH, CLEVELAND, OHIO.

89

LAST	FIRST	MIDDLE
MCCRUDDEN	RICHARD	
MCCUMBER	RAYMOND	
MCDOWELL	TOM	
MCGANN	MICHAEL	L
MCGARY	JAMES	W
MCGRANAHAN	LLOYD	
MCILRATH	IDA	E
MCINTOSH	EUGENE	
MCKIBBEN	BRANTLYB	
MCKINLEY	BERNARD	
MCKINLEY	JAMES	H
MCKINNEY	ALFRED	
MCMAHAN	ANN	
MEIER	RAY	F
MEISER	CLARENCE	G
MENGER	DALE	F
MERZDORF	EDWARD	
MESTEK JR	WILLIAM	J
METCALF	HARRY	C
METCALF	JACK	
MEYER	LOUIS	
MICHELLI	DONALD	
MILLER	MARGARET	
MILLER	MICHAEL	W
MILLER	PETER	
MILLER	RONALD	
MILLER	STEVE	R
MILLS	WAYNE	E
MINNILLO	ROSE	MARIE
MINNILLO	RUTH	
MIOZZI	ARLENE	
MIOZZI	RAMON	C
MISLEY	JAMES	J
MITCHELL	CHESTER	P
MODIC	HAROLD	
MOHR	GARY	D
MOKLEY	JOHN	W
MOLNAR	ROBERT	L
MOLNAR	RONALD	
MONTAGNO	SALVATORE	
MOORE	MAY	C
MORGAN	CHARLES	
MORGAN	EDWARD	
MORGANTI	JOSEPH	L
MORIARTY	DANIEL	E
MORRISON	JOHN	
MOSCOVITZ	ABRAHAM	
MOSIER	FREDA	V
MOUGHAN	TERRY	
MOYNAN	JAMES	
MULCAHY	JOHN	E
MULLIN	LAURA	P
MUMAW	SUSAN	B
MURPHY	JAMES	R
MYERS	DONALD	E
NACHTIGAL	JAMES	A
NAIDA	ALEX	L
NEFF	ARTHUR	C
NEIBERT	GILBERT	W
NEKICH	ROBERT	
NELSON	JAMES	V
NESBETT	NORVIN	
NEUBAUER	KATHRYN	MELVIN
NEWMAN	NATHAN	
NICOLL	ROBERT	G
NIDO	WALLACE	J
NOLAN	PETE	TOM
NOLES	FRANK	D
NORRIS	CASEY	T
OAKLEY	ADDIE	ELLEN
OLIVER	IRENE	B
ORNE	HARRY	D
OSBORNE	JAMES	R
OTT	GEORGE	S
OWENS	CLARENCE	D
OZOB	CHESTER	V
PACHINGER	FRED	J
PARK	DONALD	
PARKER	WILLIAM	
PASHKO	HOLMES	
PATRICK	EDDIE	V
PATRICK	JOHN	
PATRICK	WILLIAM	J
PATTON	HARRY	
PATTON	HARVEY	R
PATTON	HENRY	E
PECK	ROBERT	D
PEPPLE	VIRGINIA E	
PERROTI	JOSEPH	
PERSELL	CHARLES	
PETERS	ARNOLD	L
PETRO	SUE	JULIAN
PETRUNO	FRANK	D
PEVSNEK	MELVIN	E
PHILLIPS	JAMES	F
PIORKOWSKI	JOSEPH	S
PITKIN	CHARLES	E
PITT	DANIEL	A
PLANTSCH	MICHAEL	
POHL	EDWARD	
POLING	DOYT	E
PONSART	HENRY	G
POOLE	HERBERT	C
POPSON	JOSEPH	J
PRICE	FLOYD	L
PRICE	HARRY	R
PRIEST	LAURENCE	
PUINNO	NICHOLAS	J
QUICK	WILLIAM	H
RAIMONDO	ALFRED	
RAINES	OSCAR	
RANKER	WILBUR	
RASH	VERNER	W
RASH	WARRENE	M
RAWSON	MYRON	A
REARDON	ROGER	
REBMAN	CLIFFORD	
REDDICK	BESSIE	N
REDFIELD	CUYLER	I
REED	MARIE	E
REEVES	SAM	
REEVES	WILLIE	
REID	ROBERT	C
REILLY	ETHEL	L
REINHARD	GEORGE	MARTIN
REPELLA	THOMAS	
REPIC	EDWARD	M
REUBLIN	ROBERT	
REYNARD	CLARENCE	
RICE	ESTHER	
RICHARDS	JAMES	M

LAST	FIRST	MIDDLE
RICHARDS	LEWIS	J
RICHARDSON	ARNOLD	J
RICHARDSON	JAMES	W
RICHARDSON	ROBERT	D
RICHENS	JAMES	F
RIDER	GILBERT	L
RITCHEY	ALVIN	B
RIVERS	EVELYN	
ROBINSON	ALFONSO	
ROBINSON	EARL	D
ROBINSON	HARRY	G
ROBINSON	JOSEPH	H
ROBINSON	L	
ROBINSON	WILLIAM	
ROBISON	CARL	L
ROCCHI	FRANK	P
ROCKWELL	STILLMAN	
RODERICK	ALICE	M
ROGERS	CHARLES	
ROHL	JOE	
ROHR	ROBERT	
ROLAND	RUTH	
ROLLAND		ROBERT
ROLLINS	ERNEST	L
RONIGER	H	E
ROSS	FRANK	
ROSS	ROGER	F
ROSS	WILLIAM	
ROSSODIVITA	DAVID	
ROSSODIVITA	MARY	
ROWELL	ERNEST	
ROZANC	THOMAS	
RYAN	JOHN	
RYDMAN	ELIZABETH	
SALERO	LAWRENCE	
SANDERS	PAUL	S
SANDOW	LEWIS	
SANTORELLI	VINCENT	
SANZO	MICHAEL	
SATOR	ALICE	MARGARET
SAXE	HENRY	C
SCARANO	ANGELA	MARY
SCHAEFFER	WILLIAM	C
SCHAFFER	ALLAN	A
SCHEFFEL	HENRY	
SCHEU	LOUISE	B
SCHNEIDER	JOHN	
SCHULTZ	ELMER	A
SCHUSCHU	ROBERT	
SCHWALL	GRACE	
SCOTT JR	WALTER	W
SCOTT SR	WALTER	W
SEMAN	JOHN	
SEMAN	WILLIAM	
SHANK JR	CHARLES	
SHANNON	FLOYD	A
SHANNON	HARRIS	COOPER
SHARP	JOHN	
SHARP	WILLIAM	G
SHEEHAN	JAMES	D
SHEEHAN	MABELLE	B
SHEPPARD	THEODORE	R
SHIBLEY	BURLEIGH	
SHILLIDAY	EVERETT	P
SHILLIDAY	JOHN	B
SHOENBERGER	EDITH	A
SHORT	ERNWAY	
SHORT	EVERETT	
SHORT	VERTIS	
SICKELS	EARL	F
SIGLER	JACK	
SIMA	JAMES	J
SIMPSON	EDWARD	
SISSON	GEORGE	
SKENDER	RAYMOND	
SKERL	THOMAS	H
SKODA	JOSEPH	J
SKOMROCK	FRANCIS	E
SLEE	HENRY	
SMITH	ALAN	H
SMITH	ALICE	
SMITH	ALVIN	W
SMITH	CHARLES	L
SMITH	FRED	C
SMITH	HARRY	
SMITH	IRENE	MARIE
SMITH	MARY	C
SMITH	ROBB	S
SMITH	WALLACE	
SMITH	WILLIAM	P
SNYDER	GEORGE	
SNYDER	RALPH	
SOLTIS	MARCELLA	
SOMMERS	ERNESTINE	
SONNIE	PETER	E
SONNIE	WALTER	O
SPEES	JAMES	I
SPIRES	BERYLE	E
SPUZZILLO	JOHN	A
SPUZZILLO	MARY	
SPUZZILLO	MICHAEL	A
SPUZZILLO	MICHAEL	F
ST CLAIR	JULIUS	J
STAKICH	DANIEL	D
STANCZYK	BRUNO	C
STANG JR	GEORGE	H
STANICIC	PAUL	
STANISKIS	JULIUS	
STANKOVITCH	CHARLES	
STAPP	PAUL	R
STARINA	FLORENCE	
STARK	RAYMOND	A
STARMAN	LOUIS	
STAVOLE	ANGELO	J
STEELE	ARTHUR	W
STEELE	RICHARD	W
STEHR	DONALD	L
STEIGNER	ISA	
STEIN	ALBERT	F
STELTER	ELIZABETH	H
STEPP	CLARICE	P
STEPP	GENUS	W
STERBA	NORMAN	R
STEVENS	VALENTINE	G
STEVERDING	VINCENT	E
STIELAU	EDWARD	R
STONE	ARNOLD	T
STONEBACK	HOWARD	DETWEILER
STONEMAN	DELBERT	
STOVALL	WILLIAM	N
STRIKLAND	AMOS	
STRONG	ELEANOR	
STUART	JAMES	MILLER
STUMPF	CORA	MARIE

37:—Camping Grounds, Euclid Beach Park, Cleveland, Ohio.

LAST	FIRST	MIDDLE
SULLIVAN	JAMES	D
SULLIVAN	JOHN	H
SULLIVAN	TIMOTHY	
SUMPTER	MILTON	M
SUTTON	DALE	L
SWANEY	WILLIAM	E
SWINEHART	LOWELL	
SWINEHART	PEARLE	
SZEMPLENSKI	WALTER	
TABOR	ALBERT	
TAGLIA	AMELIO	J
TALLETT	MARGARET	R
TATE	BLANE	
TAYLOR	EVANS	
TAYLOR	JAMES	E
TAYLOR	THOMAS	
TELLJOHN	THOMAS	F
TEODOSIO	VINCENT	
THOMAS	PAUL	EDWARD
THOMAS	PRYSE	
THOMAS	STEVEN	
THOMPSON	IRVIN	C
THOMPSON	WALTER	E
THUNHURST	EDYTHE	
TILLER	GRAHAM	K
TIMMS	DAVID	R
TIMPERIO	JOHN	C
TIZZANO	MARIA	
TOIGO	JOSEPH	D
TOLHURST	ROBERT	G
TOLL	ORLIN	O
TOLL	RUTH	
TORRENCE	CARL	R
TRANCHITO	JOSEPH	
TRIVISON	CHARLES	A
TRIVISONNO	GRACE	
TRIVISONNO	JOHN	
TRIVISONNO	MARY	
TRMBLE	ARTHUR	
TRUSKOLAWSKI	PETER	
TUGER	JOHN	J
TURNER	DONALD	R
URBAN	THOMAS	S
USSAI	JORDAN	J
VACCARIELLO	JOSEPHINE	
VACCARIELLO	URBAN	
VALENTINE	JAMES	
VALENTINE	LEWIS	
VALETICH	MARTIN	
VATH	CHARLES	R
VEGNEY	EUGENE	T
VODVARKA	JAMES	F
VOGEL	ROBERT	
VRANEKOVIC	JOS	D
WAID	ROBERT	C
WAITE	JOHN	W
WALKER	HARRY	G
WALKER	ROBERT	
WALTON	RICHARD	A
WARREN	PAUL	E
WATERS	LULU	M
WATERS	WILLIAM	
WEBER	ARCHIBALD	AUGUST
WEBER	LLOYD	R
WEISSENBURGER	CHARLES	
WENIG	LEONARD	
WHITE	JOHN	M
WHITE	JONES	C
WHITE	MARY	E
WICKER	DAN	
WICKER	MELVIN	
WICKES	JAY	E
WILEY	RICHARD	
WILLIAMS	CHARLES	
WILLIAMS	GRACE	WYATT
WILLIAMS	JACK	M
WILLIAMS	MARGARET	
WILLIAMS	MARY	
WILLIAMS	ROGER	A
WILLIAMS	WALTER	DEWEY
WILMOT	RICHARD	C
WILSON	FRANK	A
WILSON	LLOYD	PAUL
WILSON	PAUL	
WILSON	RICHARD	G
WILSON	WILLIAM	S
WILT	MINNIE	
WINTERS	EARL	R
WISE	MELVIN	
WITZKE	EMIL	D
WOLF	ADA	G
WOOD	ROBERT	
WOODS	FRANK	HENRY
WOODY	ANNIE	
WRIGHT	WILLIAM	
WYGAL	HOWARD	L
YOUNG	JAMES	A
YOUNG	LLOYD	WESLEY
YOUNG	MOSES	
YOUNG	ROBERTA	
YOXALL	LESLIE	J
ZAGAR	FRANK	
ZAGAR	STANLEY	
ZARLINSKY	EDWARD	J
ZELL	ALBERT	H
ZIBERT	RUDOLPH	
ZIMMERMAN	ED	
ZIMMERMAN	F	E
ZIMMERMAN	GEORGE	C
ZIMMERMAN	NORMAN	
ZUPANCIC	FRANK	

1952 EMPLOYEE ROSTER

LAST	FIRST	MIDDLE
ABBEY	MAE	L
ALBRIGHT	GEORGE	H
ALDRICH	PAUL	B
ALLEN	CHARLES	R
ALLEN	HAROLD	A
ANDERSON	JOHN	ANDREW
ANDERSON	MAE	
ANDERSON	WAYNE	
ANDREWS	EUGENE	
ANGELORO	CARMELA	M
ANGERMANN	J	WILLIAM
ANGERMANN	PEARL	
ANJESKEY	JAMES	
ANTOLICK	FRANCIS	J
APLIN	CLIFFORD	
ASBECK	HELEN	
AUSTIN	CHARLES	
AUSTIN	CLIFFORD	R
AUSTIN	JOSEPH	G
AVERILL	EDGAR	N
AXE	PAUL	E
BAIRD	MARY	E
BAIRD	WALTER	
BAITT	EDWARD	
BAJKOVEC	JOHN	P
BAKER	CILLIUS	M
BAKER	JAMES	C
BALDWIN	DANIEL	
BALDWIN	FRANK	D
BALDWIN	HAROLD	E
BALOGH	DONALD	
BANKER	RAYMOND	
BARBA	SAM	
BARBO	FRANK	
BARELLA	RUBY	M
BARON	GEORGE	A
BARRY	DOROTHY	MARIE
BARRY	HENRY	JOHN
BARTO	ALOYSIUS	FRANK
BASTARDO	THOMAS	
BECK	GEORGE	R
BEDA	JULIUS	J
BEHRA	VERNON	
BENNETT	HENRY	R
BENZ	FRANKLIN	L
BERKOVITZ	MEYER	
BIBB	LEON	
BILLETT	DAVID	L
BITTEL	ROGER	A
BIXLER	PAUL	A
BLACK	DAVID	J
BLADE	GILBERT	A
BLAKEMORE	WILLIAM	
BLEAM	C	M
BLEAM	LULA	ELSIE
BLOEDE	WILLIAM	CARL
BLOUGH	ROBERT	E
BONATO	VALERIA	A
BONOMO	DONALD	J
BORSOS	MINNIE	
BOWHALL	ELMER	F
BOYETTE JR	ROY	B
BRADFORD	JOHN	
BRAMBOR	NICK	
BRANCH	ANNA	ROSE
BRANCH	WILFRED	
BRAND	JOHN	F
BRIGGS	THOMAS	LEE
BRISLINGER	JOHN	A
BRODNIK	CHARLES	
BROWDER	LUDIE	
BROWN	DAISY	MAY
BROWN	JACK	
BROWN	PAUL	K
BROWN	WILLIAM	H
BRYJA	MARGARET	
BRYLAND	VIRGINIA	
BUCKEYE	MARY	RUTH
BURNS	RICHARD	L
BURTON	JAMES	
BURTON	WILLIE	D
BUSDIECKER	RONALD	
BUTLER	JOSEPH	HACKNEY
CALLAGHAN	BEATRICE	BERNICE
CAMERON	GORDON	
CAMPBELL	ALEXANDER	ROBERT
CAMPBELL	FRED	
CAMPOLI	NANCY	
CANADA	JAMES	W
CAPRA	LOUISA	
CARLTON	GERALD	G
CARROLL	GERTRUDE	M
CARROLL	HAZELLE	BELLE
CARROLL	ROY	SANFORD
CARROLL	URSULA	F
CARTWRIGHT	LYLE	B
CARWARDINE	EDWARD	
CASON	ALMA	
CASSERLY	JAMES	
CASTLE	JOSEPH	C
CATANIA	FRANK	
CATEY	EARL	W
CAVER	LULA	MAE
CHAMBERLAIN	CAROL	
CHAMBERLAIN	OMA	
CHAMBERLAIN	VERN	
CHRISTIAN	PHILIP	H
CHRISTOPHER	JOSEPH	
CIPITI	THOMAS	F
CIRINO	JOHN	
CIRINO	THERESA	M
CLARK	TERRY	J
COCITA	JOSEPH	
COGHLAN	MARY	E
COMBS	CHRISTINE	M
CONNAVINO	JOSEPHINE	
CONRAD	BRUCE	
CONTENTO	GUISEPPE	
CONWAY	LENA	
CONWAY	STELLA	M
COOK	ORIENE	
CORRIGAN	BOB	
CORRIGAN	WALTER	T
COX	HERBERT	M
COX	JAMES	E
COX	RAYMOND	
COX	ROBERT	G
CRALL	JAMES	E
CRAMER	GERALD	
CRANDALL	HOWARD	
CRAWLEY	CHARLES	P
CREDICO JR	JOHN	
CROOKS	EMIL	
CROOKS	EMMA	
CROTTY	LAURENCE	A
CRUMYE	RICHARD	

LAST	FIRST	MIDDLE
CURRY	ANDREW	
DAMES	LAWRENCE	E
DANIELS	RAYMOND	P
DARBY	ERNESTINE	
DAVIS	DOROTHY	E
DAVIS	PAULINE	
DAWSON	THOMAS	J
DAY	WILLIAM	J
DAY	WILLIAM	R
DAY	WILLIAM	V
DEGNER	NORMAN	C
DELAMBO	FRANK	FRANCIS
DELONG	STANLEY	
DEMARCHEK	LOUIS	
DEROSSETT	THELMA	L
DEVINE	HOWARD	H
DICARPO	CHARLES	
DILIBERTO	DONALD	R
DIRUSSO	SARAH	
DONAHUE	DAVID	
DOSSA	STEVE	
DOTSON	MARY	R
DOTY	LONE	C
DOUGLAS	JAMES	W
DRAVES	JOSEPH	F
DRESSLER	H	C
DUCCA	ALEXANDER	
DUCKER	CHARLES	E
DUNCAN	GLENN	E
DURHAM	CHARLES	M
DYKES	WILLIE	H
EDDY	BYRON	W
EDDY	VIRGINIA	K
EDWARDS	EARL	L
EDWARDS	GEORGE	E
ELLISON	MERRITT	
EPSTEIN	SAMUEL	
EVANCIC	FRANK	G
EVERETT	CLIFF	E
EWING	JEANNETTE	
FAGAN	EDWARD	J
FARNER	GEORGE	H
FAUNCE	RICHARD	C
FAWCETT	HENRY	B
FEDERICO	ANTHONY	
FEDERICO	CARMELA	
FEDERICO	CATHERINE	
FEDERICO	JOSEPHINE	
FERRERI	GEORGE	
FIELDS	HERBERT	J
FINNIGAN	THOMAS	F
FISCUS	HARRY	
FITZGERALD	TROY	
FITZPATRICK	WALTER	
FLEMING	GEORGE	
FLETCHER	ENOCH	E
FOCARETO	LOUIS	N
FOLLAS	HERBY	
FONDA	FRANK	W
FOSTER	MATTIE	
FOUSER	THOMAS	C
FOX	CLARA	M
FRANCIS	JOHN	S
FRANTZ	LORETTA	
FREDERICKS	ELIZABETH	
FREDERICKS	FRANK	J
FROHMBURG	ROSE	
FUNK	ROBERT	N
GAERTNER	JAMES	L
GALIK	JOHN	
GAMBATESE	FLORENCE	
GANNET	WILLIAM	J
GARDNER	JOSEPH	D
GENSERT	GEORGE	M
GENTILE	JOHN	R
GERTZ	CHARLES	R
GETZIEN	HARRY	
GIBBONS	WILLIAM	E
GIBSON	OLLIE	J
GIERMAN	WILLIAM	F
GIFFOR JR	NICHOLAS	J
GILMORE	RALPH	E
GIOVANETTI	REGO	F
GLIEBE	KARL	
GLOWE	LEONARD	
GLOWE	MARGARET	
GODWIN JR	THOMAS	W
GOLDMAN	JOAN	
GOOD	RUTH	
GRADY	RICHARD	T
GRANT	WILLIAM	C
GREEN	ARNETT	
GREEN	ROBERT	D
GREEN	WILLIAM	A
GREENWA	FRED	STUDER
GREGORY JR	ROBERT	
GRIGGS	WILHELMINA	
GRON	PHILIP	M
GRUBB	JACK	F
GUNTERMAN	MARY	
GUNTON	WINIFRED	L
GUTKOWSKI	LEONARD	
HAGER	CARL	H
HAGY JR	WILLIAM	H
HALADUS	JAMES	D
HALL	EUGENE	
HAMILTON	GLADYS	M
HAMILTON	ROBERT	
HAMMOND	RICHARD	
HANEY	EDWARD	W
HANLON	MARY	
HANNAN	WILLIAM	
HANTZ	JACK	M
HARMON	CHARLES	J
HARRIS	ROBERT	O
HARRISON	WILLIE	MAE
HASLEM	JAMES	
HASSELBACH	ROGER	N
HEALY	PATRICIA	A
HECKMAN	EDWARD	
HEFNER	CHARLES	D
HEGLAW	ROBERT	
HENDERSON	RAYMOND	
HERMAN	ALBERT	W
HIGLEY	MAUD	M
HINES	ROY	C
HINKO	JERRY	E
HINTON	ELSIE	L
HINTON	THOMAS	B
HINTON	WILLIAM	R
HIRSCHAUER	ROGER	D

CONCRETE COTTAGE, EUCLID BEACH PARK, CLEVELAND, OHIO.

LAST	FIRST	MIDDLE
HODNIK	JOHN	
HOFFMAN	MARY	
HOFMAN	ROBERT	C
HOLLAND	JOHN	
HOLLEY	NOVELLA	
HOLMAN	FRED	H
HOLMES	FLEMING	C
HOLMES	ROBERT	
HOLSTINE	RALPH	
HOWARD	EUGENE	
HUBBARD	HARRY	R
HUDSON	EDWIN	P
HUGHES	ELLA	M
HUNT	BLANCHE	L
HUNTER	ROY	CHARLES
HURLESS	JACKIE	
HURLEY	ELIZABETH	E
HURLEY	ROBERT	
IAFELICE	FRANK	
IAMMATTEO	MICHELINA	
IAUS	JOHN	F
ISHEE	ROBERT	
IVES	ROBERT	L
JACKLITZ	ALBERT	
JACKSON	ALVA	L
JACKSON	NORTON	C
JAMES	JOHN	W
JANCIN	NANCY	M
JANES	PETER	J
JARMON	MAXINE	
JAYNES	TED	R
JAZBEC	ROBERT	
JEHLICKA	LOUIS	W
JENKINS	MARION	E
JOHNS JR	NICK	
JOHNSON	CHARLES	
JOHNSON	HATTIE	
JOHNSON	RALPH	
JOHNSON	RICHARD	L
JOHNSON	SOL	
JOHNSTON	CLIFFORD	BASIL
JOHNSTON	DONALD	HUMPHREY
JOHNSTON	GEORGE	W
JOHNSTON	RICHARD	W
JORNDT	FLORENCE	
JUDD	ESTL	LEE
JULIAN	MARGARET	H
KACSMAR	FRANCES	
KAJFEZ	MATTHEW	J
KAMINSKI	RAYMOND	
KARTUNE	ALBERT	W
KASPAR	MAY	E
KATCHER	WILLIAM	
KAVARAS	HELEN	
KEELER	ELIZABETH	
KELLER	SAM	
KELLEY	CHARLES	H
KELLEY	JAMES	J
KELLY	WILLIAM	JAMES
KELLY	WILLIAM	L
KEMTER	AGNES	
KENDALL	KATHERINE	
KENNEDY	ROBERT	
KENT	EDWARD	C
KERN	SYLVESTER	
KIKOLI	FRANK	
KINSELLA	TOM	J
KIRKPATRICK	CATHERINE	
KLEIN	SAM	
KOCZAK	WILLIAM	C
KOERNER	EMIL	J
KOLEZAR	RICHARD	
KORTA	WALTER	A
KOTABISH	RICHARD	
KOVACH	JOSEPH	
KOVACH	ROBERT	
KOVACIC	VICTOR	E
KOVELL	ALFRED	J
KRAFT	FRANK	
KRAMER	JANE	
KRAUSE	HARRY	
KREBS	JOHN	P
KREUGER	RICHARD	P
KRIVDO	JEROME	M
KRUEGER	JOSEPH	
KUSHLAN	LOUIS	C
LACY	CHARLES	E
LANIER	LOUIS	
LAPINSKAS	ROLAND	P
LATKOVICH	ELEANOR	
LAUNDERS	PRISCILLA	
LAWRENCE	JIMMIE	
LAWSON	ALFRED	
LEARY	ROSE	B
LEESBERG	ANNA	L
LEESON	HARVEY	
LEFFEL	JOHN	
LENEHAN	JOHN	P
LICKER	MANNIE	R
LIVINGSTON	RAYMOND	
LOMBARDO	DON	
LONGFIELD	TESS	EVANS
LONGWELL	JOSEPH	
LOPATICH	ANTON	
LORBER	MIKE	
LOSAK	RALPH	
LOVE	HEZEKIAH	
LOVE	J	C
LUDEMAN	ANN	AMRIE
LUDEMAN	KENNETH	
LUND	WILLIAM	W
LUPTON	ROBERT	
LYNN	EVELYN	M
LYNN	H	A
LYNN	SHIRLEY	E
LYNN	STEWART	G
LYONS	ELIZABETH	C
MAC DONALD	JOHN	
MACHAK	STEVE	F
MACKENZIE	ELLEN	
MACQUEEN	CHARLES	
MAGRUDER	JARVIS	D
MAHONE	WILLIE	B
MAHONE JR	JOHN	
MANDEL	IDA	
MANLEY	HAROLD	J
MANNING	DOUGLAS	L
MANNING	EDWARD	
MANNING	HOOD	E
MANNING	JOHN	H
MANNION	JOHN	B
MANSELL	JAMES	

LAST	FIRST	MIDDLE
MANTEL	ANNA	
MARGHERET	GENE	
MARINARO	VERNE	
MARKEL	JOSEPH	
MARKENS	EDNA	BROWNING
MARRA	ANTHONY	
MARSH	THOMAS	E
MARSHALL	CADDIE	
MARTIN	KIERAN	J
MARTIN	WILLIAM	
MARTUCCI	ROBERT	M
MASON	VERA	
MASSIE	GEORGE	T
MATTERN	WESLEY	R
MAURER	MICHAEL	G
MAVINS	JEWELL	
MAYS	EVELYN	
MCCARSKY	EDW	J
MCCARTHY	JAMES	D
MCCREADY	HOWARD	
MCCRUDDEN	ROBERT	
MCDOWELL	FLORENCE	
MCGANN	MICHAEL	L
MCHENRY	JAMES	M
MCILRATH	IDA	E
MCKENNA	JOHN	W
MCKENZIE	LEWIS	
MCKINLEY	BERNARD	
MCKINLEY	JAMES	H
MCKINNEY	WILLIAM	
MCLEOD	MERVIN	
MCNIECE	RAYMOND	
MCVEEN	JERRY	M
MENGER	DALE	F
MERZDORF	EDWARD	
METHOD	FRANCES	S
MEYER SR	LOUIS	
MICHELLI	DONALD	
MILES	ARCHIE	
MILLER	CARL	H
MILLER	CLOYD	L
MILLER	FRED	W
MINARIK	JAMES	
MINNILLO	RUTH	
MIOZZI	ARLENE	
MISLEY	JAMES	J
MODIC	HAROLD	
MOORE	SUSIE	TAYLOR
MORGAN	CATHERINE	
MORGAN	EUGENE	F
MORGAN	HUGH	
MORIARTY	DANIEL	E
MORRISON	MAURINE	
MOSCOVITZ	ABRAHAM	
MOSIER	FREDA	V
MOYNAN	JAMES	
MRAMER	EDWARD	
MULLINS	JAMES	
MUMAW	SUSAN	B
MURRAY	JOHN	
NACHTIGAL	JAMES	A
NAGLE	CLINTON	
NAJARIAN	RICHARD	
NEER	CLARK	E
NEFF	ARTHUR	C
NEUBAUER	KATHRYN	MELVIN
NEWMAN	ARTHUR	E
NICOLL	ROBERT	G
NORRIS	CASEY	T
NOYER JR	JOHN	A
OAKLEY	ADDIE	ELLEN
OBOCZKY	JOHN	G
OBRICK	LEONARD	A
ODEN	SUSIE	B
OKROS	MARIE	H
ONEIL	JOHN	C
OPALK	MARTIN	
ORNE	HARRY	D
ORZECH	OTTO	
OTCASEK	THEODORE	
OWENS	CLARENCE	D
PAPES	MARTIN	
PARK	DONALD	
PARKER	JAMES	A
PARKER	WILLIAM	
PATFIELD	RONALD	
PATRICK	EDDIE	V
PAULIN	JOHN	C
PAYNE	GLADYS	M
PECK	ROBERT	D
PERROTI	JOSEPH	
PERSELL	CHARLES	
PERUSEK	RICHARD	J
PETERSON	LLOYD	G
PETRARCA	ANTHONY	
PETRECCA	JOSEPH	
PHELPS	JOSEPH	C
PIORKOWSKI	JOSEPH	S
PITTOCK	ARTHUR	J
PITTOCK	RON	ELMER
PLANTSCH	MICHAEL	
PLESCIA	LUCY	
POCH	NICK	
PODLOGER	FRANK	M
POE	JACKSON	LEE
POLING	DOYT	E
PONTAU	WARNER	A
POWELL	CHARLES	H
PRATT	RICHARD	
PRICE	FLOYD	L
PRICE	JACK	W
PUINNO	NICHOLAS	J
RADEL	ROBERT	J
RAIMONDO	ALFRED	
RAINES	OSCAR	
REARDON	ROGER	
REDFIELD	CUYLER	I
REED	MARIE	E
REESE	DONALD	J
REESE	GEORGE	J
REESE	VERA	
REEVES	JIMMIE	
REEVES	WILLIE	
REINER	JOSEPH	
REINHARD	GEORGE	ANDREW
REINHARD	GEORGE	MARTIN
REPELLA	THOMAS	
REPIC	EDWARD	M
RICE	WILLIAM	D
RICHARD	JACK	A
RICHARD	PHILLIP	E
RICHARDSON	ARNOLD	J

40:—AERIAL VIEW OF EUCLID BEACH PARK, CLEVELAND, OHIO.

LAST	FIRST	MIDDLE
RIDGEWAY	HOWARD	
RILEY	NEVIN	M
RIO	RICHARD	J
ROBINSON	JOSEPH	H
ROCCHI	FRANK	P
ROCCO	MARY	ANN
ROCKER	LEONARD	
RODERICK	ALICE	M
ROLLINS	ERNEST	L
RONIGER	H	E
ROOME JR	HOWARD	
ROSE	EUGENE	
ROSS	EDGAR	E
ROSS	FRANK	
ROSS	HOWARD	J
ROSSODIVITA	DAVID	
ROSSODIVITA	MARY	
ROUSCH	EDWARD	R
ROWELL	ERNEST	
ROZANC	THOMAS	
ROZINSKY	FRANCIS	
RUCKER	CARTER	H
RUSS	STANLEY	G
RYDMAN	ELIZABETH	
RYON	JAMES	J
SALINGER	RUSSELL	
SANTUCCI	MELINA	A
SAPUTKA	JERRY	
SCHAFER	FRANCES	
SCHIEVE	ARTHUR	E
SCHINKO	JULIA	
SCHNEIDER	JOHN	
SCHREIBER	ARTHUR	
SCOTT	KENNETH	R
SCOTT JR	WALTER	W
SCOTT SR	WALTER	W
SEMAN	JOHN	
SEMAN	WILLIAM	
SETSER	GEORGE	C
SHANK	CHARLES	
SHANK JR	CHARLES	
SHANNON	HARRIS	COOPER
SHARP	WILLIAM	G
SHEA	JOSEPH	W
SHEEHAN	JAMES	D
SHEEHAN	MABELLE	B
SHEPPARD	THEODORE	R
SHIBLEY	BURLEIGH	
SHILLIDAY	EVERETT	P
SHILLIDAY	JOHN	B
SILVEROLI	JOHN	
SIMA	JAMES	J
SIMMONS	KERMIT	L
SKERL	THOMAS	H
SKOLL	ALBERT	H
SKOMROCK	FRANCIS	E
SMERDEL	COLINA	J
SMITH	ALLAN	
SMITH	ALVIN	W
SMITH	CHARLES	L
SMITH	FRED	C
SMITH	HARVEY	E
SMITH	IRENE	MARIE
SMITH	MARY	C
SNELLING	FRANCES	
SNYDER	GEORGE	
SNYDER	RALPH	
SODEN	JAMES	E
SOMMERS	ERNESTINE	
SONNIE	WALTER	O
SOUSA	EUGENE	
SPUZZILLO	JOHN	A
SPUZZILLO	MARY	
SPUZZILLO	MICHAEL	A
SPUZZILLO	MICHAEL	F
SQUIRE	JACK	E
ST CLAIR	JULIUS	J
STACKHOUSE	JOHN	R
STADLER	RAY	
STAHNKE	JOHN	F
STANCZYK	BRUNO	C
STANKOVITCH	CHARLES	
STAPE	WILLIAM	H
STAPP	CLYDE	K
STARINA	FLORENCE	
STARMAN	LOUIS	
STEELE	ARTHUR	W
STEELE	HARRY	
STEELE	RICHARD	W
STEIGNER	ISA	
STEIN	ALBERT	F
STEPP	GENUS	W
STERLING	DOHRMAN	S
STEVENS	VALENTINE	G
STONEBACK	HOWARD	DETWEILER
STONEMAN	DELBERT	
STOVALL	WILLIAM	
STRONG	ELEANOR	
STRONG	HAROLD	E
STUMPF	CORA	MARIE
SUAREZ	ANNA	
SULLIVAN	TIMOTHY	
SWINEHART	LELAN	L
SWINEHART	LOWELL	
SWINEHART	PEARLE	
TALLETT	MARGARET	R
TARSITANA	JOSEPH	
TAYLOR	THOMAS	
TAYLOR	VERBIE	
TEODOSIO	VINCENT	
THACKER	DARWIN	S
THACKER	HOMER	J
THOMAS	PAUL	EDWARD
THOMAS	PRYSE	
THOMAS	ROSIA	
THOMAS	STEVEN	
THOMPSON	IRVIN	C
THUNHURST	EDYTHE	
TIZZANO	MARIA	
TOLBERT	VIRGELLA	
TOLL	ORLIN	O
TOLL	RUTH	
TOPOLY	JOSEPH	A
TOWELL	ERNEST	
TRIVISON	CHARLES	A
TRIVISONNO	GRACE	
TRIVISONNO	JOHN	D
TRIVISONNO	MARY	
TROUT	WILLIAM	H
TROVATO	DONALD	
TUEMMER	HENRY	
TURNER	GEORGE	
TUTAY	MAE	

34:—Miniature Golf, Euclid Beach Park, Cleveland, Ohio.

97

LAST	FIRST	MIDDLE
USSAI	JORDAN	J
VACCARIELLO	FRANK	J
VACCARIELLO	JOSEPHINE	
VACCARIELLO	URBAN	
VALENTINE	LEWIS	
VALENTINO	MARY	D
VARGO	STANLEY	
VATH	CHARLES	R
VEGNEY	EUGENE	T
VOGEL	ROBERT	
VRANEKOVIC	JOS	D
VRH	JOSEPH	T
WADE	PATRICIA	ANN
WALDSMITH	DOROTHY	
WALK	EDDIE	F
WALLACE	JOHN	C
WALLINGFORD	HARRY	
WALSH	THOMAS	J
WALTMAN	FRIEDA	
WARE	MARY	J
WARREN	PAUL	E
WATERS	LULU	M
WATERS	WILLIAM	
WATERSON	ALLAN	F
WATHEN	WILLIAM	O
WATSON	PAUL	E
WEBER	ARCHIBALD	AUGUST
WEIR	THOMAS	J
WEISSERT	ROBERT	
WELLS	ROBERT	B
WELSH	PATRICK	
WHITE	HARRY	J
WHITE	JOHN	M
WHITEHOUSE	CLIFFORD	E
WICKES	JAY	E
WICKES	MABLE	ETTA
WILLIAMS	GRACE	WYATT
WILLIAMS	MARGARET	
WILLIAMS	ROGER	A
WILLIAMS	WALTER	DEWEY
WILLIAMSON	MICHAEL	A
WILMOT	RICHARD	C
WILSON	AUSTIN	E
WILSON	LLOYD	PAUL
WILSON	PAUL	
WILSON	RICHARD	G
WILSON	WILLIAM	S
WINGLE	DESMOND	A
WINTERBOTTOM	EDWIN	
WISE	MELVIN	
WISSMAN	CHARLES	
WITWER	PAUL	J
WITZKE	EMIL	D
WOLTMAN	RAYMOND	
WOODS	FRANK	HENRY

LAST	FIRST	MIDDLE
YAGER	FRED	H
YAMBOR	RICHARD	
YERICH	ANTHONY	
YORK	CLARENCE	
YOUNG	JAMES	A
YOUNG	LLOYD	WESLEY
YOUNG	MOSES	
YOUNG	ROBERTA	
YOXALL	LESLIE	J
ZAGAR	FRANK	
ZAMBONI SR	M	J
ZENOBY	JESSE	V
ZICARELLI	FLORENCE	
ZIMMERMAN	GEORGE	C
ZNIDARSIC	ADOLPH	
ZUCHELLI	RICHARD	

1953
EMPLOYEE ROSTER

LAST	FIRST	MIDDLE
ACHIN	JAMES	
ADAMS	JAMES	W
AGNEW	JAMES	
ALBRIGHT	GEORGE	H
ALLEN	CHARLES	R
ALLEN	LOUIS	B
AMBROSE	FRANK	
ANDERSON	JOHN	ANDREW
ANDREWS	EUGENE	
ANDREWS	GEORGE	E
ANGERMANN	J	WILLIAM
ANGERMANN	PEARL	
ARCARO	DONALD	
ASHBY	SARA	P
AXE	PAUL	E
BAILEY	GEORGE	K
BAIRD	MARY	E
BAKER	CILLIUS	M
BALDWIN	DANIEL	
BALDWIN	HAROLD	E
BALL	ROBERT	
BARBA	SAM	
BARBAGIOVANNI	CATERINA	
BARELLA	RUBY	M
BARNHARD	RICHARD	W
BARRY	DOROTHY	MARIE
BARRY	HENRY	JOHN
BARTO	ALOYSIUS	FRANK
BEAM	GORDON	R
BECK	GEORGE	R
BECK	J	K
BECKMEYER	GEORGE	F
BEGY	ALBERT	J
BEHREND	WILLIAM	J
BELLER	ROBERT	
BERRY	CLYDE	
BERTRAM	ROBERT	L
BIACOFSKY	ROBERT	J
BIBB	LEON	
BIRKENHEAD	THOMAS	
BITTEL	ROGER	A
BIXLER	PAUL	Z
BLACK	ROBERT	L
BLAKEMORE	WILLIAM	
BLEAM	C	M
BLEAM	LULA	ELSIE
BLOEDE	WILLIAM	CARL
BLYTH	DON	R
BORTZ	PAUL	D
BOTTOMLEY	GEORGE	
BOWHALL	ELMER	F
BOWHALL	HOWARD	
BRADFORD	JOHN	
BRADSHAW	JESSIE	
BRANCH	WILFRED	
BREEN	GROVER	J
BREWSTER	JOHN	
BRICE	MARY	
BROCKWAY	RUTH	M
BROWN	DAISY	MAY
BROWN	ESTILL	J
BROWN	PAUL	K
BRUMFIELD	BETTY	
BRYANT	JAMES	L
BUCKEYE	MARY	RUTH
BURKHART	FREDERICK	L
BURNS	RONALD	G
BURTON	JAMES	
BUTLER	JOSEPH	HACKNEY
CALLAGHAN	BEATRICE	BERNICE
CAMPBELL	ALEXANDER	ROBERT
CAMPBELL	FRED	
CAPASSO	JENNIE	
CAPPELLETTI	JENNIE	
CAPRA	LOUISA	
CARLTON	GERALD	G
CARROLL	DIANE	
CARROLL	GERTRUDE	M
CARROLL	HAZELLE	BELLE
CARROLL	URSULA	F
CASSERLY	JAMES	
CATALANO	FRANK	A
CHAMBERLAIN	CAROL	
CHAMBERLAIN	OMA	
CHAMBERLAIN	VERN	
CHAMBERLAIN	WAYNE	L
CHESLEY	ARTHUR	E
CHILDERS	PRICE	
CHRISTOPHER	JOSEPH	
CHRUSZCZAK	WALTER	
CIPITI	THOMAS	F
CIRINO	JOHN	
CIRINO	THERESA	M
CISTERNINO	LOUISE	
CODOSKY	MARY	
CONNAVINO	JOSEPHINE	
CONTENTO	GUISEPPE	
CONWAY	HARRY	F
CONWAY	LENA	
CONWAY	STELLA	M
COPELAND	JAMES	T
CORRIGAN	WALTER	T
COSTA	RICHARD	J
COX	ROBERT	G
CRAMER	GERALD	
CRAMER SR	GEORGE	
CRANDALL	HOWARD	
CRANE	WESLEY	E
CRAWFORD	HERMAN	
CRAWLEY	CHARLES	P
CROOKS	EMMA	
CROOKS	ROBERT	L
CROTTY	LAURENCE	A
CULVER	DALE	R
DARLIN	WILLIAM	F
DAVIS	CHARLES	E
DAVIS	KATIE	MAY
DAY	WILLIAM	R
DEDICH	MARY	C
DEERE	RICHARD	
DELAMBO	FRANK	FRANCIS
DEVANNA	ELEANOR	
DEVRIES	RICHARD	
DEWEY	DAVE	
DILIBERTO	DONALD	R
DIRUSSO	SARAH	
DONAHUE	DAVID	
DORRIS	CARL	
DOSSA	STEVE	
DOUGHERTY	MICHAEL	J
DOUGLAS	JAMES	W
DRAGONETTE	CATHERINE	
DUCCA	ALEXANDER	
DUFEK	JAMES	K
DUNCAN	EARL	E

LAST	FIRST	MIDDLE
DUPONT	LUCIEN	
EDWARDS	GEORGE	E
EGER	WILLIAM	J
EGGLESTON	DANIEL	
EGGLESTON	MAUDE	M
EISLEY	THOMAS	
ELERSON	RUBY	
ELY	HOWARD	M
EVERETT	AUGUST	R
FARNER	GEORGE	H
FAWCETT	HENRY	B
FEDERICO	CATHERINE	
FEDERICO	JOSEPHINE	
FERRERI	GEORGE	
FINUCAN	NEIL	T
FISCUS	HARRY	
FITZWORTH	JAMES	
FLYNN	ALFRED	
FOLLAS	HERBY	
FOSTER	ISABEL	
FOUSER	THOMAS	C
FOWLER	CHARLES	C
FOX	CLARA	M
FRANTZ	LORETTA	
FREDERICKS	ELIZABETH	
FREDERICKS	FRANK	JOSEPH
FRIEDEN	ROBERT	
FROHMBURG	ROSE	
GAERTNER	JAMES	L
GALIK	JOHN	
GAMBATESE	FLORENCE	
GARDNER	HOWARD	
GARDNER	JOSEPH	D
GATT	VINCYNNE	
GEIGER	LOUIS	J
GENTILE	JOHN	R
GENTRY	LUTHER	
GETZIEN	HARRY	
GIBBONS	WILLIAM	E
GIBSON	OLLIE	J
GIBSON	R	
GIERARD	STANLEY	F
GILBERT	MARY	
GILMORE	RALPH	E
GOOD	RUTH	
GOODSON	ALATHEA	
GOODSON	WILLIAM	D
GORE	WILLIAM	D
GORNIK	FRANK	
GRANT	DON	S
GRAVES	JAMES	L
GREENWAY	FRED	STUDER
GRON	PHILIP	M
GRUBB	JACK	F
GRUBB	URSULA	M
GUE	ROBERT	J
GUNTON	WINIFRED	L
HAGER	CARL	H
HAGY JR	WILLIAM	H
HALADUS	JAMES	D
HALE	KENNETH	W
HAMILTON	PAUL	R
HANLON	MARY	
HANNAN	WILLIAM	
HARMON	CHARLES	J
HARMONIC	ROBERT	
HARRAND	JOE	
HARRIS	GEORGE	K
HATTER	RONALD	M
HAYDEN	MAXINE	
HEALY	PATRICIA	A
HEARN	ROBERT	
HEGLAW	ROBERT	
HEISLER	WILBUR	
HEMPHILL	BRUCE	A
HENDERSON	RAYMOND	
HIGLEY	MAUD	M
HILL	ART	
HINTON	DAVID	R
HINTON	ELSIE	L
HINTON	WILLIAM	R
HIVELY	CHARLES	K
HODGES	BERT	
HOLLAND	JOHN	
HOLLEN	ARTHUR	F
HOLLEN	WILMER	
HOLMES	FLEMING	C
HOLMES	ROBERT	
HORRIGEN	LAWRENCE	A
HOWARD	EUGENE	
HOYT	JIM	
HUBBARD	HARRY	R
HUDSON	EDWIN	P
HUDSON	STEVE	P
HUGHES	ELLA	M
HUMMEL	CHRIS	C
HUNT	BLANCHE	L
HUNTER	ROY	CHARLES
HURLESS	JACKIE	
HUSTON	JOHN	P
IAFELICE	FRANK	
IAMMATTEO	MICHELINA	
ICE	GEORGE	V
JACKLITZ	ALBERT	
JACKSON	ALVA	L
JACKSON	ANDREW	I
JACKSON	MARY	H
JACKSON	WILLIAM	
JAMES	JOHN	W
JANES	PETER	J
JAYNES	TED	R
JOHNSON	CARL	
JOHNSON	JUSTIN	B
JOHNSON	MARTIN	G
JOHNSTON	CLIFFORD	BASIL
JOHNSTON	DONALD	HUMPHREY
JOHNSTON	GEORGE	W
JONES	JEANNE	M
JUBACH	GEORGE	R
JUDD	ESTIL	LEE
JUSKO	JOHN	
KAJFEZ	MATTHEW	J
KAMINSKI	RAYMOND	
KARKOSKA	JOSEPH	
KASPAR	MAY	E
KAST	HOWARD	R
KAZIMOUR	JAMES	F
KEARNEY	ROBERT	
KEELER	RUSSELL	J
KELLER	SAM	

LAST	FIRST	MIDDLE
KELLY	WILLIAM	L
KEMTER	AGNES	
KENNEDY	MERLE	
KENT	EDWARD	C
KERN	SYLVESTER	
KIKOLI	FRANK	
KIRKPATRICK	CATHERINE	
KISTHARDT	MARIE	L
KLAUS JR	WILLIAM	
KLEIN	MATHEW	W
KLEMPAN	ED	
KLESS	WILBERT	
KMETT	KENNETH	J
KOSS	RICHARD	A
KRAFT	FRANK	
KRAMER	JANE	
KRAUSE	HARRY	
KRAUSE	RONNIE	
KRESNYE	CHARLES	
KRIVDO	JEROME	M
KUBANCIK	THOMAS	
KUBAT	MARY	
KUNDROT	EDWARD	A
KURTZ	GRACE	M
LACEY	ROBERT	E
LAMAR	DAULTON	L
LANIER	LOUIS	
LANTA	RAYMOND	
LAWRENCE	DONALD	J
LENARCIC	JOHN	
LESSO JR	ANDREW	
LICKER	MANNIE	R
LINDER JR	ROBERT	
LIVINGSTON	RAYMOND	
LONGFIELD	TESS	EVANS
LONGWELL	JOSEPH	
LOPATICH	ANTON	
LORBER	MIKE	
LORD	RICHARD	K
LUPTON	JAMES	ALFRED
LUPTON	ROBERT	
LUZIER	GLADYS	L
LUZIER	ROBERT	F
LYNN	H	A
MAC DONALD	JOHN	
MACEK	RONALD	
MACHAK	STEVE	F
MACKENZIE	ELLEN	
MACQUEEN	CHARLES	
MAHONE	WILLIE	B
MAHONE JR	JOHN	
MAHONEY	RALPH	G
MALAKAR	AGNES	V
MANCHESTER	EDWARD	
MANDEL	IDA	
MANNARINO	LOUIS	A
MANNING	DOUGLAS	L
MANNING	EDWARD	
MANNING	JOHN	H
MANNING	PATRICK	J
MARCHESANO	RONALD	
MARFONGELLA	ELIZABETH	
MARGHERET	GENE	
MARIANO	MIKE	
MARINARO	VERNE	
MARKEL	JOSEPH	
MARKHAM	THOMAS	G
MARRA	ANTHONY	
MARSHALL	CADDIE	
MASSIE	GEORGE	T
MATSON	THOMAS	A
MCANDREWS	JOHN	
MCBRIDE	CALVIN	P
MCCLOSKY	JAMES	J
MCCOY	MARY	
MCCRUDDEN	ROBERT	
MCCUMBER	RAYMOND	
MCDOWELL	FLORENCE	
MCGANN	MICHAEL	L
MCGOWAN	DONALD	
MCKENNA	JOHN	W
MCKINLEY	BERNARD	
MCKINLEY	JAMES	H
MCKINNEY	CLEOPHUS	
MCKINNEY	WILLIAM	
MCLEAN	CATHERINE	
MCLEAN	JACK	H
MCNIECE	RAYMOND	
MEADOWS	JOHN	R
MERZDORF	EDWARD	
METHOD	FRANCES	S
METZGER	CARL	
MEYER	JOSEPH	A
MIERKE	ALBERT	H
MILES	ARCHIE	
MILLER	RONALD	
MIOZZI	ARLENE	
MISLEY	JAMES	J
MITCHELL	CHESTER	
MONTAGNO	SALVATORE	
MOORE	MAY	C
MORGAN	EUGENE	F
MORGAN	RAYMOND	F
MORRIS	ISABELLE	
MOSCOVITZ	ABRAHAM	
MOSIER	FREDA	V
MOYNIHAN	FRANCIS	
MRAMER	EDWARD	
MULLINS	HERBERT	
MULLINS	VIRGIL	
MUMAW	SUSAN	B
MURRAY	JOHN	
MUSIC	MARY	ELLEN
MYERS	LEO	E
NACHTIGAL	JAMES	A
NAGLE	CLINTON	
NAGLE	ROBERT	
NEUBAUER	KATHRYN	MELVIN
NEWCOMB	WILLIAM	E
NICOLL	ROGERT	G
NORRIS	CASEY	T
OAKLEY	ADDIE	ELLEN
OBOYLE	JOHN	J
OKROS	MARIE	H
OLINZOCK	FRANK	R
ONEIL	JOHN	C
OPALK	MARTIN	
OSBORNE	CHARLES	
OSRETKAR	WILLIAM	J
OWENS	CLARENCE	D
PAGE	GEORGE	S

LAST	FIRST	MIDDLE
PAONSHA	ANTON	
PARKER	WILLIAM	
PARKER	WILLIAM	FORMAN
PARMELEE	CHARLES	W
PATFIELD	ROBERT	
PATTON	HARRY	V
PATTON	PATRICK	B
PATTON	THOMAS	
PAVLINA	EDWARD	
PEARCE	AGNES	
PEARSON	ROY	A
PELL	KENNETH	
PERROTI	JOSEPH	
PERSELL	CHARLES	
PETERKA	LIBBIE	M
PETRECCA	JOSEPH	
PETRECCA	NICHOLAS	
PETRO	SUE	JULIAN
PICCOLI	VICTOR	R
PITTOCK	RON	ELMER
PLANTSCH	MICHAEL	
POCH	NICK	
PODMORE	GOLDIA	
POMPLAS	ALBERT	
POWELL	CHARLES	
PREWITT	ELDON	J
PRICE	FLOYD	L
PRICE	HARRY	R
PRIMUS	FELIX	
PURDY	MARGARET	
PURO	EDWARD	
RAFFERTY	PETER	
RAMSEY	DON	C
RANEY	LUTHER	
RAYER	WILLIAM	D
REARDON	ROGER	
REDFIELD	CUYLER	I
REED	MARIE	E
REESE	GEORGE	J
REINHARD	GEORGE	MARTIN
RENTON	STEPHEN	
RHODES	ROBERT	C
RICHARDS	FLORENCE	
RICHARDSON	BRAD	
RITTWAGE	GEORGE	W
ROBINSON	JOSEPH	H
ROCEWICKY	MIKE	
ROCKER	LEONARD	
RODERICK	ALICE	M
ROGERS	HORACE	
ROGERS	ROBERT	
ROLLINS	ERNEST	L
RONIGER	H	E
ROONEY	LAWSON	F
ROSS	EDGAR	E
ROSS	FRANK	

LAST	FIRST	MIDDLE
ROSSODIVITA	DAVID	
ROSSODIVITA	MARY	
ROUSCH	EDWARD	R
ROZANC	THOMAS	
RUBRIGHT	EDWARD	
RUCK	JAMES	A
RUCKER	CARTER	H
RUGGLES	ROBERT	B
RUTH	JAMES	
RYAN	JOHN	
RYDMAN	ELIZABETH	
SALAMONE	DOMINIC	A
SANTUCCI	MELINA	A
SARGENT	EVELYN	M
SBROCCO	SAM	
SCARANO	ANGELA	MARY
SCARDELLETTI	GENEVA	
SCHAEFER	MARTHA	L
SCHAFER	FRANCES	
SCHEU	LOUISE	B
SCHILL	EUGENE	
SCHILL JR	AUGUST	J
SEDLAK	ELEANOR	
SHANK	CHARLES	
SHANK JR	CHARLES	
SHANNON	HARRIS	COPPER
SHEEHAN	JAMES	D
SHEEHAN	MABELLE	B
SHELTON	CHARLES	
SHERMAN	MARGARET	
SHILLIDAY	EVERETT	P
SHIVELY	GEORGE	G
SHOTWELL	ALMA	M
SICKELS	LESTER	F
SICKELS	LESTER	G
SICKELS	MARJORIE	
SIELAFF	ROLLIN	G
SILVERMAN	JEROME	
SILVEROLI	JOHN	
SIMA	JAMES	J
SIMMS	LOU	CARSON
SIMON	ALVIN	F
SISNEY	ALLENE	
SMITH	ALAN	H
SMITH	ALVIN	W
SMITH	CHARLES	L
SMITH	EDDIE	
SMITH	FRED	C
SMITH	IRENE	MARIE
SMITH	MARY	C
SMITH	MARY	LOUISE
SMITH	WILBERT	
SMRDEL	STANLEY	J
SNELLING	FRANCES	
SNOW	RICHARD	L
SOMMERS	ERNESTINE	
SOUSA	EUGENE	
SPINOS	JAMES	
SPRINGBORN	ROBERT	
SPUZZILLO	JOHN	A
SPUZZILLO	MARY	
SPUZZILLO	MICHAEL	A
ST CLAIR	JULIUS	J
STAMM	J	G
STAMM	JOHANNA	M
STAMM	MOLLY	M
STANG JR	GEORGE	H
STANKOVITCH	CHARLES	
STAPE	WILLIAM	H
STAPLES JR	EDWARD	W
STAPP	PAUL	R
STARINA F	LORENCE	
STEELE	ARTHUR	W
STERLING	THOMAS	
STEVENS	VALENTINE	G
STINEDURF JR	LEROY	
STONEBACK	HOWARD	DETWEILER

47. OPEN AIR DANCE FLOOR, EUCLID BEACH PARK, CLEVELAND, OHIO.

102

LAST	FIRST	MIDDLE
STOVALL	WILLIAM	N
STRATTON	ROLLAND	A
STRAUSS	CHARLES	
STRONG	ELEANOR	
STRONG	HAROLD	E
STRONG	PEARL	
SUHAR	JOSEPH	
SUSONG	FLOYD	
SWEENEY	JOHN	
SWEIGERT	CHARLES	
TALLETT	MARGARET	R
TATE	BLANE	
TATUM	MASO	
TAYLOR	ROY	C
TAYLOR	THOMAS	N
TEODOSIO	VINCENT	
THOMAS	PRYSE	
THOMAS	STEVEN	
THOMPSON	IRVIN	C
THORNTON	JOHN	
THUNHURST	EDYTHE	
TIZZANO	MARIA	
TONNEAS JR	FRED	W
TRIPOLI	BART	J
TRIVISON	CHARLES	A
TRIVISONNO	GRACE	
TROUT	WILLIAM	H
TROVATO	JOSEPH	
TUEMMER	HENRY	
VACCARIELLO	JOSEPHINE	
VACCARIELLO	URBAN	
VALENTINE	LEWIS	
VALENTINO	MARY	D
VAN STRAATEN	JAMES	
VERHOVEC	CAROL	
VILLANI	RAYMOND	
VLACH	ANTHONY	
VOGEL	ROBERT	
VRANEKOVIC	JOS	D
VRH	JOSEPH	T
WACKER	WILLIAM	
WADE	PATRICIA	ANN
WALDSMITH	DOROTHY	
WALK	EDDIE	F
WALKER	SIDNEY	L
WALSINGER	RICHARD	
WARD	DONNA	MAE
WARD	JEROME	D
WATERS	LULU	M
WATERS	WILLIAM	
WATERSON	ALLAN	F
WATSON	DAVIS	
WATTS	EPHRAIM	
WEBB	JOHN	R
WEBER	RICHARD	L
WEED	CHARLES	E
WEISBARTH	PHILIP	
WEISS	ED	
WEISSBERG	BELLE	
WEISSERT	ROBERT	
WHITMORE	GUS	
WICKES	JAY	E
WICKES	MABLE	ETTA
WIEGAND	RAY	
WILEY	GEORGE	H
WILLIAMS	BERTHA	
WILLIAMS	GRACE	WYATT
WILLIAMS	KENNETH	A
WILLIAMS	MARGARET	
WILLIAMS	ROGER	A
WILLIAMS	WALTER	DEWEY
WILSON	AUSTIN	E
WILSON	LLOYD	PAUL
WILSON	PAUL	
WILSON	RICHARD	G
WILSON	WILLIAM	S
WINTERBOTTOM	EDWIN	
WOLFE	JAMES	
WOODS	FRANK	HENRY
WOOTTON	GENELLA	M
WRAY	ATHALYNE	
YAGER	FRED	H
YELCHO	JACOB	
YOUNG	HILDA	
YOUNG	JAMES	A
YOUNG	LLOYD	WESLEY
YOUNG	MOSES	
YOUNG	ROBERTA	
YOUNG	WILLIAM	E
ZABLOTNY	WALTER	S
ZAGAR	FRANK	
ZAMBONI	HAZEL	
ZERAVICA	NICK	
ZIMMERMAN	GEORGE	C

Engineer Mike Mower demonstrates how to lubricate the train. ca.1941

Euclid Beach Park garage and maintenance building.

1954
EMPLOYEE ROSTER

LAST	FIRST	MIDDLE
ACHIN	JAMES	
AGNEW	JAMES	
AKINS	RONALD	
ALBRIGHT	GEORGE	H
ALLEMAN	RALPH	
ALLEN	LOUIS	B
ALVAREZ	EDWARD	
ANDERHALT JR	RUSSELL	
ANGERMANN	PEARL	
APPLEBEE	JULIUS	
AUSTIN	CHARLES	
AVERILL	EDGAR	N
AVERILL	NEAL	
AXE	PAUL	E
BAILEY	MABEL	
BAIRD	MARY	E
BAKER	CILLIUS	M
BALTURSHAT	ESTHER	
BARBAGIOVANNI	CATERINA	
BARBATO	JERRY	
BARINDT	HELEN	
BARON	GEORGE	A
BARRY	DOROTHY	MARIE
BARRY	HENRY	JOHN
BARRY	JOHN	C
BARTO	ALOYSIUS	FRANK
BECKER	HARTWIG	
BECKMEYER	GEORGE	
BEHREND	WILLIAM	J
BELLER	ROBERT	
BERRY	CLYDE	
BERTRAM	ROBERT	L
BETZ	BRUCE	
BIACOFSKY	ROBERT	J
BIRKENHEAD	THOMAS	
BLAKEMORE	WILLIAM	
BLEAM	C	M
BLEAM	LULU	ELSIE
BLOEDE	WILLIAM	CARL
BOEY	LEO	
BOGAN	MISSOURI	
BORDERS	EVELYN	
BOTTOMLEY	GEORGE	
BRADFORD	JOHN	
BRADSHAW	JESSIE	
BRANCH	WILFRED	
BRANDON	RICHARD	
BRAWNER	TED	
BROBST	FRANK	
BROWN	DAISY	MAY
BROWN	LIBBIE	
BROWN	RAY	
BRUMYE	RICHARD	
BURNEY	HENRY	
BURTON	JAMES	
BUSH	KATHERINE	
BUTLER	JOSEPH	HACKNEY
CALLAGHAN	BEATRICE	BERNICE
CALVERT	FRANK	
CAMP	WINFIELD	S
CAMPBELL	ALEXANDER	ROBERT
CAMPBELL	FRED	
CAPPELLETTI	JENNIE	
CAPRA	CRISTINA	
CAPRA	LOUISA	
CARLTON	GERALD	G
CARROLL	GERTRUDE	M
CARROLL	HAZELLE	BELLE
CARROLL	ROSELLE	
CASSERLY	JAMES	
CATALANO	FRANK	A
CATANESE	ANGIE	
CATEY	EARL	W
CHAMBERLAIN	CAROL	
CHAMBERLAIN	VERN	
CHRISTOPHER	JOSEPH	
CIESLAK	DANIEL	E
CIPITI	THOMAS	F
CIRINO	JOHN	
CIRINO	THERESA	M
CISCO	GEORGE	
CLARK	FRED	
CLARK	PRESTON	
CLARK	ROBERT	
COLLIDGE	ROY	
CONNAVINO	JOSEPHINE	
CONTENTO	GUISEPPE	
CONWAY	HARRY	F
CONWAY	LENA	
CONWAY	STELLA	M
COOK	PEGGY	L
COOLIDGE	CALVIN	
CORRIGAN	WALTER	
COURNOYER	LEO	
COWDEN	ALLEN	R
COX	HERBERT	W
COX	JAMES	E
COX	ROBERT	G
CRAFTON	MEVIN	
CRAMER	GERALD	
CRANDALL	HOWARD	
CRANE	WESLEY	E
CRAWFORD	HERMAN	L
CRAWFORD JR	JOHN	
CRAWLEY	CHARLES	P
CROOKS	EMMA	
CROSS	JOHN	D
CROTTY	LAURENCE	A
CROTTY JR	LAWRENCE	J
CROW JR	RALPH	
CULVER	DALE	R
CUTHILL	JOHN	
DANIELS	RAYMOND	P
DAVIS	CHARLES	E
DAVIS	KATIE	MAY
DEDICH	MARK	
DEDICH	MARY	C
DEERE	RICHARD	
DEGERIO	LOUIS	
DELAMBO	FRANK	FRANCIS
DELBALSO	ANGELINA	
DERCOLE	ANTONETTE	
DERLATKA	MATTHEW	
DICORPO	MIKE	
DILL	PAUL	A
DINERO	ANTHONY	
DOBRANIC	STEVE	
DOSSA	STEVE	
DOUGLAS	JAMES	W
DUCCA	ALEXANDER	
DUDLEY	WALTER	S
DWYER	EDWARD	
DZURENKO	FRANCIS	
EDWARDS	BERT	E

LAST	FIRST	MIDDLE
EGAN	MARY	JO
EGGERT	EDWIN	W
ELY	HOWARD	M
ERTTER	JAMES	
ERTTER	JOSEPH	
ESTEP	JUSTUS	
ESTEP	KATHLEEN	
EVERETT	AUGUST	R
FARONE	ANGELA	
FEATHER	MARY	ELLEN
FEDERICO	CATHERINE	
FEDERICO	JOSEPHINE	
FERRERI	GEORGE	
FICKE	CHARLES	H
FICKE	HERBERT	G
FICKE	ROBERT	E
FISHER	EDWARD	
FITZWORTH	JAMES	
FLYNN	HARRY	K
FOLTZ	CALVIN	
FORSYTH	RUSSELL	
FOX	DAVID	
FOX	RICHARD	M
FRANTZ	LORETTA	
FREEMAN	MELVIN	W
FROHMBURG	ROSE	
GAERTNER	JAMES	L
GALIK	JOHN	
GALL	HELEN	J
GALLA	JOHN	
GALLAGHER	JAY	
GALUSICK	DOROTHY	
GARDNER	JOSEPH	D
GATT	VINCYNNE	
GEHRING	WILLIAM	
GERVAC	STEPHEN	B
GETZIER	HARRY	
GIES	EDWARD	C
GILBERT	MARY	
GIPSON	CONNA	
GIPSON	DONNA	
GIPSON	LEVYONNE	
GLIOZZI	ANTHONY	
GOLSON	ELLA	L
GONSALVES	DONALD	G
GOOD	RUTH	
GOODSON	ALATHEA	
GOODSON	WILLIAM	D
GORNIK	FRANK	
GORSHA	PATRICIA	
GOVE	RICHARD	
GRANDE	DOMENIC	
GREENBERG	JACK	
GREENWAY	FRED	STUDER
GRIFFITH	J	C
GRON	PHILIP	M
GROSSE	GEORGE	
GUILLOZET	NOEL	
GUTTMAN	RICHARD	
HALE	WILBERTH	
HAMILTON	PAUL	R
HANLON	MARY	
HAWKINS	THELMA	
HAYDEN	MAXINE	
HEARN	ROBERT	
HECKMAN	EDWARD	
HEGLAW	ROBERT	
HEISLER	WILBUR	
HENDERSON	ELEANOR	
HENDERSON	RAYMOND	
HERKE	GEORGE	W
HERRICK	ROBERT	
HERTENSTEIN	ARWILDA	
HERTENSTEIN	EDWARD	
HEUER	ANNA	C
HICKEY	JERRY	
HILL	ARTHUR	M
HILL	HAROLD	
HINES	ROY	C
HINTON	ELSIE	L
HINTON	WILLIAM	R
HODNIK	JOHN	
HOFMAN	ROBERT	
HOLLAND	JOHN	
HOLLEN	ARTHUR	F
HOLLEN	WILMER	
HOLMES	ROBERT	
HOOD	DONALD	
HORWATH	STANLEY	
HOWARD	EUGENE	
HUBBARD	HARRY	R
HUBER	GEORGE	
HUDSON	EDWIN	P
HUGHES	ELLA	M
HUMMEL	CHRIS	C
HUNT	BLANCHE	L
HUNTER	ROY	CHARLES
HUSTON	JOHN	P
IAMMATTEO	MICHELINA	
JACKLITZ	ALBERT	
JACKSON	ALVA	L
JACKSON	ANDREW	I
JACKSON	MARY	H
JAMES	JOHN	W
JANCO	DOROTHY	
JANES	PETER	J
JARMON	MAXINE	
JARMUSCH	MARILYN	
JAYNES	TED	R
JOHNSON	JAN	P
JOHNSON	PATRICIA	
JOHNSTON	DONALD	HUMPHREY
JUDD	ESTIL	LEE
JUDD	WALLACE	S
KARKOSKA	JOSEPH	
KASPAR	MAY	E
KAST	HOWARD	R
KASTELIC	VERONICA	
KAYLOR	JOSEPH	
KEARNEY	CHARLES	V
KEARNEY	ROBERT	
KELLER	SAM	
KELLEY	PAUL	J
KELLY	ROBERT	
KEMTER	AGNES	
KENT	EDWARD	C
KESKE	WALTER	
KIFER	JOHN	J
KIKOLI	FRANK	
KING	WILLIE	F
KIRCHENER	ELROY	

LAST	FIRST	MIDDLE
KIRKPATRICK	CATHERINE	
KIRKPATRICK	LAWRENCE	K
KISTHARDT	MARIE	L
KLAUS	AUGUST	J
KLEINSHROT	BETTY	
KOVACIK	VIC	
KOVIC	JOYCE	L
KRAMER	JANE	
KRAUSE	HARRY	
KRESNYE	CHARLES	
KREVES	EDWARD	J
KRIVDO	JEROME	M
KRUCHZEK	RICHARD	
KUBEA	ROSE	
KUFER	JOHN	J
KULMACZESKI	ALVIN	
KURTZ	GRACE	M
LACEY	ROBERT	E
LANCASTER	WALLACE	
LARDIE	KENNETH	R
LARDIE	RONALD	
LATKOVICH	ELEANOR	
LENGEL	TOM	S
LENGVEL	LADISLAUS	
LESKOVEC	FRANK	
LESSO	ANDREW	M
LILLIS	LAWRENCE	E
LILLIS	STEVEN	J
LINAMEN	CARRIE	
LINDER JR	ROBERT	
LIVINGSTONE	BERTHA	
LONGAZEL	ANNA	
LONGFIELD	TESS	EVANS
LONGWELL	JOSEPH	
LORBER	MIKE	
LOWERY	THOMAS	J
LUCA	EDITH	
LUCAS	FRANK	
LUPO	RICHARD	J
LUPTON	ROBERT	
LUZIER	GLADYS	L
LUZIER	ROBERT	F
LYKES	PEARL	
LYNN	H	A
MAC DONALD	JOHN	
MACEK	RONALD	
MACHAK	STEVE	F
MACKENZIE	ELLEN	
MACZUGA	LEONARD	
MAHONE	WILLIE	B
MALLOW	ROBERT	
MANDEL	IDA	
MANNARINO	LOUIS	A
MANNING	DOUGLAS	L
MANNING	EDWARD	
MANNING	JOHN	H
MANNING	PATRICK	J
MARETTE	RALPH	
MARFONGELLA	ELIZABETH	
MARIANO	MIKE	
MARINO	JOE	
MARKEL	JOSEPH	
MARSHALL	CADDIE	
MARTENS	WALTER	J
MARTIN	BRIAN	
MARTIN	RAYMOND	
MASON	ARTHUR	L
MASSIE	GEORGE	T
MCBRIDE	CALVIN	P
MCBRIDE	MARY	E
MCCONOCHA	DANIEL	
MCCRUDDEN	ROBERT	
MCCULLEY	PRESTON	I
MCDERMOTT	CATHERINE	
MCDERMOTT	GERALD	
MCDOWELL	FLORENCE	
MCGANN	MICHAEL	L
MCKENNA	JOHN	W
MCKINLEY	JAMES	H
MCLEOD	MERVIN	
MCMAHON	HELEN	
MELBER	EDWARD	
MELI	PHILIP	
MERZDORF	EDWARD	
MICHAEL	SYLVIA	
MICHEL	GORDON	F
MICHEL	GORDON	J
MIERKE	ALBERT	H
MILES	ARCHIE	
MILLER	RONALD	
MIOZZI	ARLENE	
MISLEY	JAMES	J
MISNY	JOHN	
MITCHELL	CHESTER	
MOLITOR	ROBERT	F
MOLNAR	ROBERT	L
MOMINEY	RAY	
MOODY	CLARENCE	
MOORE	MAY	C
MORRIS	ARTHUR	
MOSIER	FREDA	V
MOUGHAN	KEVIN	
MULLINS	HERBERT	
MUMAW	SUSAN	B
MUNSEY	J	P
MURRAY	JOSEPH	
NASH	WILLIAM	J
NEUBAUER	KATHRYN	MELVIN
NICHOLS	ALAN	
NICOL	ANDREW	
NICOLL	ROBERT	G
OAKLEY	ADDIE	ELLEN
OBANNON	PEARL	
OCONNOR	JAMES	
OKROS	MARIE	H
ONEIL	BESSIE	
ONEIL	JOHN	C
ONEILL	CHARLES	R
OPALK	MARTIN	
OPPENHEIMER	LOUIS	
OSBORNE	CHARLES	
OSBORNE	EMMETT	L
OTTO	VICTOR	C
OWENS	CLARENCE	D
OWENS	DOLORES	
PARKER	WILLIAM	
PARKER	WILLIAM	FORMAN
PARMELEE	CHARLES	W
PARMELEE	EDITH	
PARRISH	BRYAN	J
PEARCE	AGNES	

THE PAVILION, EUCLID BEACH

LAST	FIRST	MIDDLE
PECK	WILBERT	J
PERSELL	CHARLES	
PETERKA	LIBBIE	M
PETRECCA	JOSEPH	
PETRECCA	NICHOLAS	
PETRICIG	JOHN	
PETRO	SUE	JULIAN
PEZDIC	ANDREW	
PITTOCK	RON	ELMER
PLANTSCH	MICHAEL	
POLING	CHARLES	E
POMPLAS	ALBERT	
PREWITT	ELDON	J
PRICE	FLOYD	L
PRICE	HARRY	R
PRIEBE	RICHARD	
PRUIETT	LILLIAN	
PURDY	MARGARET	
PURO	EDWARD	J
RAFFERTY	PETER	
RAIMONDO	ALFRED	
RANKER	WILBUR	
RAUKAR	FRANK	
RAY	PAUL	
REARDON	ROGER	
REDFIELD	CUYLER	I
REED	ELIZABETH	P
REED	MARIE	E
REESE	GEORGE	J
REIBEL	DESSIE	
REINHARD	GEORGE	MARTIN
RICHARD	PHILLIP	E
RICHARDSON	BRAD	
RICHMOND	JAMES	E
ROBINSON	SHIRLEY	
ROCKER	LEONARD	
RODERICK	ALICE	M
RODWAY	EDWARD	
ROLLINS	ERNEST	L
RONIGER	H	E
ROSS	EDGAR	E
ROSS	FRANK	
ROSSODIVITA	DAVID	
ROZANC	THOMAS	
RUCKER	CARTER	H
RYDMAN	ELIZABETH	
SBROCCO	SABASTINO	
SCHAUB	GEORGE	
SCHILL	EUGENE	
SCHILL JR	AUGUST	J
SCHNEIDER	JOHN	
SCHRAMM	HAROLD	R
SCOTT	ARTHUR	H
SHANK	CHARLES	
SHANNON	HARRIS	COOPER
SHEEHAN	JAMES	D
SHEEHAN	MABELLE	B
SHEPPARD	THEODORE	R
SHILLIDAY	EVERETT	P
SIGNS	MARIE	
SILVEROLI	JOHN	
SIMA	JAMES	J
SIMON	ALBERT	M
SIPP	RAMELLE	
SLOAN	LORA	
SMITH	ALAN	H
SMITH	ALVIN	W
SMITH	CHARLES	L
SMITH	CHARLES	T
SMITH	DAVID	
SMITH	FRED	C
SMITH	HARRY	A
SMITH	IRENE	MARIE
SMITH	KITTY	A
SMITH	LESTER	L
SMITH	MARY	C
SMITH	MARY	LOUISE
SMITH	RALPH	E
SMITH	RAYMOND	S
SMRDEL	STANLEY	J
SOLTIS	MAXINE	
SOMMERS	ERNESTINE	
SOUCHIK	WALTER	
SOUSA	EUGENE	
SPUZZILLO	MARY	
SPUZZILLO	MICHAEL	A
SPUZZILLO	MICHAEL	F
ST CLAIR	JULIUS	J
STALEY	JOHN	
STAMM	JOHANNA	M
STANKOVITCH	CHARLES	
STAPE	WILLIAM	H
STARBIRD	RICHARD	O
STEELE	ARTHUR	W
STEVENS	VALENTINE	G
STONEBACK	HOWARD	DETWEILER
STONEBACK	WILLIAM	
STRAUSS	CHARLES	W
STRONG	ELEANOR	
STRONG	LARRY	
STURGHILL	LEE	ANN
SUBIN	EDITH	
SUSONG	FLOYD	
SUSONG	JAMES	F
SWEENEY	JOHN	
SWIMER	MIKE	
TEETER	ARNOLD	
TERRIZZI	PETER	
TESTA	DONALD	
THOMAS	PRYSE	
THOMAS	ROBERT	G
THOMAS	STEVEN	
THOMPSON	CHARLES	H
THOMPSON	IRVIN	C
THUNHURST	EDYTHE	
TIDERMAN	FRANK	E
TIMKO JR	JOHN	
TIZZANO	MARIA	
TONCRE	WESLEY	
TONELLI	BENEDETTE	
TRACZYNSKI	EDWARD	
TRIPOLI	BART	J
TRIVISON	CHARLES	A
TRIVISONNO	GRACE	
TRUMBLEY	ROBERT	C
TUEMMER	HENRY	
URBANCIC	JAMES	W
VACCARIELLO	ANITA	
VACCARIELLO	JOSEPHINE	
VACCARIELLO	URBAN	
VALENTINO	MARY	D
VEGNEY	EUGENE	T

LAST	FIRST	MIDDLE
VILLANI	RAYMOND	
VOGEL	ROBERT	
VRANEKOVIC	JOS	D
VRH	JOSEPH	T
WADSWORTH	ELSIE	
WAGNER	EDWARD	
WALDRON	KATHERINE	
WALK	EDDIE	F
WALSINGER	RICHARD	
WAMBOLD	GLENNA	
WARE	DONNA	MAE
WATERS	LULU	M
WATERS	WILLIAM	
WATERSON	ALLAN	F
WEATHERLY	FRANCINE	
WEEDON	JOHN	A
WEISS	ED	
WEISSERT	ROBERT	
WEST	NETTIE	
WHITE	ROBERT	J
WICKES	JAY	E
WICKES	MABLE	ETTA
WILCOX	GERTRUDE	
WILEY	GEORGE	H
WILLARD	EUGENE	C
WILLIAMS	BERTHA	
WILLIAMS	EUGENE	
WILLIAMS	GRACE	WYATT
WILLIAMS	KENNETH	A
WILLIAMS	MARGARET	
WILLIAMS	ROGER	A
WILLIAMS	WALTER	DEWEY
WILLIS	RICHARD	
WILSON	GEORGE	W
WILSON	LLOYD	PAUL
WILSON	PAUL	
WILSON	RICHARD	G
WILSON	ROBERT	O
WILSON	WILLIAM	S
WINSLOW	GEORGE	
WINTERBOTTOM	EDWIN	
WOODS	FRANK	HENRY
WORKMAN	WILLIAM	
WRIGHT	ETHEL	M
WURM	CHARLOTTE	L
YAGER	FRED	H
YARKER	HORACE	
YERICH	JOHN	
YOUNG	LLOYD	WESLEY
YOUNG	MOSES	
YOUNG	ROBERTA	
ZAGAR	FRANK	
ZAMBONI	HAZEL	
ZIGLI	RONALD	M
ZIMMERMAN	GEORGE	C
ZINK	GEORGE	
ZORETICH	PAUL	

109

Bruce Young Collection

Photo: Robert Runyan

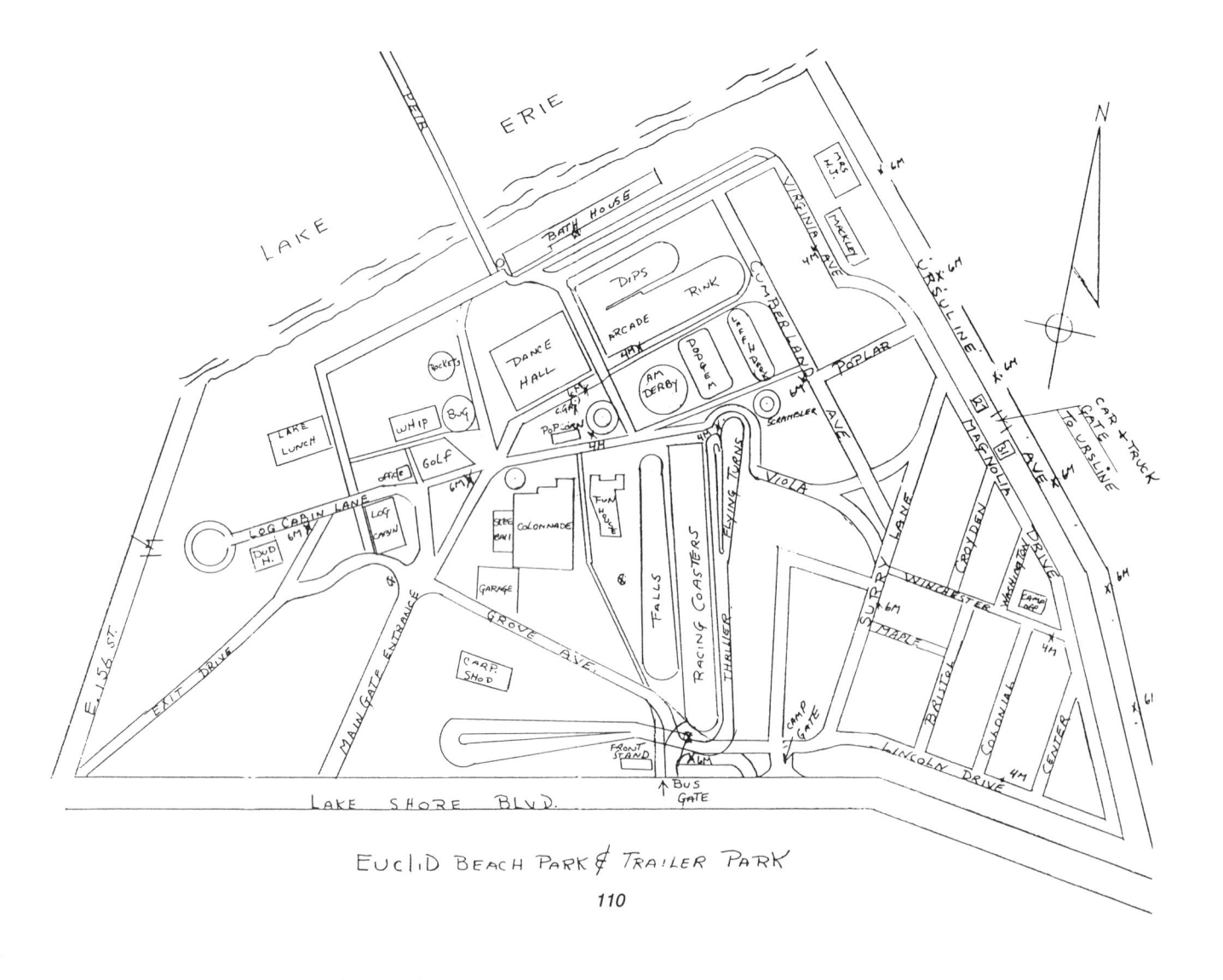

Euclid Beach Park & Trailer Park

110

1955 EMPLOYEE ROSTER

LAST	FIRST	MIDDLE
ADAMS	W	L
AGNEW	JAMES	
ALBRIGHT	GEORGE	H
ALICK	FRANK	
ALLEN	LOUIS	B
ANDERSON	GEORGE	
ANGERMANN	PEARL	
AUSTIN	IRENE	D
AVERILL	EDGAR	N
AXE	PAUL	E
BAILEY	MABEL	
BAIRD	MARY	E
BAKER	CILLIUS	M
BALAS	ROBERT	
BALCAM	GWENDOLYN	
BALTURSHAT	ESTHER	
BANDY	CHARLES	
BANISH	AUDREY	
BARBATO	JERRY	
BARINDT	HELEN	
BARON	GEORGE	A
BARRY	DOROTHY	MARIE
BARRY	HENRY	JOHN
BARTH	CAROLINE	
BARTH	GEORGE	
BARTO	ALOYSIUS	FRANK
BAUER	DAVID	
BAUMAN	WAYNE	
BEETLER	ROBERT	
BELLER	ROBERT	
BENNETT	ARVIA	
BENNETT	CHARLES	
BENNETT	HENRY	R
BERRY	CLYDE	
BERTRAM	ROBERT	L
BIACOFSKY	ROBERT	J
BIRKENHEAD	THOMAS	
BLEAM	C	M
BLEAM	LULA	ELSIE
BLOEDE	WILLIAM	CARL
BLOOM	WAYNE	
BOGAN	MISSOURI	
BOLDEN	ADMIRAL	D
BOTTOMLEY	GEORGE	
BOVINO	VINCENT	
BOWHALL	HOWARD	
BRADFORD	JOHN	
BRAWNER	TED	
BRENNAN	JAMES	E
BRILL	LEONARD	
BROWN	DAISY	MAY
BROWN	IRA	E
BRYANT	THOMAS	
BUDNICK	ALFRED	
BURNEY	HENRY	
BURTON	JAMES	
BUTLER	JOSEPH	HACKNEY
CADA	PHILIP	
CALDWELL	HOMER	
CALLAGHAN	BEATRICE	BERNICE
CAMPBELL	ALEXANDER	ROBERT
CAMPBELL	FRED	
CAMPBELL	MARLA	
CAPRA	LOUISA	
CAPRA	MARIA	
CARPENTER	SARELLA	
CARROLL	GERTRUDE	M
CARROLL	HAZELLE	BELLE
CARSON	HAMILTON	
CARSON	WALTER	
CATALANO	FRANK	A
CATALANO	JAMES	
CATEY	EARL	W
CHAMBERLAIN	HENRY	
CHAMBERLAIN	VERN	
CHAMPAGNE	GROVER	
CHANEY	GEORGE	
CHRISTY	MARGARET	
CIPITI	THOMAS	F
CIRINO	JOHN	
CIRINO	THERESA	M
CLARK	DAVID	
CLARK	EVALYN	
CLARK	FRED	
CLARKE	JOHN	G
CONARD	MINNIE	
CONNAVINO	JOSEPHINE	
CONNOR	NORMAN	
CONTENTO	GUISEPPE	
CONWAY	HARRY	F
CONWAY	LENA	
CONWAY	STELLA	M
COSTLOW	RICHARD	
COUGHLIN	GEORGE	
CRAMER	GERALD	
CRAMER SR	GEORGE	
CRANDALL	HOWARD	
CRANE	WESLEY	E
CRAWFORD	OSCAR	B
CRAWFORD JR	JOHN	
CRAWLEY	CHARLES	P
CROOKS	EMMA	
CROTTY JR	LAWRENCE	J
CULLEN	JOSEPH	
CUTHILL	JOHN	
DAMATO	MARY	
DAUGHERTY	J	A
DAVIS	ARTHUR	
DAVIS	CHARLES	E
DAWSON	DONN	
DAY	WILLIAM	R
DEDICH	MARY	C
DELAMBO	FRANK	FRANCIS
DELEMBO	LOUIS	
DELURA	FRANCES	
DEMETER	RICHARD	
DICK	ARCHIBALD	
DICKSON	VIVIAN	
DICORPO	MIKE	
DIEBOLT	FLORENCE	
DIEBOLT	WILLIAM	
DILLON	OLIVER	
DINERO	ANTHONY	
DIXON	JAMES	
DOSSA	STEVE	
DOTY	C	
DUCCA	ALEXANDER	
DUCKWORTH	SENNIE	
DUNAWAY	JASPER	
DUNCH	CARL	J
EMERICH	JOHN	
ENGLAND	FRANKLIN	
ERICKSON	BRUCE	
ERRTER	JAMES	
ERRTER	JOSEPH	

LAST	FIRST	MIDDLE
ESTEP	JUSTUS	
ESTEP	KATHLEEN	
FARNER	GEORGE	H
FEATHER	MARY	ELLEN
FEDERICO	CATHERINE	
FEDERICO	JOSEPHINE	
FEDORCHEK	ROBT	
FERRERI	GEORGE	
FITZWORTH	JAMES	
FOLIO	LOUIS	
FOLTZ	CALVIN	
FORD	AL	
FORSYTH	RUSSELL	
FOWLER	CHARLES	
FRALEY	WARREN	
FRANKOVICH	NICKOLAS	
FRANTZ	LORETTA	
GAERTNER	JAMES	L
GALIK	JOHN	
GALL	HELEN	J
GALLA	JOHN	
GARDNER	JOSEPH	D
GEARIETY	CHARLES	
GENSERT	GEORGE	
GERHARD	BARBARA	
GETZIEN	HARRY	
GIBSON	RICHARD	
GIBSON	ROBERT	
GILBERT	MARY	
GILMORE	R	D
GOOD	RUTH	
GOODSON	ALATHEA	
GOODSON	WILLIAM	D
GOODWIN	JUDITH	
GORNIK	FRANK	
GOVE	RICHARD	
GRAVES	JAMES	L
GRAVITT LE	ROY	
GREEN	AURETHA	
GREEN	JULIUS	
GREENWAY	FRED	STUDER
GRIFFIS	JOYCE	
GRIFFITH	JAHUE	C
GUE	ROBERT	
GUTTMAN	RICHARD	
HALE	WILBERT	H
HAMILTON	PAUL	R
HAYDEN	MAXINE	
HAYLES	RUTH	
HECKMAN	EDWARD	
HECTER	DENNY	
HEGLAW	ROBERT	
HENDERSON	RAYMOND	
HERKE	GEORGE	W
HERRICK	F	

HERTENSTEIN	ARWILDA	
HERTENSTEIN	EDWARD	
HESS	HARRIETT	
HICKEY	JERRY	
HINTON	ELSIE	L
HINTON	WILLIAM	R
HIVELY	CHARLES	K
HIVELY	NELLIE	F
HOBEIN	EARL	
HOFFMAN	RONALD	
HOGARTH	DONALD	
HOGARTH	MARTHA	
HOLLAND	JOHN	
HOLLEN	ARTHUR	F
HOOVER	HARRY	
HOOVER	JOHN	
HOWARD	EUGENE	
HUBER	GEORGE	
HUDSON	EDWIN	P
HUGHES	ELLA	M
HUNT	BLANCHE	L
HUNTER	ROY	CHARLES
IAFELICE	NORMAN	
IAMMATEO	MICHELINA	
JACKLITZ	ALBERT	
JACKSON	ANDREW	I
JACKSON	MARY	H
JACKSON	W	C
JAMES	JOHN	W
JANES	PETER	J
JARMUSCH	MARILYN	
JAYNES	TED	R
JEAN	RUTH	
JOHNSTON	DONALD	HUMPHREY
JUDD	ESTIL	LEE
KARKOSKA	JOSEPH	
KASPAR	MAY	E
KEELER	RUSSELL	J
KELLER	SAM	
KELLEY	PAUL	J
KELLY	W	L
KELLY	WALTER	C
KENT	EDWARD	C
KESTER	LILLIAN	
KIBBLER	HENRY	
KIDD	LEO	J
KIFER	JOHN	J
KIKOLI	FRANK	
KILLIANY	JOHN	
KINZER	JAMES	
KIRCHNER	KEITH	E
KIRKPATRICK	CATHERINE	
KISTHARDT	MARIE	L
KLAUS	AUGUST	
KLEINSHROT	BETTY	
KLEMPAN	ED	
KLESS	ARTHUR	
KOONTZ	MELVILLE	
KOPREVEC	AL	
KOROSEC	ROBERT	
KOSTORA	P	J
KOVIC	JOYCE	L
KRALL	ANGIE	
KRAMER	JANE	
KRAUSE	HARRY	
KREVES	EDWARD	J
KRIVDO	JEROME	M
KRIVDO	NELL	
KUBEA	ROSE	
KUCHARSKI	RONALD	
KUNSMAN	RAYMOND	
LACONTE	FRANK	
LANCASTER	WALLACE	
LANTZY	RICHARD	
LATHAM	O	W

127 Trailer Park at Euclid Beach Park, Cleveland, Ohio

LAST	FIRST	MIDDLE
LATKOVICH	ELEANOR	
LATKOWSKI	RICHARD	
LAVELLE	MICHAEL	
LAVO	J	
LEIBY	CHARLES	
LENGVEL	LADISLAUS	
LESSO	ANDREW	M
LICKER	MANNIE	R
LILLIS	STEVEN	J
LITTLE	FRANCES	
LIVINGSTONE	BERTHA	
LOHSE	HAL	
LONGAZEL	ANNA	
LONGFIELD	TESS	EVANS
LONGWELL	JOSEPH	
LOVE	DALE	
LUCAS	FRANK	
LUKS	DONALD	
LUPTON	ROBERT	
LYNN	H	A
MAC DONALD	JOHN	
MACKENZIE	ELLEN	
MAHONE	WILLIE	B
MALEC	FRANK	
MALENGO	LEO	
MANCINE	BENJAMIN	
MANDAU	RUDOLPH	
MANDEL	IDA	
MANNING	DOUGLAS	L
MARETTE	RALPH	
MARIANO	MIKE	
MARKEL	JOSEPH	
MARMASH	JOHN	
MARN	LOUIS	
MARRA	TONY	
MARSHALL	CADDIE	
MARTENS	WALTER	J
MARTINO	FRANK	
MASON	ARTHUR	L
MASSIE	GEORGE	T
MASTRAN	LEE	
MAUSAR	JOSEPH	
MAZZA	M	E
MCBRIDE	CALVIN	P
MCBRIDE	MARY	E
MCCARTHY	FRANK	
MCCAULEY	ISABEL	
MCCLOSKEY	EDWARD	
MCCULLEY	PRESTON	I
MCGANN	MICHAEL	L
MCGOWEN	THOMAS	
MCKINLEY	JAMES	H
MEEHAN	J	
MEISER	ROBERT	
MELBER	EDWARD	
MERZDORF	EDWARD	
MICHEL	GORDON	F
MIERKE	ALBERT	H
MIERTL	DON	
MILES	ARCHIE	
MILLER	A	D
MILLER	ALICE	
MISNY	JOHN	
MITCHELL	DORITA	
MITCHELL	EDDIE	
MITCHELL	MARY	
MOMINEY	RAY	
MOORE	MAY	C
MOUNTS	JAMES	
MUERS	HARRY	
MULLINS	HERBERT	
MUMAW	SUSAN	B
MURRAY	JOSEPH	
MYERS	GLENN	
NEUBAUER	JAY	H
NEUBAUER	KATHRYN	MELVIN
NICOL	ANDREW	
NOLAN	ROBERT	
NOVAK	ANTON	
NOVAK	RALPH	
NUTTER	WILLIAM	
OAKLEY	ADDIE	ELLEN
OBERDANK	L	
OCONNOR	JAMES	
OCONNOR	MARY	
OKROS	BETTE	JO
OKROS	MARIE	H
OLESKI	RAYMOND	
ONEIL	BESSIE	
ONEIL	JOHN	C
OPALK	MARTIN	
ORNE	HARRY	D
PAGLIA	ROSA	
PALEK	FRANK	
PAPCKE	DAVID	
PARKER	WILLIAM	
PARKER	WILLIAM	FORMAN
PARMELEE	CHARLES	W
PARMELEE	EDITH	
PARSONS	EDWARD	
PEARCE	AGNES	
PERKINS	HERBERT	
PERKINS	WILLIAM	
PERSELL	CHARLES	
PETERKA	LIBBIE	M
PETRECCA	JOSEPH	
PETRECCA	NICHOLAS	
PLANTSCH	MICHAEL	
POWELL	JAMES	
POZUN	VERA	
PRICE	FLOYD	L
PRICE	LEO	E
PRYMMER	GEORGE	
PUINNO	NICHOLAS	J
PURDY	MARGARET	
RAFFERTY	PETER	
RAUPACH	WILLIAM	
RAY	PAUL	
REARDON	ROGER	
REDFIELD	CUYLER	I
REED	MARIE	E
REIBEL	DESSIE	
REINER	JOSEPH	
REINHARD	GEORGE	MARTIN
REISING	RICHARD	
REPELLA	THOMAS	
RICHARD	PHILLIP	E
RICHARDS	FLORENCE	
RODERICK	ALICE	M
ROLLINS	ERNEST	L
RONIGER	H	E
ROSKOPH	BARBARA	

"Kiddie Hook and Ladder, Euclid Beach Park, Cleveland, Ohio"

LAST	FIRST	MIDDLE
ROSS	EDGAR	E
ROZANC	THOMAS	
RUCKER	CARTER	H
RYAN	EDWARD	
RYDMAN	ELIZABETH	
SAINTZ	RAYMOND	
SANDERS	PATRICIA	
SAUNDERS	NORMAN	
SBROCCO	SABASTINO	
SCAFIDI	JOE	
SCHAFFER	ROBERT	
SCHAUB	GEORGE	
SCHELLENTRAGER	DAVID	
SCHILL JR	AUGUST	J
SCHNELLER	ARTHUR	
SCHRAMM	HAROLD	R
SCHROEDER	WILLIAM	
SCHUESSLER	ALAN	
SCHULTZ	NORMAN	
SCOTT	LESTER	
SEDMAK	ALLAN	
SHAFFER	CLARENCE	
SHANNON	HARRIS	COOPER
SHAVER	DAVID	
SHAVER	LORETTA	
SHEEHAN	JAMES	D
SHEEHAN	MABELLE	B
SHEPPARD	THEODORE	R
SHILLIDAY	EVERETT	P
SILVEROLI	JOHN	
SIMA	JAMES	J
SKULLY	JERRY	
SMELIK	JOHN	
SMILEY	OTTO	
SMITH	ALAN	H
SMITH	ALVIN	W
SMITH	BETTY	BRUMFIELD
SMITH	CHARLES	L
SMITH	CLEVELAND	
SMITH	FRED	C
SMITH	HARRY	A
SMITH	IRENE	MARIE
SMITH	MARY	C
SMITH	RAYMOND	S
SMRDEL	STANLEYJ	
SNELLING	SANDRA	
SOLTIS	MAXINE	
SOMMERS	ERNESTINE	
SOMMERS	JOANNE	
SOUSA	EUGENE	
SPEIDEL	MELVIN	
SPRENGER	WILLIAM	
SPUZZILLO	MARLENE	
SPUZZILLO	MARY	
SPUZZILLO	MICHAEL	
SPUZZILLO	MICHAEL	A
ST CLAIR	JULIUS	J
STARE	F	
STEELE	ARTHUR	W
STEELE	RICHARD	W
STEPP	GENUS	W
STEWART	JEAROLD	
STONEBACK	HOWARD	DETWEILER
STONEBACK	WILLIAM	
STRAND	CARL	
STRONG	ELEANOR	
STUMPF	CORA	MARIE
SUSONG	FLOYD	
SWIMER	MIKE	
TAYLOR	RICHARD	
TAYLOR	ROY	C
TEETER	ARNOLD	
TEODOSIO	VINCENT	
TERRIZZI	PETER	
THOMAS	PAUL	EDWARD
THOMAS	PRYSE	
THOMAS	ROBERT G	
THOMPSON	IRVIN	C
THUNHURST	EDYTHE	
TIZZANO	MARIA	
TOMPKINS	ROSA	
TONCRE	WESLEY	
TONELLI	BENEDETTE	
TOTH	PATRICIA	
TRANCHITO	CARMEN	
TRIPLETT	SYLVESTER	
TRIPOLI	BART	J
TRIVISONNO	GRACE	
TUCKER	TOM	
TUEMMER	HENRY	
URANKAR	GERALD	
VACCARIELLO	ANITA	
VACCARIELLO	CAROL	
VACCARIELLO	JOSEPHINE	
VACCARIELLO	URBAN	
VALENTINE	LEWIS	
VALENTINO	MARY	D
VERCILLO	ROGER	
VILLANI	RAYMOND	
VOGEL	ROBERT	
VOSS	RICHARD	
VRANEKOVIC	JOS	D
WAGNER	FLORENCE	CALLAHAN
WAITE	JOHN	
WALDRON	KATHERINE	
WALLON	CLIFFORD	
WALSH	EDWARD	
WANCHISN	GEORGE	
WARD	AGNES	
WARD	CHARLES	
WARD	DONNA	MAE
WARD	IRMA	
WATERS	LULU	M
WATERS	WILLIAM	
WATERSON	ALLAN	F
WEATHERLY	FRANCINE	
WEBER	EDWARD	
WEISSERT	ROBERT	
WERBER	DONN	
WESKE	ALBERT	
WESOLOSKI	TONY	
WEST	NETTIE	
WILEY	GEORGE	H
WILKANOSKI	JEROME	
WILLIAMS	BERTHA	
WILLIAMS	GRACE	WYATT
WILLIAMS	KENNETH	A
WILLIAMS	MARGARET	
WILLIAMS	ROGER	A
WILLIAMS	WALTER	DEWEY
WILLIAMSON	JON	

Rocket Ships Euclid Beach Park Cleveland, Ohio

LAST	FIRST	MIDDLE
WILSON	LLOYD	PAUL
WILSON	RICHARD	G
WILSON	ROBERT	O
WILSON	WILLIAM	S
WINTERBOTTOM	EDWIN	
WOODS	FRANK	HENRY
WRAY	ATHALYNE	
WRIGHT	ETHEL	M
WURM	CHARLOTTE	L
YAGER	FRED	H
YAGER	LOUISE	E
YARKER	HORACE	
YOUNG	LLOYD	WESLEY
YOUNG	MOSES	
YOUNG	ROBERTA	
ZACK	ANDY	
ZAMBONI	HAZEL	
ZAMBONI SR	M	J
ZANYK	JOHN	
ZIGLI	RONALD	M
ZIMMERMAN	GEORGE	C
ZODNIK	F	
ZOMBECK	JACOB	
ZOMBECK	LEO	
ZORETICH	PAUL	

H. E. Rosniger and W. D. Williams check the train. ca. 1954.

Do you remember the French Frolic?

Or the Giant Vacuum?

1956
EMPLOYEE ROSTER

LAST	FIRST	MIDDLE
ADAMS	DAVID	
ALBRIGHT	GEORGE	H
ALLEN	LOUIS	B
ANDERSON	FRANZ	
ARISTOTILE	ERMINIA	
ARNOLD	TIMOTHY	
AUSTIN	CHARLES	
BACHER	ROBERT	
BAILEY	MABEL	
BAILEY	WILLIS	
BAIRD	MARY	E
BAKER	CILLIUS	M
BALCAM	GWENDOLYN	
BALTURSHAT	ESTHER	
BANDSUCH	RONALD	
BANDY	CHARLES	
BANISH	AUDREY	
BARBATO	JERRY	
BARINDT	HELEN	
BARON	GEORGE	A
BARONE	ANTHONY	
BARRY	DOROTHY	MARIE
BARTH	CAROLINE	
BARTH	GEORGE	
BARTO	ALOYSIUS	FRANK
BASS	ARLENE	
BATTAGLIO	JANE	
BAUER	DAVID	
BAUER	ROBERT	
BAUMAN	WAYNE	
BENNETT	ARVIA	
BENNETT	CHARLES	
BERTRAM	ROBERT	L
BIACOFSKY	ROBERT	J
BIRKENHEAD	THOMAS	
BITTEL	ROGER	
BLAIWES	ARTHUR	
BLEAM	C	M
BLEAM	LULA	ELSIE
BLOEDE	WILLIAM	CARL
BLUHM	JOHN	
BOVINO	VINCENT	
BRACCIA	MARY	
BRADFORD	JOHN	
BRADSHAW	JESSIE	
BRAXTON	LEANA	
BRENNAN	JAMES	E
BRILL	LEONARD	
BROKAW	PAUL	
BRONSTON	THEODORE	
BROWN	DAISY	MAY
BROWN	J	B
BRYANT	ANNA	
BRYANT	THOMAS	
BUCHANAN	RICHARD	
BURNEY	HENRY	
BURTON	JAMES	
BUTLER	JOSEPH	HACKNEY
CADA	PHILIP	
CAMPBELL	ALEXANDER	ROBERT
CAMPBELL	FRED	
CAMPBELL	ROBERT	
CARROLL	GERTRUDE	M
CARROLL	HAZELLE	BELLE
CATALANO	FRANK	A
CATALANO	JAMES	
CATEY	EARL	W
CHAMBERLAIN	VERN	
CHIN	VINCYNNE	
CHRISTEL	RUDOLPH	
CIERTNIK	VICTOR	
CIESLAK	DANIEL	E
CIPITI	THOMAS	F
CIRINO	JOHN	
CIRINO	LUISA	
CIRINO	THERESA	M
CLARK	FRED	
CLIFFORD	FRANCIS	
COLLINS	MABEL	
CONNAVINO	JOSEPHINE	
CONNOR	NORMAN	
CONRAD	GEORGE	
CONTENTO	GUISEPPE	
CONWAY	HARRY	F
COX	OLA	
CRAMER SR	GEORGE	
CRANDALL	HOWARD	
CRAWFORD	HERMAN	L
CRAWFORD	OSCAR	B
CRIGHTON	JOHN	
CROOKS	EMMA	
CUSTER	JOHN	
DANEWITZ	BARBARA	
DAVID	GERALD	
DAVIS	CHARLES	E
DAVIS	KATIE	
DEERING	GEORGIA	
DELAMBO	FRANK	FRANCIS
DEMETER	RICHARD	
DESPENES	PATRICIA	
DICK	ARCHIBALD	
DICORPO	MIKE	
DITIRRO	MARTA	
DOSSA	STEVE	
DOUGHERTY	JEAN	
DUCCA	ALEXANDER	
DUNAWAY	JASPER	
EVENDEN	RICHARD	
FARREN	ROBERT	
FEATHER	MARY	ELLEN
FEDERICO	CATHERINE	
FEDERICO	MARYANNE	
FERRERI	GEORGE	
FORTUNA	WILLIAM	
FRANKLIN	ALFRED	
FRANTZ	LORETTA	
FRANTZ	WILLIAM	
FRENCH	MILDRED	
FULTON	JOHN	
GALIK	JOHN	
GALL	HELEN	J
GALLA	EDWARD	
GALLA	FRED	
GALLA	JOHN	
GARDNER	JOSEPH	D
GASTON	CHARLES	
GEARIETY	CHARLES	
GETZIEN	HARRY	
GIAVONETTE	JAMES	
GIBBS	BARTON	
GIBSON	ROBERT	
GIERMAN	WILLIAM	
GIMPLEY	JOSEPH	

LAST	FIRST	MIDDLE
GOOD	RUTH	
GOODSON	ALATHEA	
GOODSON	WILLIAM	D
GORNIK	FRANK	
GRADY	DAVID	
GREENWAY	FRED	STUDER
GRISSON	CLASSIE	
GUENTZLER	RONALD	
GUHDE	CHARLES	R
HALL	MARY	
HASCAL	RUTH	
HAWKINS	EDWARD	
HECKMAN	EDWARD	
HENDERSON	RAYMOND	
HERKE	GEORGE	W
HERTENSTEIN	ARWILDA	
HERTENSTEIN	EDWARD	
HINTON	ELSIE	L
HINTON	WILLIAM	R
HITT	WILLIAM	
HOGARTH	DONALD	
HOGARTH	MARTHA	
HOLLAND	JOHN	
HOLLEN	ARTHUR	F
HOLLEN	JAMES	
HOLLEN	WILMER	
HOWARD	EUGENE	
HOWELL	ROBERT	
HUBBARD	MARY	
HUBER	GEORGE	
HUDSON	EDWIN	P
HUDSON	ERNEST	
HUDSON	WILLIAM	J
HUGHES	BEULAH	
HUGHES	ELLA	M
HUNT	BLANCHE	L
HUNTER	ROY	CHARLES
HURLESS	JACKIE	
HURLESS	JANET	
IAMMATTEO	MICHELINA	
JACKLITZ	ALBERT	
JACKSON	ANDREW	I
JACKSON	GRACE	
JACKSON	MARGARET	
JACKSON	MARY	H
JAMES	JOHN	W
JAMIESON	DAVID	
JANES	PETER	J
JAYNES	TED	R
JEAN	JOHN	L
JOHNSTON	AGNES	
JOHNSTON	DONALD	HUMPHREY
JUBACH	GEORGE	R
JUDD	ESTIL	LEE

KARKOSKA	JOSEPH	
KASELONIS	KENNETH	
KASPAR	MAY	E
KELLER	SAM	
KELLIC	JULIA	
KENT	EDWARD	
KENT	EDWARD	C
KIKOLI	FRANK	
KING	DOLORES	
KINZER	JAMES	
KIRCHNER	KEITH	E
KISTHARDT	MARIE	L
KLEINSHROT	BETTY	
KLEMENCIC	EGIDI	
KOROSEC	ROBERT	
KOVIC	JOYCE	L
KRAFT	FRANK	
KRAMER	JANE	
KRAUSE	HARRY	
KRIVDO	JEROME	M
KUCHARSKI	RONALD	
KUHTA	ROBERT	
KUNSMAN	RAYMOND	
KUNSMAN	WALTER	
LACY	CHARLES	
LANTZ	RAYMOND	
LARDIE	GIDEON	
LARDIE	RONALD	
LEECE	LEICESTER	
LEECE	MARIE	E
LEWIS	RONALD	
LICKER	MANNIE	R
LONGAZEL	ANNA	
LONGFIELD	TESS	EVANS
LONGWELL	JOSEPH	
LOVE	CORA	
LOVE	DALE	
LUCAS	FRANK	
LUPO	RICHARD	
LUPTON	ROBERT	
LYNN	H	A
MAC DONALD	JOHN	
MAHONE	JOHN	
MAHONE	WILLIE	B
MALOVRH	WILLIAM	
MANCINE	LOUIS	
MANCINI	JOSEPHINE	
MANDATO	MADONNA	
MANDAU	RUDOLPH	
MANDEL	IDA	
MANNING	DOUGLAS	L
MANSON	DAVID	
MARCUM	HENRY	
MARIANO	MIKE	
MARKEL	JOSEPH	
MARN	LOUIS	
MARRA	ANTHONY	
MARSHALL	CADDIE	
MARTENS	ALBERT	
MARTENS	GOLDA	
MARTENS	WALTER	J
MARTIN	EDWARD	
MARTINO	FRANK	
MASKUNAS	JEANETTE	
MAUSAR	JOSEPH	
MCBRIDE	CALVIN	P
MCBRIDE	MARY	E
MCCARTHY	FRANK	
MCKENZIE	MARTIN	
MCKINLEY	JAMES	H
MEADE JR	JOHN	
MEDVES	EDWARD	
MELARAGNO	ASSUNTA	
MELBER	EDWARD	
MERZDORF	EDWARD	
MIERKE	A	H
MILES	ARCHIE	

LAST	FIRST	MIDDLE
MILLER	ALICE	
MILLER	MILDRED	
MOLESKY	ROBERT	
MOORE	MAY	C
MOSES	CECIL	E
MOSES	KIM	E
MOUNTS	JAMES	
MULLINS	HERBERT	
MUMAW	SUSAN	B
NARDINI	JOHN	
NEUBAUER	JAY	H
NEUBAUER	KATHRYN	MELVIN
NICOL	ANDREW	
OAKLEY	ADDIE	ELLEN
OCONNOR	MARY	
OHARA	PATRICK	
OKER	DONALD	R
OKROS	MARIE	H
OLESKI	DONALD	
OLESKI	RAYMOND	
OPALK	MARTIN	
OSBORNE	CHARLES	
PAGLIA	ROSA	
PAKOSH	MICHAEL	
PALEK	FRANCIS	
PARKER	CLYDE	
PARKER	WILLIAM	
PARMELEE	CHARLES	W
PARMELEE	EDITH	
PARMER	LUVADIA	
PASSALACQUA	JOHN	
PATRICK	EDDIE	
PEARCE	AGNES	
PEPPLE	VIRGINIA	
PERSELL	CHARLES	
PESTA	EDWARD	
PETERKA	LIBBIE	M
PETRECCA	JOSEPH	
PETRECCA	NICHOLAS	
PHILLIPS	ERNEST	
PLANTSCH	MICHAEL	
PLANTSCH	MIKE	
POWELL	MAXINE	
PRICE	FLOYD	L
PURDY	MARGARET	
RADOVANIC	EVA	
RAUPACH	WILLIAM	
RAY	PAUL	
REARDON	ROGER	
REDFIELD	CUYLER	I
REED	MARIE	E
REINHARD	GEORGE	A
REINHARD	GEORGE	MARTIN
REYNOLDS	JAMES	
ROBINSON	PERCY	
ROBINSON	WILLIAM	
ROCKFORD	JAMES	
RODERICK	ALICE	M
ROGERS	ROBERT	
ROHLOFF	WALTER	
ROLLINS	ERNEST	L
RONIGER	H	E
ROSS	FRANK	
ROSSODIVITO	MARIE	
ROWLEY	HARRY	
RUCKER	CARTER	H
RYDMAN	ELIZABETH	
SANDERS	PATRICIA	
SARTAIN	KENNETH	
SBROCCO	SABASTINO	
SCHABRONI	GERALD	
SCHAUB	GEORGE	
SCHILL JR	AUGUST	J
SCHNELLER	ARTHUR	
SCHORSTEN	PAULINE	
SCHRAMM	HAROLD	R
SENNISH	GEORGE	
SHANNON	HARRIS	COOPER
SHAVER	DAVID	
SHAVER	LORETTA	
SHEEHAN	JAMES	D
SHEEHAN	MABELLE	B
SHEEHAN	SYLVESTER	
SHEPPARD	THEODORE	R
SHILLIDAY	EVERETT	P
SISTEK	MAE	
SMELIK	JOHN	
SMITH	ALAN	H
SMITH	ALVIN	W
SMITH	CHARLES	L
SMITH	FRED	C
SMITH	GATRAL	
SMITH	HARRY	A
SMITH	IRENE	MARIE
SMITH	MARY	C
SMITH	RAYMOND	S
SODEQUIST	CHERIE	
SOUSA	EUGENE	
SPRENGER	WILLIAM	
SPRINGER	CATHERINE	
SPUZZILLO	MICHAEL	A
ST CLAIR	JULIUS	J
STARCHER	ORIS	
STARE	F	
STARE	FRANK	
STEELE	ARTHUR	W
STELMACH	EVELYN	
STEWART	JAY	T
STIRES	CARL	
STOKES	THEODORE	
STONEBACK	HOWARD	DETWEILER
STRAND	CARL	
STRONG	ELEANOR	
STURM	ANTON	
SUGLIA	MARIE	
SUSONG	FLOYD	
SUTKUS	ALICE	
SVETINA	JOHN	
SWARTZWELDER	JOHN	
SWEENEY	ANGELA	
TAYLOR	RICHARD	
TAYLOR	ROY	C
TAYLOR	THOMAS	
TEMPLIN	RONALD	
TERRIZZI	PETER	
THACKER	CHARLES	
THACKER	OTTO	
THOMAS	JAMES	P
THOMAS	ROBERT	G
THUNHURST	EDYTHE	

Flying Scooter — Euclid Beach Park, Cleveland, Ohio

LAST	FIRST	MIDDLE
TIZZANO	MARIA	
TONCRE	WESLEY	
TOTH	PATRICIA	
TRANCHITO	CARMEN	
TRENT	FRANK	
TRIVISONNO	CHARLES	A
TUCKER	TOM	
TUEMMER	HENRY	
TURNER	MAXINE	
ULRICH	FRANK	
UNDERWOOD	LEAMAND	
URANKAR	GERALD	
URGELEIT	LEONARD	
VACCARIELLO	JOSEPHINE	
VACCARIELLO	URBAN	
VALENTINO	MARY	D
VEGNEY	EUGENE	T
VERARDI	GIULIO	
VERBA	ROBERT	
VOGEL	ROBERT	
VRANEKOVIC	JOS	D
WAGNER	FLORENCE	CALLAHAN
WALKUSKI	CHESTER	
WANDA	CHARLES	
WARD	CHARLES	
WATERS	LULU	M
WATERS	MATILDA	
WATERS	WILLIAM	
WEBER	EDWARD	
WEISSERT	ROBERT	
WELLING	PETER	
WERBER	DONN	
WESKE	ALBERT	
WHEATCROFT	DAVE	
WHITE	ALDONA	
WILEY	GEORGE	H
WILLARD	EUGENE	C
WILLIAMS	GRACE	WYATT
WILLIAMS	KENNETH	A
WILLIAMS	MARGARET	
WILLIAMS	ROGER	A
WILLIAMS	WALTER	DEWEY
WILSON	ANNA	R
WILSON	LLOYD	PAUL
WILSON	PAUL	
WILSON	RICHARD	G
WILSON	ROBERT	O
WILSON	WILLIAM	S
WINTERBOTTOM	EDWIN	
WOODS	FRANK	HENRY
WRAY	ATHALYNE	
WRIGHT	LULA	
WRIGHT	WILLIAM	N
WURM	CHARLOTTE	L

YAGER	FRED	H
YARKER	HORACE	
YOCUM	DAVID	
YOUNG	ANN	
YOUNG	LLOYD	WESLEY
YOUNG	MOSES	
YOUNG	REGINALD	
YOUNG	ROBERTA	
ZAMBONI	HAZEL	
ZAMBONI SR	M	J
ZANYK	JOHN	

Harris C. Shannon, Park Manager 1901-1959.

A slow day on the Thriller.

120

1957-58
EMPLOYEE ROSTER

AMERICAN DERBY

LAST	FIRST	MIDDLE
BRADSHAW	J	
BRADSHAW	JOHN	
DEMETER	RICHARD	
EDDY	TOM	
JONES	DAVID	
KEEFER	LEE	
LANGE	DON	
MALENGO	LEO	
MANNING	DOUG	
MELBER	ED	
NICOL	ANDREW	
OCONNOR	JIM	
PAUL	FRED	
PEGORARO	JIM	
PHILLIPS	JERRY	
ROADS	KENNETH	
ROSS	FRANK	
SMITH	ALVIN	
SMITH	FRED	
SMITH	HARRY	
THOMPSON	IRVIN	
WILSON	LLOYD	
WILSON	PAUL	
WILSON	RICHARD	
WILSON	ROBERT	
WILSON	WM	S
WOLF	THOMAS	

BALLOON STAND

LAST	FIRST	MIDDLE
COVOLTA	MARILYN	
FIER	RHODA	
MUMAW	SUSAN	
PEARCE	AGNES	
PETERKA	LIB	
ROBBINS	KATHLYN	
KROEGER	LOUISE	
MCCAULEY	ISABELLE	
SCHMIDT	ALBERTA	
STELNIAK	EVELYN	
BENNETT	ARVIA	
LUIKART	CAROLYN	

BIG 4 STAND

LAST	FIRST	MIDDLE
ADAMS	DAVID	
BARGHOLT	NORMAN	
BARON	GEO	
BENNETT	CHARLES	
BLODGETT	RICHARD	
BROWN	ALLEN	
CHOVANEC	ANDREW	
COLBERT	CHARLES	
COLLINS	EDWARD	
COWIE	TOM	
ESTEP	JUSTUS	
FANSLAU	HOLDIGARD	
GALLA	JOHN	
HERKE	GEORGE	
HOGARTH	DONALD	
HOLLEN	JAMES	
KAST	HOWARD	

KELLER	SAM	
KILBECK	MIKE	
MANDAU	RUDY	
MARTENS	AL	
MAUSER	JOSEPH	
MECHTENSIMER	HARRY	
MIERKE JR	ALBERT	
MIERKE SR	ALBERT	
PARMELEE	C	W
RUCKER	CARTER	
SAWITKE	DALE	
SMITH	ALAN	
SMITH	CHARLES	
TIMKO	EUGENE	
TONELLI	GIOVANNI	
WATERS	WILLIAM	
WATTS	RAYMOND	

BOULEVARD STAND

LAST	FIRST	MIDDLE
ADKINS	BARBARA	
BALCAM	GWENDOLYN	
BANDSUCH	RONALD	
BANISH	AUDREY	
BEALL	CHARLES	
BLACK	MARCIA	MARIE
BROOKS	JOY	
CAST	MIKE	
CATALANO	JIM	
CIRINO	JOHN	
CLEVENGER	JEAN	
COSTANZO	JOSEPHINE	
FEATHERS	MARY	ELLEN
FEDERICO	MARYANNE	
GOGOLIN	HENRY	
GOODWIN	JUDY	
GRAY	JIM	
HURLESS	JACKIE	
JUDD	ESTIL	
KAST	DALE	
KAUFMAN	RICHARD	
KLEINSMITH	RONALD	
KNOWLES	SHARI	
KRIVDO	NELL	
KUCHARSKI	RONALD	
KUHTA	ROBERT	
LATKOVICH	ELEANOR	
LERZ	ANGIE	
MEYER	BLONDELL	
MILLER SR	PETER	
MOORE	HAROLD	
NARDINI	JOHN	
OSWALT	VIRGINIA	
PERKOVICH	JOHN	
PETRECCA	NICK	
PRICE	FLOYD	
ROSSA	ARLENE	
SAID	NORINE	
SCHUBERT	ROY	
SELLARDS	RIBECCA	
SELLERS	ESSIE	
SHIMKO	STEVE	
SMITH	JEANNETTE	
TETTELBASH	BARBARA	
TOMINO	DONALD	
TRIVISONNO	CHARLES	
TURK	CAROLINE	

LAST	FIRST
URANKAR	GERALD
WHITE	ALDONA
WILSON	DOROTHY
WISE	LAVERNE
WISE	PHYLLIS
WRAY	ATHALYN
ZADELL	DOLORES

CAMP OFFICE

LAST	FIRST	MIDDLE
BRENNAN	JAMES	
CREIGHTON	JOHN	
DAVIS	KATIE	
DUKE	EFFIE	
FANSLOW	HERMAN	
KLEMENCIC	EGIDI	
LONGWELL	JOE	
MATHEWSON	CHAS	
MORICA	LARRY	
MRS HUNT		
MULLINS	DAVE	
MULLINS	HERB	
PLANTSCH	MIKE	
YOUNG	REGI	
YOUNG	ROBERTA	

CANDY KISS STAND

LAST	FIRST	MIDDLE
ADKINS	BARBARA	
AUSTIN	CHUCK	
BANDSUCH	RONALD	
BANISH	AUDREY	
BEALL	CHARLES	
CAST	MIKE	
CATALANO	JIM	
CIRINO	JOHN	
CRIMI	ANGELO	
DANIELS	RAY	
EAST	MIKE	
FEATHERS	MARY	
GOGOLIN	HENRY	
GRAFIO	ENRICO	
GRAY	JIM	
JUDD	ESTIL	
KAST	DALE	
KAUFMAN	RICHARD	
KLEINSMITH	RONALD	
KORDICK	DAVID	
KUCHARSKI	RONALD	
KUHTA	ROBERT	
LARUE	GLEN	
LECLAIR	FRANK	
MANNING	DENNIS	
MOORE	HAROLD	
NARDINI	JOHN	

LAST	FIRST
PERKOVICH	JOHN
PERNUS	PAT
PETRECCA	JOE
PETRECCA	NICK
PRICE	FLOYD
RAY	DICK
RICHMOND	DENNIS
SCHUBERT	ROY
SPINELLI	JIM
TOMINO	DONALD
TRIVISONNO	CHARLES
URANKAR	GERALD
WHITE	ADONA
WILSON	DOROTHY
WISE	LAVERNE
ZADELL	DOLORES

CAROUSEL

LAST	FIRST	MIDDLE
BLODGETT	RICHARD	
CHAPMAN	PETE	
GALA	FRED	
GUHDE	CHAS	
KNAUSS	RAY	
MANTOOTH	PAUL	
MCCREIGHT	GEORGE	
MOLESKY	BOB	
MONROE	BILL	
NICOL	ANDREW	
RONKE	BERT	
RUCK	JIM	
SMITH	CHAS	
SMITH	RAY	
STACKS	JOE	
THOMAS	ROGER	
WILLARD	EUGENE	

CARPENTERS

LAST	FIRST	MIDDLE
HUDSON	E	P
KLEMENCIC	EGIDI	
LONGWELL	JOE	
MATRKA	FRANK	
MCKINELY	JIM	
MULLINS	HERB	
OCONNOR	JIM	
OCONNOR	JIM	
OCONNOR	MIKE	
OCONNOR	MIKE	
PLANTSCH	MIKE	
STONEBACK	H	D
VALENTINE	ED	
VOGEL	BOB	
WHEATCROFT	DAVID	
YELCHO	JACOB	

CASHIERS

LAST	FIRST	MIDDLE
BAIRD	MARY	
BARRY	DOROTHY	
BLEAM	LULU	
BOLINGER	RUTH	
BRADSHAW	JESSIE	
BRAUN	ESTELLE	
CARROLL	HAZEL	
COLLINS	MABEL	
GOOD	RUTH	
KISTHARDT	MARIE	
KRAMER	JANET	
KUHELAVIC	KATHRYN	
LEECE	MARIE	

Dancing Pavilion, Euclid Beach Park, Cleveland, Ohio

LAST	FIRST	
LONGFIELD	TESS	
MARTENS	GALDA	
MCMILLEN	LUCILLE	
MOORE	MAY	
PARMELEE	EDITH	
RODERICK	ALICE	
SHAVER	LORETTA	
SHEEHAN	MABEL	
STELMACH	EVELYN	
STONEBACK	ELVIRA	
THUNHURST	EDYTHE	
VALENTINE	GARNETT	
WAGNER	FLORENCE	
WATERS	LULU	
WRIGHT	ETHEL	
WURM	CHARLOTTE	

CHOPPERS

LAST	FIRST	MIDDLE
BENNETT	ARVIA	
BROWN	DAISY	
CONWAY	LENA	
CRAIG	RUTH	
ESCHEN	MARION	
HUBBARD	MARY	
HUGHES	ELLA	
JONES	KATHRYN	
KIRKPATRICK	MARY	
MANDEL	IDA	
MARSHALL	CADDIE	
OKROS	BETTE	JO
PAPES	JOSEPHINE	
PAPES	VALERIA	
REED	MARIE	
RYDMAN	ELIZABETH	
SHANNON	PEARL	
SISTEK	MAE	
WRIGHT	LULU	
ZIEL	LENA	

COLLONNADE

LAST	FIRST	MIDDLE
BALTURSHOT	ESTHER	
BRYANT	ANN	
CAPRA	MARIA	
CARTER	OTIE	
CIRINO	THERESA	
ESTEP	KAY	
FRENCH	MILDRED	
GALL	HELEN	
HOGARTH	MARTHA	
IAMMATTEO	MICHELINA	
KRALL	ANGIE	
LATKOVICH	ELEANOR	
LONGAZEL	ANNA	
MANNING	DENNIS	
MARMASH	JACK	
MCBRIDE	MARY	
NEUBAUER	JAY	
OKROS	MARY	
OLESKI	DONALD	
PALEK	FRANCIS	
PURDY	MARGARET	
RADOVANIC	EVA	
ROCK	JOHN	
SANTUCCI	MELINA	
SMITH	IRENE	
TIMPERIO	FILOMENA	
VALENTINO	MARY	
WICKS	GERALDINE	
WRAY	ATHALYNE	

DODGEM

LAST	FIRST	MIDDLE
BRADSHAW	JOHN	
EDDY	TOM	
FERGUSON	JIM	
KEEFER	LEE	
KNAUSS	RAY	
MALENGO	LEO	
MANNING	DOUG	
MELBER	ED	
MELBER	ED	
NICOL	ANDREW	
OCONNOR	JIM	
PAUL	FRED	
PEGORARO	RUDOLPH	
ROSS	FRANK	
SMITH	ALVIN	
SMITH	CHAS	
SMITH	HARRY	
TAYLOR	TOM	
THOMPSON	IRVIN	
WETMORE	ROBERT	
WHEATCROFT	DAVE	
WILSON	PAUL	
WILSON	RICHARD	
WILSON	ROBERT	
WOLF	THOMAS	
YOUNG	MOSES	

DRINK STAND

LAST	FIRST	MIDDLE
BLOEDE	WILLIAM	
BROWN	GEORGE	
COLBERT	CHARLES	
CRESS	JOHN	
HENDERSON	RAY	
HILL	JERRY	
MULLINS	HERBERT	
WILLIAMS	KENNETH	
YOUNG	REGINALD	

FLYING TURNS

LAST	FIRST	MIDDLE
BRENNAN	JAMES	
CATEY	EARL	
DOBLER	G	ROLAND
GETZIEN	HARRY	
HUBER	GEORGE	
MARRA	ANTHONY	
MERZDORFF	EDWIN	
MONROE	WILLIAM	
PETTIT	HAROLD	
REARDON	RODGER	

LAST	FIRST	MIDDLE
RONKE	BERT	
SWIMER	MYRON	

FROZEN WHIP

LAST	FIRST	MIDDLE
BARRETT	FRANK	
BROWN	GEORGE	
CARLTON	JERRY	
CRAMER	GEORGE	
HILL	JERRY	
KINZER	ALLEN	
KUNSMAN	WALTER	
MIZLEY	JAMES	
MIZLEY	PAUL	
MULLINS	DAVID	
MULLINS	HERBERT	
PESTA	EDWARD	
RAY	PAUL	
SCHUTT	KENNETH	
TRENT	FRANK	

GATEMEN

LAST	FIRST	MIDDLE
BLUHM	J	
DOEDDERLEIN	M	
DVORAK	R	
LEECE	L	
PARKER	C	H
PINCHBECK	H	
REDFIELD	C	
STAVE	F	
TUEMMER	H	

KIDDIE LAND

LAST	FIRST	MIDDLE
ADAMS	TOM	
ALLEN	LOU	
APANITAS	J	
BARKER	BOB	
BARRETT	RICHARD	
BLEAM	C	
BROWN	AL	
BURKHARDT	OTTO	
BURNS	ROBT	
BURTON	RICHARD	
BUSHMAN	FRANK	
CLARK	THOMAS	
CONLEY	T	
CONNORS	NORMA	
COOLEY	L	
CUTLIP	HARRY	
DAVENBORT	SID	

LAST	FIRST	MIDDLE
DAVIS	CHAS	
DICK	ARCH	
DROBNICK	KEN	
DUCCA	R	
DUKE	BILL	
FIGIEL	F	
FLYNN	JOE	
FRATE	DON	
FULTON	JOHN	
GEARIETY	CHAS	
GEDEON	LEO	
GREGER	RICHARD	
GUHDE	CHAS	
GUNTZLER	RON	
HACKMAN	R	
HALE	C	
HALL	C	H
HALL	H	
HARKERNA	P	
HARPSTER	H	H
HAUPT	T	
HECKMAN	ROBERT	
HERBST	CHAS	
HERTENSTEIN	ED	
KAMINSKAS	RICHARD	
KELLER	JOHN	
KRAUSE	KEN	
LARYEL	RICHARD	
LAWSON	CHAS	
LUTZ	RON	
MILSTOVIC	JOHN	
MUELLER		BOB
PARKER	JIM	
PETELINKAR	R	
PRICE	RAY	
RAUPACH	WILLIAM	
REILLY	JAMES	
RONKE	B	
ROTHAUNEL	ROY	
RUSCSAK	STEVE	
RUTKOUSKI	CHAS	
SCOTT	AL	
SEMAN	JIM	
SENNISH	GEO	
SHEEHAN	JAMES	D
SMITH	L	
SNYDER	JOHN	
SPICA	DON	
STEWART	JEFF	
STEWART	JIM	
SWANKHAMER	W	
SWANSON	F	H
TEKUNTZ	C	
THOMAS	ROGER	
TIANILLO	BILL	
TINKER	E	
TINKER	NEIL	
TURICK	GEO	
VERBERTER	CLIFFORD	
VRYELIST	LIN	
WALKER	DUDLEY	
WILEY	GEO	
ZALEWSKI	TED	

LABORERS

LAST	FIRST	MIDDLE
BARBATO	JERRY	
BERTRAM	LAWRENCE	
BRADFORD	JOHN	
BRYANT	THOMAS	
BUTLER	JOE	
CERIO	ANTONIO	
COERTNIK	VICTOR	
CONTENT	JOE	
DELAMBO	FRANK	

Entrance to Euclid Beach, Cleveland

LAST	FIRST	
DICORPO	MIKE	
DIPITI	THOMAS	
FERRERI	GEORGE	
GORNIK	FRANK	
HUDSON	ERNEST	
JACKSON	ANDREW	
JAMES	JOHN	
KIKOLI	FRANK	
KRIVDO	JERRY	
LENARCIC	JOHN	
MAHONE	JOHN	
MAHONE	WILLIE	
MARIANO	MIKE	
MARN	LOUIS	
MCCOWAN	HERMAN	
MICHAELS	ERVAN	
NEIBERT	GILBERT	
PASALACQUA	JOHN	
PATRICK EDDIE		
POINTER	FRANCIS	
POKRANT	LEONARD	
POKRANT JR	LEONARD	
RICHARDSON	JAMES	
SCHAFFER	WALTER	
SCHILL JR	AUGUST	
SCRUGGS	CECIL	
SILVEROLI	JOHN	
SPUZZILLO	MIKE	
TIRABASSI	CAMILLO	
TRIVISONNO	RAYMOND	
WALKER	LOUIS	
YOCUM	DAVID	
YOUNG	MOSE	
YOUNG	REGINALD	

LADIES ROOM

LAST	FIRST	MIDDLE
STRONG	ELEANOR	
GRISSON	CLASSIE	
BRITTION	JAMIE	
JACKSON	RUBY	M
BROUGHTON	ETHEL	
LEWIS	FRANCES	
BOYD	BARBARA	
DEAN	OLLIE	
RIVERS	MILLIE	
HARRIS	FLORENCE	
JACKSON	WILLADA	
SMITH	MARY LOUISE	
COTTINGHAM	MARGARITE	
JACKSON	MAMIE	
JAMISON	LEUDELLA	
REDRICK	ALDA	M
WASHINGTON	LORENA	
LANIER	FRANCES	
CRAWFORD	FRANCES	
RAINEY	MALISIA	

LAFF IN THE DARK

LAST	FIRST	MIDDLE
BRADSHAW	JOHN	
BRYANT	THOMAS	
DELONG	VANCE	
EDDY	TOM	
JAYNES	TED	
JONES	DAVE	
MALENGO	LEO	
MANNING	DOUG	
NICOL	ANDY	
OCONNOR	JIM	
PEGORARO	RUDOLPH	
PERSELL	CHAS	
ROCKFORD	JIM	

LAST	FIRST	
ROSS	FRANK	
RUCK	JAMES	
SMITH	CHAS	
SMITH	FRED	
SMITH	RAY	
TAYLOR	TOM	
THOMPSON	IRVIN	
TONELLI	BEN	
WILSON	PAUL	
WILSON	RICHARD	
WILSON	ROBERT	
WOLF	TOM	

MACHINE SHOP

LAST	FIRST	MIDDLE
ADAMS	WM	
CAMPBELL	FRED	
JACKLITZ	AL	
KRAUSE	H	
KRIVDO	J	
LUPTON	BOB	
PARKER	BILL	
RONNIGER	H	E
VOGEL	BOB	
WILLIAMS	ROGER	
WILLIAMS	WALTER	
WOODS	F	

MAIN LUNCH

LAST	FIRST	MIDDLE
BERES	ANNA	
CAPRA	CHRISTINA	
CARTER	KENNETH	
CIRINO	LOUISA	
CONNAVINO	JOSEPHINE	
FEDERICO	CATHERINE	
FRENCH	MILDRED	
GALL	HELEN	
JACKSON	MARY	
KRALL	ANGIE	
MARDIS	MARGARET	
MARMASH	JACK	
MCBRIDE	MARY	
NEUBAUER	JAY	
OKROS	MARY	
PLESCIA	LUCY	
SISTEK	MAE	
SMITH	IRENE	
VACCARIELLO	JOSEPHINE	
VALENTINO	MARY	
WICKS	GERALDINE	

Picnickers, Euclid Beach Park, Cleveland, Ohio

OVER THE FALLS

LAST	FIRST	MIDDLE
ALLER	DENNIS	
BARON	GEORGE	
BIRD	GEORGE	
BRANEKOVIC	JOSEPH	
BROOKS	PHILLIP	
CHOVANEC	ANDREW	
CRAMER	GERALD	
DAVIES	LEE	
DEVRIES	DONALD	
GIBBARD	CHRISTOPHER	
HECKMAN	EDWARD	
KRAUSE	HARRY	
LONCELLA	TEDDY	
LOWMAN	DAVID	
MARRA	ANTONIO	
MITCHELL	CHESTER	
MULLINS	HERBERT	
MULLINS	WILLIAM	
PAKOSH	MICHAEL	
PETKOFF	MARKO	
RASH	VERN	
RICHARD	PHILIP	
RICHARDSON	JAMES	
RINALDI	NICK	
RUTLEDGE	ROBERT	
SCHILL JR	AUGUST	
SCHILL SR	AUGUST	
SLAPER	RONALD	
SOUSA	EUGENE	
TURNER	P	B
WHEATCROFT	DAVE	
WISE	JAMES	
WOERTH	LEONARD	

POLICE

LAST	FIRST	MIDDLE
BARBERIO	F	
BRACALE	E	
CAMPBELL	A	R
CIESLAK	D	
DUCCA	A	
GLICKER	W	
JANES	P	
LAUDIE	G	
MARTINO	F	
MILES	A	
ROLLINS	E	
TRIVISON	C	
WEISSERT	R	
WOODROW	B	
YAGER	F	
ZIMMERMANN	J	

PORTERS

LAST	FIRST	MIDDLE
BANDY	CHARLES	
BRACCIA	MARY	
BRADFORD	JOHN	
BRONSON	THEODORE	
BUTLER	JOE	
CIRINO	LOUISA	
CONNAVINO	JOSEPHINE	
CUSTER	JOHNNIE	
DITIRRO	MARIA	
FARINACCI	FILOMENA	
FEDERICO	CATHERINE	
FORILLO	ANNA	
HOWARD	EUGENE	
JACKSON	GRACE	
JAMES	JOHN	
LEWIS	CEDRIC	
MELARAGNO	ASSUNTA	
PAGLIA	ROSE	
PATRICK	EDDIE	
ROBINSON	WILLIAM	
SBROCCO	GISBERTA	
SHEPPARD	T	R
ST CLAIR	JULIUS	
STRUNA	JIM	
SUGLIA	MARIA	
SUSONG	FLOYD	
TROTTA	JOHN	
VACCARIELLO	JOSEPHINE	
YOCUM	DAVE	

RACING COASTERS

LAST	FIRST	MIDDLE
ALLAR	DENNIS	
BARNES	DONALD	
BROOKS	PHILLIP	
CADA	PHILIP	
COLEMAN	WILLIAM	
DEVRIES	DONALD	
HEGLAW	ROBERT	
HOLLAND	JOHN	
MARRA	ANTONIO	
MOREHOUSE	DONALD	
MULLINS	WILLIAM	
NACHTIGAL	JAMES	
PETKOFF	MARKO	
PEVARNICK	ANTHONY	
REICHERT	LAWRENCE	
RICHARD	PHILLIP	
RINALDI	NICK	
SCHAUB	GEORGE	
SCHILL	AUGUST	
SCHILL JR	ROBERT	
SOUSA	EUGENE	
STANTON	SAMUEL	
STAPP	PAUL	
STOCKTON	VERL	
TURNER	P	B
VRANEKOVIC	JOSEPH	
WESKE	ALBERT	
WHEATCROFT	DAVE	
WISE	EDWARD	

RESTAURANT

LAST	FIRST	MIDDLE
COOK	PAT	
DANIELSON	OPAL	
FORMICK	SUE	
HETH	PAUL	
HETH	PAUL	CLINTON
HOFFMAN	O'JEAN	

Pulling Taffy at Humphrey's Candy & Popcorn Stand — Euclid Beach Park — Cleveland, Ohio

LAST	FIRST	MIDDLE
JOHNSTON	CLARA	
KATZ	RONALD	
KOLENCE	PAULINE	
KRIVEC	DIANE	
LITTLE	PAT	
MEDPACK	GRACE	
MILLER	MARY	LOU
MONROE	BETTY	
PENN	BETTY	
POZERELLI	ELIZABETH	
SCOTT	DOROTHY	
SCOTT	DOROTHY	H
SEBOCK	VIRGINIA	
SHAT	FRANK	J
STAUER	ED	
SULLIVAN	JOHN	
SVETINA	FAYE	
WHIGHAM	RUTH	

ROCK-O-PLANE

LAST	FIRST	MIDDLE
DEMETER	RICHARD	
EDDY	TOM	
FERGUSON	JIM	
KNAUSS	RAY	
MALENGO	GEO	
NICOL	ANDREW	
OCONNOR	JIM	
PAUL	FRED	
PEGORARO	JIM	
RUCK	JIM	
SMITH	CHAS	
SMITH	FRED	
SMITH	RAY	
TEASTER	HOWARD	
WILSON	PAUL	
WOLF	EDWARD	

ROTOR

LAST	FIRST	MIDDLE
ALLER	DENNIS	
BARKHAUER	EDWIN	
BIRKENHEAD	THOMAS	
BROOKS	PHILLIP	
CADA	PHILIP	
DAVIES	LEE	
GIBBARD	CHRISTOPHER	
HEGLAW	ROBERT	
HEMPLEWIS	CHARLIE	
HIVELY	CHARLES	
HOLLAND	JOHN	
KRAUSE	HARRY	
MARINO	TONY	
MARKELONIS	RALPH	
MASSENGLE	GREENE	
MONAGHAN	WILLIAM	
MONROE	WILLIAM	
PAKOSH	MICHAEL	
RINALDI	NICK	
RUSS	JOSEPH	
SCHAUB	GEORGE	
SCHILL	EUGENE	
SCHILL JR	AUGUST	
SOUSA	EUGENE	
VRANEKOVIC	JOSEPH	
WESKE	ALBERT	
WISE	EDWARD	
WOERTH	LEONARD	

SOUVENIR STAND

LAST	FIRST	MIDDLE
COVOLTA	MARILYN	
FIER	RHODA	
MUMAW	SUSAN	
PEARCE	AGNES	

SURPRISE HOUSE

LAST	FIRST	MIDDLE
BURMEISTER	TED	
EASTON	JAMES	
EVERETS	NICHOLAS	
KRIVDO	JERRY	
LOGAN	LAMONT	
MCCONOCHA	DANIEL	
MOLESKY	BOB	
WILLARD	EUGENE	
WISE	CHARLES	

THRILLER

LAST	FIRST	MIDDLE
BAUER	DAVID	
CASSERLY	JAMES	
CATEY	EARL	
CHAMBERLAIN	VERN	
EHRKE	RUSSELL	
GOODSON	WILLIAM	
MCBRIDE	CALVIN	
MEADE JR	JOHN	J
OPALK	MARTIN	
STEELE	ARTHUR	
MEADE	JOHN	J
GIMPLEY	JOSEPH	
DOBLER	J	ROLAND
MERZDORFF	EDWIN	
BRENNAN	JAMES	
SOUSA	EUGENE	
MARRA	ANTHONY	

WEINER STAND

LAST	FIRST	MIDDLE
SMITH	MARY	C
GOODSON	ALTHEA	
FRANTZ	LORETTA	
HERTENSTEIN	ARWILDA	
FRENCH	MILDRED	

* INCOMPLETE

UNDERWATER GAMES

promoted by
The Cleveland Skin Divers Club

SUNDAY, SEPTEMBER 27, 1959

All Skin & Scuba Divers Invited

TO BE HELD AT EUCLID BEACH PARK

4. Relay Race

Bob Lupton checks divers' passes.

1959-60
EMPLOYEE ROSTER

LAST	FIRST	MIDDLE
ADAMS	DAVID	
ADAMS	TOM	
ADAMS	WILLIAM	
ADKINS	BARBARA	
ALLEN	LOU	
ALLER	DENNIS	
AMOS	HAROLD	
ANDERSON	PAUL	
ANTHONY	JANE	
APANITAS	J	
ARMSTRONG	WILLIAM	
ATTWOOD	E	
AUFDENHAUS	M	
AUSTIN	CHUCK	
AUSTIN	IRENE	
BAILEY	WILLIAM	
BAIRD	MARY	
BAKER	RUSS	
BALCAM	GWENDOLYN	
BALTURSHOT	ESTHER	
BANDSUCH	RONALD	
BANDY	CHARLES	
BANISH	AUDREY	
BARBATO	JERRY	
BARBERIO	F	
BARGHOLT	NORMAN	
BARKER	BOB	
BARKHAUER	EDWIN	
BARNES	DONALD	
BARNES	GERTRUDE	
BARNES	R	
BARON	GEORGE	
BARRETT	FRANK	
BARRETT	RICHARD	
BARRY	DOROTHY	
BAUDY	MARY	
BAUER	DAVID	
BAUER	JIM	
BAUGHMAN	JOANN	
BEALL	CHARLES	
BECK	JUDY	
BELL	CLARENCE	
BELL	MARGARET	
BENNETT	ARVIA	
BENNETT	CHARLES	
BERTRAM	LAWRENCE	
BIRD	GEORGE	
BIRKENHEAD	THOMAS	
BLACK	MARCIA MARIE	
BLEAM	C	
BLEAM	LULU	
BLODGETT	RICHARD	
BLOEDE	WILLIAM	
BLUHM	J	
BOLINGER	RUTH	
BOYD	BARBARA	
BRACALE	E	
BRACCIA	MARY	
BRADFORD	JOHN	
BRADSHAW	JESSIE	
BRADSHAW	JOHN	
BRANEKOVIC	JOSEPH	
BRAUN	ESTELLE	
BRENNAN	JAMES	
BRITTION	JAMIE	
BROD	WILLARD	
BRODNICK	JOHN	C
BRONSON	THEODORE	
BROOKS	JOY	
BROOKS	PHILLIP	
BROUGHTON	ETHEL	
BROWN	ALLEN	
BROWN	DAISY	
BROWN	GEORGE	
BRYANT	ANN	
BRYANT	GENEVA	
BRYANT	THOMAS	
BUHDE	CHARLES	
BURKHARDT	OTTO	
BURMEISTER	TED	
BURNS	ROBERT	
BURTON	JAMES	
BURTON	RICHARD	
BUSHMAN	FRANK	
BUTLER	JOE	
CABOT	WILLIAM	
CADA	PHILIP	
CAMPBELL	A	R
CAMPBELL	CHARLES	
CAMPBELL	FRED	
CAPRA	CHRISTINA	
CAPRA	MARIA	
CARLSON	FRANK	
CARLSON	JEFF	
CARLSON	LELIA	
CARLTON	JERRY	
CARNEY	ROGER	
CARRELL	JOSEPHINE	
CARRIG	B	
CARROLL	GERTRUDE	
CARROLL	HAZEL	
CARTER	KENNETH	
CARTER	OTIE	
CASSERLY	JAMES	
CAST	MIKE	
CATALANO	FRANK	
CATALANO	JIM	
CATEY	EARL	
CAVANAUGH	TOM	
CERANOWICZ	LORETTA	
CERCEK	DOUG	
CERCEK	G	
CERIO	ANTONIO	
CHAMBERLAIN	VERN	
CHAMPS	ETHEL	
CHAPMAN	PETE	
CHERNICKY	RON	
CHIPRANI	THOMAS	
CHOVANEC	ANDREW	
CIESLAK	D	
CIPITI	THOMAS	
CIRINO	JOHN	
CIRINO	LOUISA	
CIRINO	PATRICK	
CIRINO	THERESA	
CLARK	EVELYN	
CLARK	FRED	
CLARK	NATHANIEL	
CLARK	THOMAS	
CLARKE	DAVID	
CLEVENGER	JEAN	
CLIFFORD	JAMES	
COERTNIK	VICTOR	
COLBERT	CHARLES	
COLDREN	ELLEN	
COLEMAN	WILLIAM	
COLLINS	EDWARD	M

LAST	FIRST	MIDDLE
COLLINS	MABEL	
CONLEY	T	
CONNAVINO	JOSEPHINE	
CONNORS	NORMA	
CONTENTO	JOE	
CONWAY	EDWARD	J
CONWAY-DAHLER	LENA	
COOLEY	L	
COPE	WILLIAM	
CORNELIOUS	WILLIAM	
COSTANZO	JOSEPHINE	
COTTER	JAMES	
COTTINGHAM	MARGARITE	
COVOLTA	MARILYN	
COWIE	TOM	
CRAIG	RUTH	
CRAMER	GEORGE	
CRAMER	GERALD	
CRAWFORD	FRANCES	
CREIGHTON	JOHN	
CRESS	JOHN	
CRICKARD	PETE	
CRIMI	ANGELO	
CROCKER	MIKE	
CUSTER	JOHNNIE	
CUTLIP	HARRY	
DANIELS	RAY	
DAVENPORT	SID	
DAVIES	LEE	
DAVIS	CHARLES	
DAVIS	KATIE	
DAWSON	JOHN	
DEADWYLER	JEROME	
DEADWYLER	MILTON	
DEAN	OLLIE	
DEBBY	G	
DEBBY	J	
DELAMBO	FRANK	
DEMETER	RICHARD	
DENHOLM	DAVE	
DEONG	VANCE	
DEVRIES	DONALD	
DICK	ARCH	
DICORPO	MIKE	
DIETRICH	R	
DIETZ	RONALD	
DIETZEL	ROBERT	
DIETZEL	THOMAS	
DILIBERTO	ALAN	
DIPITI	THOMAS	
DITIRRO	MARIA	
DITZEL	EDWARD	
DOBLER	J	ROLAND
DOEDDERLEIN	M	
DORT	GEORGE	
DOSSA	STEVE	

LAST	FIRST	MIDDLE	
DRAHEIM	EDWARD		
DROBNICK	KEN		
DUCCA	A		
DUCCA	RONALD		
DUKE	BILL		
DUKE	EFFIE		
DUNCAN	EDNA		
DUNKER	ROD		
DUNKIN	VIOLET		
DUVAL	BERNICE		
DVORAK	R		
EAST	MIKE		
EASTON	JAMES		
EDDY	TOM		
EDWARDS	EVELYN		
EHRKE	RUSSELL		
ELIE	CARRIE		
EMCH	JAMES		
EMERSON	TIMOTHY		
ERICKSON	JAMES		
ERSKIN	BARBARA		
ERSKIN	CLARENCE		
ESCHEN	MARION		
ESEP	JUSTUS		
ESTEP	KAY		
EUTY	MILDRED		
EVERETS	NICHOLAS		
FANSLAU	HERMAN		
FANSLAU	HOLDIGARD		
FARINACCI	FILOMENA		
FARRAR	R		
FEATHERS	EDWARD		
FEATHERS	MARY ELLEN		
FEDERICO	CATHERINE		
FEDERICO	MARYANNE		
FERGUSON	JIM		
FERRERI	GEORGE		
FIER	ADA		
FIER	F		
FIER	RHODA		
FIGER	R		
FIGIEL	F		
FIRESTONE	DR L	H	
FLETCHER	G	W	
FLETTERICH	JESSIE		
FLORAN	M		
FLYNN	JOE		
FOGARTY	G		
FOLTZ	CALVIN		
FONICK	JIM		
FORILLO	ANNA		
FRANTZ	LORETTA		
FRATE	DON		
FREDLE	LOTTIE		
FRENCH	MILDRED		
FULDAUER	FRED		
FULTON	JIM		
FULTON	JOHN		
FULTON	LARRY		
GALA	FRED		
GALL	HELEN		
GALLA	JOHN		
GEARIETY	CHARLES		
GEDEON	LEO		
GETZIEN	HARRY		
GIBBARD	CHRISTOPHER		
GILMOUR	DON		
GIMPLEY	JOSEPH		
GLICKER	W		
GOGOLIN	HENRY		
GOLDSTEIN	S		
GOOD	RUTH		
GOODSON	ALTHEA		
GOODSON	WILLIAM		
GOODWIN	JUDY		
GORDON	D		

Miniature Train in Sleepy Hollow Village, Euclid Beach Park, Cleveland, Ohio

LAST	FIRST	MIDDLE
GORNIK	FRANK	
GRAFIO	ENRICO	
GRASSELL	LOUIS	
GRASSI	M	
GRAY	JIM	
GRAY	M	
GRDINA	VICTOR	
GRDOLNIK	N	
GREEN	CURTIS	
GREENE	JOHN	
GREENWAY	FRED	
GREGER	RICHARD	
GRIFFON	N	
GRISSON	CLASSIE	
GRONERT	HERMAN	
GUENTHER	HIRAM	
GUENTZLER	DAVE	
GUHDE	CHARLES	
GUHDE	DONALD	
GUILIANO	MIKE	
GUNTZLER	RON	
GUSTIN	G	
GUTEKUNST	ADAM	
GUTEKUNST	ADELE	
GUTEKUNST	BARBARA	
HABE	JOHN	
HACKMAN	R	
HALE	C	
HALL	C	H
HALL	H	
HAMILTON	MARGARET	
HANZEL	R	
HARKEMA	P	
HARPSTER	H	H
HARRIS	FLORENCE	
HAUPT	T	
HAZARD	STONEWALL	
HECKMAN	EDWARD	
HECKMAN	ROBERT	
HEGLAW	ROBERT	
HEINTZ	RICHARD	
HEITMAN	DANIEL	
HEMPLEWIS	CHARLIE	
HENDERSON	RAY	
HENDRICKS	JUDY	
HERBST	CHARLES	
HERKE	GEORGE	
HERTENSTEIN	ARWILDA	
HERTENSTEIN	ED	
HEUER	B	
HILBERG	ELMER	
HILL	JERRY	
HIRSCH	TOM	
HIVELY	CHARLES	
HIXSON	ART	
HIXSON	ELECTA	
HOERNIG	MARY	
HOFFART	LOUIS	
HOGARTH	DONALD	
HOGARTH	MARTHA	
HOINSKI	THOMAS	
HOKAVA	DENNIS	
HOLLAND	JOHN	
HOLLEN	JAMES	
HOLLIS	H	P
HOLT	GERTRUDE	
HOUSEHOLDER	DOROTHY	
HOWARD	EUGENE	
HOWLE	E	P
HUBBARD	MARY	
HUBER	GEORGE	
HUDSON	ERNEST	
HUGHES	ELLA	
HUMPHREY	GERTRUDE	
HURD	CARL	
HURLESS	JACKIE	
IAFELICE	RICHARD	
IAMMATEO	MICHELINA	
IVEY	JOHN	
JACKLITZ	ALBERT	
JACKSON	ANDREW	
JACKSON	GRACE	
JACKSON	MARY	
JACKSON	RUBY	M
JACKSON	WILLADA	
JAFFRAY	D	
JAKOVAC	IVAN	
JAMES	JOHN	
JAMISONLE	UDELLA	
JANES	P	
JARVIS	JERRY	
JASULATIS	P	
JAYNES	TED	
JOHNSON	DOUGLAS	
JOHNSON	RHODA	
JONES	DAVID	
JONES	KATHRYN	
JONES	W	O
JUDD	BARBARA	
JUDD	ESTIL	
KAMINSKAS	RICHARD	
KANIS	JOHN	
KANJOR	CARL	
KARKOSKA	JOE	
KARNAK	WILLIAM	
KAST	DALE	
KAST	HOWARD	
KAUFMAN	RICHARD	
KEEFER	LEE	
KEKIC	MARJORIE	
KEKIC	THOMAS	
KELLER	JOHN	
KELLER	SAM	
KELLIC	JULIA	
KELLOGG	S	
KENT	EDWARD	
KIDD	RICHARD	
KIKOLI	FRANK	
KILBECK	MIKE	
KILLEEN	K	
KINZER	ALLEN	
KINZER	ELWOOD	
KINZER	WILMA	
KIRK	E	
KIRKPATRICK	MARY	
KISTHARDT	MARIE	
KLEINSMITH	RONALD	
KLEMENCIC	EDWARD	
KLEMENCIC	EGIDI	
KLEMENS	LEON	
KNAUSS	JIM	
KNAUSS	RAY	

"Kiddieland," Euclid Beach Park, Cleveland, Ohio

LAST	FIRST	MIDDLE
KNECHT	R	
KNOWLES	SHARI	
KONRAD	EDWARD	
KORDIC	ELIZABETH	
KORDICK	DAVID	
KORDIK	E	
KOVACH	JOSEPH	
KOVIC	JUDY	
KRALL	ANGIE	
KRALL	ANTHONY	
KRAMER	EDWARD	
KRAMER	JANE	
KRAUSE	DOROTHY	
KRAUSE	HARRY	
KRAUSE	KEN	
KRAUSE	LINDA	
KRIVDO	JERRY	
KRIVDO	NELL	
KRIVOY	D	
KROEGER	LOUISE	
KUBELAVIC	KATHRYN	
KUCHARSKI	RONALD	
KUHTA	ROBERT	
KUNSMAN	WALTER	
KURZEIKA	J	
LALAK	JOHN	
LANDERS	JOSEPH	
LANESE	TONY	
LANGE	DON	
LANIER	FRANCES	
LARUE	GLEN	
LATKOVICH	ELEANOR	
LAUDIE	G	
LAWSON	CHARLES	
LAYREL	RICHARD	
LECLAIR	FRANK	
LEECE	L	
LEECE	MARIE	
LENARCIC	JOHN	
LERZ	ANGIE	
LEWIS	CEDRIC	
LOGAN	LAMONT	
LOMBARDO	T	
LONCELLA	TEDDY	
LONGAZEL	ANNA	
LONGFIELD	TESS	
LONGWELL	JOE	
LOURY	MAJOR	
LOWMAN	DAVID	
LUCE	HOMER	
LUCIANO	ANTHONY	
LUIKART	CAROLYN	
LUPTON	ROBERT	
LUPTON	RUTH	
LUTZ	RON	
LUTZ	VERNA	

MACEK	GERTRUDE	
MADISON	DAN	
MAHONE	JOHN	
MAHONE	WILLIE	
MAIN	H	
MALENGO	GEORGE	
MALENGO	LEO	
MANDAU	RUDY	
MANDEL	IDA	
MANNING	DANIEL	
MANNING	DENNIS	
MANNING	DOUG	
MANTOOTH	PAUL	
MARDIS	MARGARET	
MARIANO	MIKE	
MARINO	TONY	
MARINONI	PETE	
MARKELONIS	RALPH	
MARMASH	JACK	
MARN	LOUIS	
MARRA	ANTONIO	
MARSHALL	CADDIE	
MARTENS	AL	
MARTENS	GOLDA	
MARTENS	WALTER	
MARTIN	WILLIAM	
MARTINO	F	
MASSENGLE	GREENE	
MATHEWSON	CHARLES	
MATRKA	FRANK	
MATUNA	PETER	
MAURER	JOSEPHINE	
MAUSER	JOSEPH	
MCBRIDE	CALVIN	
MCBRIDE	MARY	
MCCAULEY	ISABELLE	
MCCONOCHA	DANIEL	
MCCOWAN	HERMAN	
MCCREIGHT	GEORGE	
MCDONALD	BARBARA	
MCGRATH	PHILIP	
MCINTYRE	JAMES	
MCKELVEY	DAVE	
MCKINLEY	JIM	
MCMASTER	WILLIAM	
MCMILLEN	LUCILLE	
MCNEELY	MARY	
MCNEILL	ALLAN	
MEADE	JOHN	J
MEADE JR	JOHN	J
MECHTENSIMER	HARRY	
MELARAGNO	ASSUNTA	
MELBER	ED	
MERKULOFF	GEORGE	
MERZDORF	EDWARD	
MEYER	BLONDELL	
MICHAELS	ERVAN	
MIERKE JR	ALBERT	
MIERKE SR	ALBERT	
MIKLICH	A	
MILATOVIC	J	
MILES	A	
MILES	MARY	
MILLER	LARRY	
MILLER SR	PETER	
MILSTOVIC	JOHN	
MISLEY	RAY	
MITCHEL	CHESTER	
MIZLEY	JAMES	
MIZLEY	PAUL	
MOG	DAVID	
MOLESKY	BOB	
MOLNAR	DOROTHY	
MONAGHAN	WILLIAM	
MONROE	WILLIAM	
MOORE	HAROLD	
MOORE	MAY	
MORAN	ROBERT	

GENERAL VIEW OF EUCLID BEACH, CLEVELAND, OHIO.

LAST	FIRST	MIDDLE
MOREHOUSE	DONALD	
MORICA	LARRY	
MORLEY	MARIE	
MUELLER	BOB	
MULLINS	DAVID	
MULLINS	HERBERT	
MULLINS	WILLIAM	
MUMAW	SUSAN	
NACHTIGAL	JAMES	
NARDINI	JOHN	
NASH	SAM	
NEBLETT	J	
NEELD	D	
NEIBERT	GILBERT	
NEUBAUER	JAY	
NEWMAN	D	
NEWTON	FRANCIS	
NICOL	ANDREW	
OAKLEY	ADDIE	E
OCONNOR	JIM	
OCONNOR	MIKE	
OKROS	BETTE	JO
OKROS	MARY	
OLESKI	DONALD	
OLESKI	ROBERT	
OPALK	CAROLINE	
OPALK	MARTIN	
OSBORNE	J	
OSBORNE	MILDRED	
OSTROWSKI	JERRY	
OSWALT	VIRGINIA	
PAGES	R	
PAGLIA	ROSE	
PAKOSH	MICHAEL	
PALEK	FRANCIS	
PALEY	NORMAN	
PAPES	JOSEPHINE	
PAPES	VALERIA	
PARKER	BILL	
PARKER	C	H
PARKER	JIM	
PARKER	WILLIAM	
PARMELEE	C	W
PARMELEE	EDITH	
PASALACQUA	JOHN	
PATRICK	EDDIE	
PAUL	FRED	
PAVELLA	DANIEL	
PAVELLA	THOMAS	
PEARCE	AGNES	
PECK	LOIS	
PEGORARO	JIM	
PEGORARO	RUDOLPH	
PELKO	JOHN	
PERKOVICH	JOHN	
PERNUS	PAT	
PERSELL	CHARLES	
PERUSEK	B	
PESTA	EDWARD	
PETELINKAR	R	
PETERKA	LIB	
PETKOFF	MARKO	
PETRECCA	JOE	
PETRECCA	NICK	
PETRICK	DAN	
PETTIT	HAROLD	
PEVARNICK	ANTHONY	
PHILLIPS	C	
PHILLIPS	JERRY	
PHILLIPS	R	D
PHIPPS	EARL	
PINCHBECK	H	
PLANTSCH	MIKE	
PLESCIA	LUCY	
POCKAR	JERRY	
POE	GARLAND	
POINTER	FRANCIS	
POKRANT	LEONARD	
POKRANT JR	LEONARD	
PRAYNER	EUGENE	
PRICE	FLOYD	
PRICE	RAY	
PRITCHARD	JOHN	
PUINNO	NICK	
PURDY	MARGARET	
PUSTARE	R	
RADOVANIC	EVA	
RAINEY	MALISIA	
RAMSEY	RONALD	
RANALLO	DANIEL	
RASH	VERN	
RAUPACH	RONALD	
RAUPACH	WILLIAM	
RAY	DICK	
RAY	PAUL	
RAY	RICHARD	
REARDON	ROGER	
REDFIELD	C	
REDRICK	ALDA	M
REED	GARY	
REED	MARIE	
REEDER	NORMA	
REICHENBACH	ERICH	
REICHERT	LAWRENCE	
REILLY	JAMES	
REINHARD	GEORGE	
REINHART	DONALD	
REINKE	ROBERT	
RICHARD	PHILLIP	
RICHARDS	ANDREW	
RICHARDSON	JAMES	
RICHMOND	DENNIS	
RILEY	GEORGE	
RINALDI	NICK	
RIVERS	MILLIE	
ROADS	KENNETH	
ROBBINS	KATHLYN	
ROBERTS	RUSSELL	
ROBERTS	RUTH	
ROBINSON	WILLIAM	
ROCK	JOHN	
ROCKFORD	JAMES	
ROCKFORD	JIM	
RODERICK	ALICE	
ROLLINS	E	
RONKE	BERT	
RONNIGER	H	E
ROSS	FRANK	
ROSSA	ARLENE	
ROTH	CARL	
ROTHAUNEL	ROY	
RUCK	JAMES	

The Thriller, Euclid Beach Park, Cleveland, Ohio

133

LAST	FIRST	MIDDLE
RUCKER	CARTER	
RUSCSAK	STEVE	
RUSS	JOSEPH	
RUTKOUSKI	CHARLES	
RUTLEDGE	ROBERT	
RYAN	ALAN	
RYDMAN	ELIZABETH	
SAID	NORINE	
SANDERS	CAL	
SANTUCCI	MELINA	
SAUVAIN K		
SAWITKE	DALE	
SCHAFFER	WALTER	
SCHAUB	GEORGE	
SCHILL	EUGENE	
SCHILL JR	AUGUST	
SCHILL JR	ROBERT	
SCHILL SR	AUGUST	
SCHMIDT	ALBERTA	
SCHOTT	DENNIS	
SCHOTT	I	W
SCHUBERT	ROY	
SCHUTT	KENNETH	
SCOLARO	URBAN	
SCOTT	AL	
SCOTT	DAVID	
SCOTT	PHILLIP	
SCRUGGS	CECIL	
SELJAN	R	
SELLARDS	REBECCA	
SELLER	ESSIE	
SEMAN	JAMES	
SENNISH	GEORGE	
SHANNON	PEARL	
SHAVER	LORETTA	
SHEEHAN	JAMES	D
SHEEHAN	MABEL	
SHEPPARD	T	R
SHILLIDAY	KATHRYN	
SHIMKO	STEVE	
SHULZ	R	
SIKO	THOMAS	
SILVEROLI	JOHN	
SINGLETON	WALKER	
SIPPOLA	W	
SISTEK	MAE	
SIVIK	VIOLET	
SKINNER	JESSIE	
SLAPER	RONALD	
SLATER	DONALD	
SMITH	ALAN	
SMITH	ALVIN	
SMITH	CHARLES	
SMITH	FRED	
SMITH	HARRY	
SMITH	IRENE	
SMITH	J	D
SMITH	JEANNETTE	
SMITH	L	
SMITH	MARY	LOUISE
SMITH	RAY	
SMITH	T	
SMRDEL	S	
SNYDER	JOHN	
SOLTIS	ALBERT	
SOUSA	EUGENE	
SPICA	DON	
SPINELLI JIM		
SPUZZILLO	MIKE	
SROCCO	GISBERTA	
ST CLAIR	JULIUS	
STACKS	JOE	
STAIDUHAR	G	
STANTON	SAMUEL	
STAPP	PAUL	
STARE	FRANK	
STARMAN	LOUIS	
STAVE	F	
STEELE	ARTHUR	
STEFANCIC	WILLIAM	
STELNIAK	EVELYN	
STEPP	R	
STEWART	JEFF	
STEWART	JIM	
STIH	NORMAN	
STOCKMAN	RUTH	
STOCKTON	VERL	
STONEBACK	ELVIRA	
STONEBACK	H	D
STREETER	GARY	
STRONG	ELEANOR	
STRUNA	JIM	
STUCK	RICHARD	
STUPICA	R	
SUGLIA	MARIA	
SUMPH	G	
SUSONG	FLOYD	
SUSTARIC	STEVE	
SWANEY	KEN	
SWANKHAMER	W	
SWANSON	F	H
SWEENEY	ANGIE	
SWEENEY	J	
SWIMER	MYRON	
TAKACS	J	
TATE	JOSEPHINE	
TAUKERSLEY	JERRY	
TAUSCH	BERTHA	
TAYLOR	FREDDIE	
TAYLOR	TOM	
TEASTER	HOWARD	
TEKUNTZ	C	
TERABASSI	CAMILLO	
TETTELBASH	BARBARA	
THOMAS	ROGER	
THOMAS	SAM	
THOMAS	WAYNE	
THOMPSON	IRVIN	
THUNHURST	EDYTHE	
TIANILLO	BILL	
TIMKO	EUGENE	
TIMKO	J	
TIMPERIO	FILOMENA	
TINCK	B	
TINKER	E	
TINKER	NEIL	
TKATCH	J	
TOMALSKI	F	
TOMARO	V	
TOMINO	DONALD	
TOMINO	NICK	
TONELLI	BEN	
TONELLI	GIOVANNI	

134

LAST	FIRST	MIDDLE
TOPOLY	BARB	
TRAUB	NORMAN	
TRAVERS	B	
TREM	TOM	
TRENT	FRANK	
TRIVISON	C	
TRIVISONNO	CHARLES	
TRIVISONNO	GRACE	
TRIVISONNO	RAYMOND	
TROTTA	JOHN	
TROYER	BELLE	
TUEMMER	H	
TUFTS	ROBERT	
TURICK	GEORGE	
TURK	CAROLINE	
TURK	RAY	
TURNER	P	B
TURNER	VERNON	
ULMAN	LILLIAN	
UNETIC	AL	
URANKAR	GERALD	
URSUL	GEORGE	J
VACCARIELLO	ANITA	
VACCARIELLO	JOSEPHINE	
VACCARIELLO	URBAN	
VALENTINE	EDWARD	
VALENTINE	GARNETT	
VALENTINO	MARY	
VEGNEY	EUGENE	
VERBERTER	CLIFFORD	
VIDUGERIS	WILLIAM	
VOGEL	ROBERT	
VOGRIN	G	
VRANEKOVIC	JOSEPH	
VRYELIST	LIN	
WAGNER	FLORENCE	
WALKER	DUDLEY	
WALKER	LOUIS	
WALTERS	LEN	
WALTERS	MARK	
WARD	R	
WASHINGTON	LORENA	
WATERS	LULU	
WATERS	WILLIAM	
WATTS	RAYMOND	
WEISSERT	R	
WERBER	DOUGLAS	
WESKE	ALBERT	
WEST	NETTIE	
WETMORE	ROBERT	
WHEATCROFT	DAVID	
WHITE	ALDONA	
WHITFIELD	E	
WHITMAN	BETTY	
WICKES	J	
WICKS	GERALDINE	
WILDERS	J	
WILEY	FRED	
WILEY	GEORGE	
WILLARD	EUGENE	
WILLIAMS	DORA	
WILLIAMS	GLORIA	
WILLIAMS	IVORY	
WILLIAMS	KENNETH	
WILLIAMS	LINDA	
WILLIAMS	ROBERT	
WILLIAMS	ROGER	
WILLIAMS	WALTER	
WILSON	DOROTHY	
WILSON	LLOYD	
WILSON	PAUL	
WILSON	RICHARD	
WILSON	ROBERT	
WILSON	WILLIAM	S
WISE	CHARLES	
WISE	EDWARD	
WISE	JAMES	
WISE	LAVERNE	
WISE	PHYLLIS	
WOERTH	LEONARD	
WOHLGEMUTH	TERRY	
WOLF	A	
WOLF	EDWARD	
WOLF	THOMAS	
WOLFF	LESTER	
WOODROW	B	
WOODS	F	
WRAY	ATHALYNE	
WRIGHT	ETHEL	
WRIGHT	LULU	
WURM	CHARLOTTE	
YAGER	F	
YAKOS	JAMES	
YELCHO	JACOB	
YOKUM	DAVID	
YORK	JAMES	
YOUNG	MOSE	
YOUNG	REGINALD	
YOUNG	ROBERTA	
YUHASZ	RAY	
ZADELL	DOLORES	
ZALEWSKI	TED	
ZANYK	AL	
ZANYK	JOHN	
ZAPOL	R	
ZAUDER	R	
ZEITZ	W	
ZIEL	LENA	
ZIFKO	M	
ZIMMERMANN	J	
ZOBEC	F	
ZUST	JIM	
ZUZEK	L	

* INCOMPLETE

"Over the Falls," Euclid Beach Park, Cleveland, Ohio

Dudley Humphrey Jr. test drives the Turnpike and Antique Cars with friends and family.

1961-62
EMPLOYEE ROSTER

LAST	FIRST	MIDDLE
ADAMS	DAVID	
AMOS	HAROLD	
ANDERSON	PAUL	
ANTHONY	JANE	
ARMSTRONG	WILLIAM	
ATTWOOD	E	
AUFDENHAUS	M	
AUSTIN	IRENE	
BAILEY	WM	
BAKER	RUSS	
BANISH	AUDREY	
BARBATO	JERRY	
BARBERIO	F	
BARGHOLT	NORMAN	
BARNES	GERTRUDE	
BARNES	R	
BARRY	DOROTHY	
BAUDY	MARY	
BAUER	JIM	
BAUGHMAN	JOANN	
BECK	JUDY	
BELL	CLARENCE	
BELL	MARGARET	
BENNETT	ARVIA	
BENNETT	CHARLES	
BERTRAM	LARRY	
BIRKENHEAD	TOM	
BLEAM	C	
BLEAM	LULU	
BLOEDE	WILLIAM	
BRACALE	E	
BRADFORD	JOHN	
BRADSHAW	J	
BROD	WILLARD	
BRYANT	GENEVA	
BRYANT	THOMAS	
BURKHARDT	O	
BURTON	JAMES	
CABOT	WILLIAM	
CAMPBELL	A	R
CAMPBELL	CHARLES	
CAMPBELL	FRED	
CARLSON	FRANK	
CARLSON	JEFF	
CARLSON	LELIA	
CARNEY	ROGER	
CARRELL	JOSEPHINE	
CARRIG	B	
CARROLL	GERTRUDE	
CARROLL	HAZEL	
CASSERLY	JAMES	
CATALANO	FRANK	
CATALANO	J	
CAVANAUGH	TOM	
CERANOWICZ	LORETTA	
CERCEK	DOUG	
CERCEK	G	
CHAMPS	ETHEL	
CHAPMAN	PETE	
CHERNICKY	RON	
CHIPRANI	THOMAS	
CIPITI	THOMAS	
CIRINO	JOHN	
CIRINO	PATRICK	
CLARK	EVELYN	
CLARK	FRED	
CLARK	NATHANIEL	
CLARK	THOMAS	
CLARKE	DAVID	
CLIFFORD	JAMES	
COLDREN	ELLEN	
COLLEY	L	
COLLINS	EDWARD	M
COLLINS	MABEL	
CONNOR	N	
CONTENTO	JOE	
CONWAY	EDWARD	J
CONWAY	LENA	
CONWAY-DAHLER	LENA	
COPE	WILLIAM	
CORNELIOUS	WILLIAM	
COTTER	JAMES	
COWIE	TOM	
CRAMER	GEORGE	
CRICKARD	PETE	
CROCKER	MIKE	
DAWSON	JOHN	
DEADWYLER	JEROME	
DEADWYLER	MILTOM	
DEBBY	G	
DEBBY	J	
DELAMBO	FRANK	
DENHOLM	DAVE	
DICK	A	
DICORPO	MIKE	
DIETRICH	R	
DIETZ	RONALD	
DIETZEL	ROBERT	
DIETZEL	THOMAS	
DILIBERTO	ALAN	
DITZEL	EDWARD	
DOEDDERLEIN	M	
DORT	GEORGE	
DOSSA	STEVE	
DRAHEIM	EDWARD	
DUCCA	A	
DUCCA	RONALD	
DUNCAN	EDNA	
DUNKER	ROD	
DUNKIN	VIOLET	
DUVAL	BERNICE	
DVORAK	R	
EASTON	JAMES	
EDWARDS	EVELYN	
ELIE	CARRIE	
EMCH	JAMES	
EMERSON	TIMOTHY	
ERICKSON	JAMES	
ERSKIN	BARBARA	
ERSKIN	CLARENCE	
EUTY	MILDRED	
FARRAR	R	
FEATHERS	EDWARD	
FIER	ADA	
FIER	F	
FIER	RHODA	
FIGER	R	
FIRESTONE	DR L	H
FLETCHER	G	W
FLETTERICH	JESSIE	
FLORAN	M	
FOGARTY	G	
FOLTZ	CALVIN	
FONICK	JIM	

LAST	FIRST	MIDDLE
FREDLE	LOTTIE	
FRENCH	MILDRED	
FULDAUER	FRED	
FULTON	JIM	
FULTON	LARRY	
GALLA	JOHN	
GEARIETY	C	
GEDEON	L	
GETZIEN	HARRY	
GILMOUR	DON	
GOLDSTEIN	S	
GOOD	RUTH	
GOODSON	A	
GOODSON	WILLIAM	
GORDON	D	
GORNIK	FRANK	
GRASSELL	LOUIS	
GRASSI	M	
GRAY	M	
GRDINA	VICTOR	
GRDOLNIK	N	
GREEN	CURTIS	
GREENE	JOHN	
GREENWAY	FRED	
GRIFFON	N	
GRONERT	HERMAN	
GUENTHER	HIRAM	
GUENTZLER	DAVE	
GUHDE	DONALD	
GUILIANO	MIKE	
GUSTIN	G	
GUTEKUNST	ADAM	
GUTEKUNST	ADELE	
GUTEKUNST	BARBARA	
HABE	JOHN	
HALL	C	
HAMILTON	MARGARET	
HANZEL	R	
HARKEMA	P	
HARPSTER	H	
HAZARD	STONEWALL	
HEINTZ	RICHARD	
HEITMAN	DANIEL	
HENDRICKS	JUDY	
HERTENSTEIN	A	
HERTENSTEIN	E	
HEUER	B	
HILBERG	ELMER	
HIRSCH	TOM	
HIXSON	ART	
HIXSON	ELECTRA	
HOERNIG	MARY	
HOFFART	LOUIS	
HOINSKI	THOMAS	
HOKAVAR	DENNIS	
HOLLIS	H	P

HOLT	GERTRUDE	
HOUSEHOLDER	DOROTHY	
HOWARD	EUGENE	
HOWLE	E	P
HUBBARD	MARY	
HUBER	GEORGE	
HUDSON	E	P
HUMPHREY	GERTRUDE	
HURD	CARL	
IAFELICE	RICHARD	
IVEY	JOHN	
JACKLITZ	ALBERT	
JACKSON	ANDREW	
JACKSON	GRACE	
JAFF	RAY	D
JAKOVAC	IVAN	
JAMES	P	
JARVIS	JERRY	
JASULATIS	P	
JOHNSON	DOUGLAS	
JOHNSON	RHODA	
JONES	D	
JONES	W	O
JUDD	BARBARA	
JUDD	ESTIL	
KANIS	JOHN	
KANJOR	CARL	
KARKOSKA	JOE	
KARNAK	WILLIAM	
KAST	HOWARD	
KEKIC	MARJORIE	
KEKIC	THOMAS	
KELLIC	JULIA	
KELLOGG	S	
KENT	EDWARD	
KIDD	RICHARD	
KIKOLI	FRANK	
KILLEEN	K	
KINZER	ALLEN	
KINZER	ELWOOD	
KINZER	WILMA	
KIRK	E	
KLEMENCIC	EDWARD	
KLEMENS	LEON	
KNAUSS	JIM	
KNAUSS	R	
KNECHT	R	
KONRAD	EDWARD	
KORDIC	ELIZABETH	
KORDIK	E	
KOVACH	JOSEPH	
KOVIC	JUDY	
KRALL	ANTHONY	
KRAMER	EDWARD	
KRAMER	JANE	
KRAUSE	DOROTHY	
KRAUSE	HARRY	
KRAUSE	LINDA	
KRIVOY	D	
KUBELAVIC	KATHRYN	
KUHTA	R	
KURZEIKA	J	
LALAK	JOHN	
LANDERS	JOSEPH	
LANESE	TONY	
LEECE	MARIE	
LITTLE	DAVID	
LOMBARDO	T	
LOURY	MAJOR	
LUCE	HOMER	
LUCIANO	ANTHONY	
LUPTON	ROBERT	
LUPTON	RUTH	
LUTZ	VERNA	

LAST	FIRST	MIDDLE
MACEK	GERTRUDE	
MADISON	DAN	
MAHONE	WILLIE	
MAIN	H	
MALENGO	L	
MANNING	DANIEL	
MARIANO	MIKE	
MARINONI	PETE	
MARN	LOUIS	
MARTENS	A	M
MARTENS	GOLDA	
MARTENS	WALTER	
MARTIN	BILL	
MARTIN	WILLIAM	
MARTINO	F	
MATUNA	PETER	
MAURER	JOSEPHINE	
MCDONALD	BARBARA	
MCGRATH	PHILIP	
MCINTYRE	JAMES	
MCKELVEY	DAVE	
MCMASTER	WILLIAM	
MCNEELY	MARY	
MCNEILL	ALLAN	
MEADE	JOHN	
MELARAGNO	A	
MELBER	E	
MELBER	ED	
MERKULOFF	GEORGE	
MERZDORF	EDWARD	
MIERKE JR	A	H
MIERKE SR	A	
MIKLICH	A	
MILATOVIC	J	
MILES	A	
MILES	MARY	
MILLER	LARRY	
MISLEY	JAMES	
MISLEY	RAY	
MOG	DAVID	
MOLESKY	R	
MOLNAR	DOROTHY	
MORAN	R	
MORAN	ROBERT	
MORLEY	MARIE	
MUELLER	B	
MULLINS	DAVID	
NASH	SAM	
NEBLETT		J
NEELD	D	
NEWMAN	D	
NEWTON	FRANCIS	
NICOL	A	
OAKLEY	ADDIE	E
OLESKI	ROBERT	
OPALK	CAROLINE	
OPALK	MARTIN	
OSBORNE	J	
OSBORNE	MILDRED	
OSTROWSKI	JERRY	
PAGES	R	
PALEK	FRANCES	
PALEY	NORMAN	
PAPES	JOSEPHINE	
PAPES	VALERIA	
PARKER	C	H
PARKER	WILLIAM	
PAVELLA	DANIEL	
PAVELLA	THOMAS	
PEARCE	AGNES	
PECK	LOIS	
PEGORARO	J	
PEGORARO	R	
PELKO	JOHN	
PERSELL	CHARLES	
PERUSEK	B	
PETKOFF	MARK	O
PETRECCA	JOE	
PETRICK	DAN	
PHILLIPS	C	
PHILLIPS R	D	
PHIPPS	EARL	
PLANTSCH	MIKE	
POCKAR	JERRY	
POE	GARLAND	
PRAYNER	EUGENE	
PRITCHARD	JOHN	
PUINNO	NICK	
PUSTARE	R	
RAMSEY	RONALD	
RANALLO	DANIEL	
RAUPACH	RONALD	
RAY	RICHARD	
REARDON	RODGER	
REED	GARY	
REED	MARIE	
REEDER	NORMA	
REICHENBACH	ERICH	
REINHARD	GEORGE	
REINHART	DONALD	
REINKE	ROBERT	
RICHARDS	ANDREW	
RILEY	GEORGE	
ROBERTS	RUSSELL	
ROBERTS	RUTH	
ROCKFORD	JAMES	
RODERICK	ALICE	
ROTH	CARL	
RUCKER	CARTER	
RUTLEDGE	ROBERT	
RYAN	ALAN	
RYDMAN	ELIZABETH	
SANDERS	CAL	
SAUVAIN	K	
SCHAUB	GEORGE	
SCHILL JR	AUGUST	
SCHOTT	DENNIS	
SCHOTT	I	W
SCOLARO	URBAN	
SCOTT	DAVID	
SCOTT	PHILLIP	
SELJAN	R	
SELLARDS	REBECCA	
SEMAN	JAMES	
SENNISCH	G	
SHAVER	LORETTA	
SHILLIDAY	KATHRYN	
SHULZ	R	
SIKO	THOMAS	
SINGLETON	WALKER	
SIPPOLA	W	
SIVIK	VIOLET	

"High Rides", Euclid Beach Park, Cleveland, Ohio

LAST	FIRST	MIDDLE
SKINNER	JESSIE	
SLATER	DONALD	
SMITH	CHARLES	
SMITH	F	
SMITH	IRENE	
SMITH	J	
SMITH	J	D
SMITH	JEAN	
SMITH	MARY	C
SMITH	T	
SMRDEL	S	
SOLTIS	ALBERT	
SOUSA	EUGENE	
STAIDUHAR	G	
STARE	FRANK	
STARMAN	LOUIS	
STEELE	ARTHUR	
STEFANCIC	WILLIAM	
STEPP	R	
STIH	NORMAN	
STOCKMAN	RUTH	
STREETER	GARY	
STRUNA	E	
STUCK	RICHARD	
STUPICA	R	
SUMPH	G	
SUSTARIC	STEVE	
SWANEY	KEN	
SWEENEY	ANGIE	
SWEENEY	J	
TAKACS	J	
TATE	JOSEPHINE	
TAUKERSLEY	JERRY	
TAUSCH	BERTHA	
TAYLOR	FREDDIE	
TAYLOR	TOM	
THOMAS	SAM	
THOMAS	WAYNE	
THUNHURST	EDYTHE	
TIMKO	J	
TIMPERIO	FILOMENA	
TINCK	B	
TKATCH	J	
TOMALSKI	F	
TOMARO	V	
TOMINO	NICK	
TONELLI	G	
TOPOLY	BARB	
TRAUB	NORMAN	
TRAVERS	B	
TREM	TOM	
TRENT	FRANK	
TRIVISON	C	
TRIVISONNO	GRACE	
TRIVISONNO	RAY	
TROYER	BELLE	

LAST	FIRST	MIDDLE
TUFTS	ROBERT	
TURK	RAY	
TURNER	VERNON	
ULMAN	LILLIAN	
UNETIC	AL	
URSUL	GEORGE	J
VACCARIELLO	ANITA	
VACCARIELLO	J	
VACCARIELLO	JOSEPHINE	
VACCARIELLO	URBAN	
VALENTINE	EDWARD	
VALENTINE	G	
VALENTINO	MARY	
VEGNEY	EUGENE	
VIDUGERIS	WILLIAM	
VOGEL	ROBERT	
VOGRIN	G	
WALTERS	LEN	
WALTERS	MARK	
WARD	R	
WATERS	WILLIAM	
WERBER	DOUGLAS	
WEST	NETTIE	
WHITFIELD	E	
WHITMAN	BETTY	
WICKES	J	
WICKS	GERALDINE	
WILDERS	J	
WILEY	FRED	
WILLIAMS	DORA	
WILLIAMS	GLORIA	
WILLIAMS	IVORY	
WILLIAMS	KENNETH	
WILLIAMS	LINDA	
WILLIAMS	ROBERT	
WILLIAMS	WALTER	
WILSON	L	P
WILSON	PAUL	
WILSON	RICHARD	
WILSON	ROBERT	
WILSON	WILLIAM	
WISE	EDWARD	
WISE	JAMES	
WOHLGEMUTH	TERRY	
WOLF	A	
WOLFF	LESTER	
WRIGHT	ETHEL	
WURM	CHARLOTTE	
YAGER	F	
YAKOS	JAMES	
YOKUM	DAVID	
YORK	JAMES	
YOUNG	MOSE	
YOUNG	REGINALD	
YOUNG	ROBERTA	
YUHASZ	RAY	
ZANYK	AL	
ZANYK	JOHN	
ZAPOL	R	
ZAUDER	R	
ZEITZ	W	
ZIFKO	M	
ZOBEC	F	
ZUST	JIM	
ZUZEK	L	

"Carrousel, Euclid Beach Park, Cleveland, Ohio"

1963-64
EMPLOYEE ROSTER

LAST	FIRST	MIDDLE
ADAMS	MARY	C
AIKEN	DAVID	V
ANDERSON	DAVID	L
ANDERSON	PAUL	L
ANDRUS	CHARLES	E
ANTHONY	JANE	H
APAL	RICHARD	H
ATTWOOD	ERNEST	J
AUNE JR	GEORGE	E
BACHER	LAWRENCE	G
BAILEY	REUBEN	T
BAILEY	WILLIAM	H
BANKO	JACK	E
BANKS	JERRY	
BARBATO	JERRY	M
BARBERIO	FRANK	L
BARRY	DOROTHY	M
BATES	EDNA	M
BELL	CLARENCE	A
BENNETT	ARVIA	P
BENNETT	CHARLES	D
BIDELMAN	RICHARD	L
BIRKENHEAD	THOMAS	
BLACKBURN	ROBIE	
BLAUCH	DAVID	C
BLEAM	CLARENCE	M
BLEAM	LULU	E
BLOEDE	WILLIAM	C
BONGIVONNI	JOSEPH	
BONNER	ADDISON	L
BOSSY	DOROTHY	M
BRACALE	EARL	J
BRADFORD	JOHN	
BRADLEY	SAMUEL	D
BRADSHAW	JOHN	
BRIGANTIO	ANTHONY	J
BRODNICK	ROBERT	D
BROWN	DONALD	C
BRYANT	THOMAS	E
BURICH	JAMES	A
BURKE	FLORENCE	
BURTON	JAMES	
BUTTERFIELD	AGNES	E
CABOT	WILLIAM	J
CAMPBELL	ALEXANDER	R
CAMPBELL	CHARLES	R
CAMPBELL	FRED	
CARLSON	FRANK	A
CARLSON	JEFFREY	
CARLSON	LELIA	E
CARNEY	ROGER	C
CARRIG	WILLIAM	
CARROLL	GERTRUDE	M
CASSERLY	JAMES	
CATALANO	FRANK	A
CERCEK	DOUGLAS	J
CERCEK	GARY	M
CHAL	THOMAS	
CIPITI	THOMAS	
CIRINO	JOHN	
CIRINO	PATRICK	M
CLARK	BRUCE	
CLARK	EVELYN	E
CLARK	FRED	
CLARK	ROBERT	E
CLARK	THOMAS	
COLDREN	ELLEN	
COLLINS	EDWARD	M
COLLINS	MABEL	M
CONLEY	MARY	
CONNOR	NORMAN	
CONTENTO	GUISEPPE	
CONWAY	EDWARD	J
COOK	ALICE	S
COOLEY	LAWRENCE	
COPE	WILLIAM	J
CORBERT	MICHAEL	
CORNELIOUS	WILL	CHARLES
COSTA	JOSEPH	
COWIE	THOMAS	F
COWIE	THOMAS	J
CROSS	FREDERICK	
DAGG	STEPHANIE	A
DAY	JENNIE	L
DEADWYLER	MILTON	
DEBBY	JAMES	E
DEBBY	JOHN	G
DEBELAK	MILTON	S
DELAMBO	FRANK	F
DELIBERTO	ALAN	J
DENHAM	GLENN	W
DENHOLM	DAVE	
DENHOLM	WILLIAM	
DIETRICH	ROBERT	A
DIETZ	RONALD	
DIETZEL	EDWARD	A
DIETZEL	ROBERT	J
DIETZEL	THOMAS	J
DISCENZA	ANTHONY	J
DORT	GEORGE	
DOSSA	STEVE	
DRAHEIM	EDWARD	
DREHER	LINDA	M
DROHEIM	EDWARD	W
DUCCA	ALEXANDER	
DUCCA	THOMAS	J
DUNCAN	EDNA	T
DUNKER	RALPH	O
EASTON	JAMES	R
ECKENRODE	JOSEPH	R
EDWARDS	EVELYN	P
ELBRECHT	CLYDE	
ELBRECHT	WALTER	K
ELIE	CARRIE	
EMCH	JAMES	
ENTY	MILDRED	
EVANS	LAVERNE	
FANSLAU	HOLDIGARD	
FEATHER	EDWARD	
FIER	ADA	
FIER JR	FRANK	J
FIGER	RICHARD	
FIRESTONE	DR LOUIS	H
FIRESTONE	LOUIS	H
FLETCHER	GEORGE	W
FLETTERICH	JESSIE	M
FOGARTY	GENEVIEVE	
FOLTZ	CALVIN	
FONICK	JAMES	
FRICKY	DENNIS	V
FRIER JR	FRANK	
FULTON	JOHN	E
GAMBLE	HENRY	L

LAST	FIRST	MIDDLE
GASSER	THEODORE	P
GEARIETY	CHARLES	H
GEDEON	LEO	
GEORGE	LEVI	
GEZANN	GARY	
GIANFAGNA	TERESA	M
GIBSON	MARY	
GILLESPIE	JAMES	L
GOLDSTEIN	SAM	
GOODSON	ALATHEA	
GOODSON	WILLIAM	D
GORDON	DUNCAN	
GORNIK	FRANK	
GRASSELL	LOUIS	
GRAUL	CAROL	L
GRAY	JOHN	M
GRDOLNIK	NEIL	V
GREENE	WILLIE	G
GREENWAY	FRED	S
GREGORC	MOLLY	
GRIFFIN JR	FLOYD	J
GRONERT	HERMAN	
GROSSELL	LOUIS	A
GUENTHER	HIRAM	G
GUENTZLER	DAVID	A
GUILIANO	MICHAEL	
GUTEKUNST	ADAM	
GUTEKUNST	ADELE	H
HALGASH	KENNETH	G
HALL	CLIFTON	A
HAMILTON	MARGARET	R
HANNAM	TIMOTHY	J
HANZEL	RICHARD	J
HARKEMA	PETER	
HARKEMA SR	PETER	
HARPSTER	HARRY	H
HAWKINS	JOHN	W
HEITMAN	DANIEL	C
HELMINK	JAMES	E
HEREFORD	RAYWELL	
HERTENSTIEN	ARIVILDA	
HEUER	WILLIAM	A
HIGHLAND	WILLIAM	A
HIRSCH	THOMAS	A
HIXON	ARTHUR	L
HIXSON	ELECTRA	E
HOERNIG	MARY	
HOKAVAR	DENNIS	
HOLGASH	KENNETH	
HORTENSTEIN	EDWARD	
HOUSEHOLDER	DOROTHY	
HOWARD	EUGENE	
HUBBARD	MARY	
HUBBARD	ROBERT	J
HUDSON	C	P
HUDSON	EDWIN	P

HUFFMAN	SARA	L
HUMPHREY	GERTRUDE	
IAFELIECE	RICHARD	A
IANNETTA	RICHARD	
JACKLITZ	ALBERT	
JACKSON	ANDREW	
JACKSON	GRACE	
JACKSON	JANELL	
JANES	PETER	J
JANIS	WILLIAM	
JASULAITIS	PIUS	
JENKINS	GRACE	W
JERNEJCIC	FRANK	A
JONES	GARY	D
JONES	LYNN	C
JONES	WILLIAM	O
JONES JR	EDWARD	
JUDD	BARBARA	J
JUDD	ESTIL	L
KARKOSKA	JOSEPH	J
KAST	HOWARD	R
KEKIC	MARJORIE	H
KELLOGG	SHELLY	C
KERN	MARY	R
KIBLER	MICHAEL	K
KILLEEN	KENNETH	L
KINZER	ALLEN	O
KINZER	JAMES	
KINZER	WILMA	
KIRK JR	EMERY	
KLEMENCIC	FREDRIC	A
KNAUSS	JAMES	
KNECHT	RICHARD	J
KOENIG	WILLIAM	R
KOPLOW	PHILIP	J
KORDIC	ELIZABETH	C
KOVACH	JOSEPH	
KRALL	ANTHONY	E
KRAMER	EDWARD	A
KRAMER	JANE	F
KRATOHVIL	GLADYS	
KRAUSE	DOROTHY	L
KRAUSE	HARRY	
KRAUSE	LINDA	M
KRIVOY	DANIEL	G
KUBELAVIC	KATHRYN	
LAKATOSH	JOSEPH	J
LALAK	JOHN	A
LANDEN	JOSEPH	R
LANESE	ANTHONY	D
LANGHAM	ROBERT	H
LANIAUSKAS	VICTOR	
LAPE	DOROTHY	R
LAZZARO	MILDRED	
LEAGAN JR	JAMES	E
LEE	WILLIAM	N
LEECE	MARIE	E
LEEDERS	ROBERT	A
LEFFEL	JOHN	C
LEVAR	FRANK	A
LIPTON	MARGARET	R
LOMBARDO	THOMAS	J
LUCE	HOMER	H
LUMANNICK	ALLAN	G
LUPTON	HOLLY	
LUPTON	ROBERT	
LUPTON	RUTH	M
MACADLO	PAUL	E
MACEK	GERTRUDE	L
MADEYA	ARTHUR	
MAHONE	WILLIE	B
MALENGO	LEO	
MANNING	DANIEL	
MANNING	DENNIS	

The Pier, Euclid Beach Park, Cleveland, Ohio

LAST	FIRST	MIDDLE
MARIANO	MIKE	
MARIMONI	PETER	
MARN	LOUIS	
MARTENS	ALBERT	M
MARTENS	GOLDA	E
MARTENS	WALTER	J
MARTINO	FRANK	L
MATKO	MARY	
MATUNA	PETER	
MAURER	JOSEPHINE	E
MAZZA	VINCENT	F
MCDONALD	MARVIN	J
MCGOWAN	RALPH	W
MCGRATH	PHILLIP	T
MCINTYRE	JAMES	G
MCKELVEY	DAVID	N
MCKINLEY	JAMES	R
MCMASTER	WILLIAM	G
MCNALLY	TIMOTHY	T
MEADE JR	JOHN	J
MELBER	EDWARD	
MERCHANT	ROBERT	
MERKULOFF	GEORGE	
MERVO	RONALD	
MERZDORF	EDWARD	
MICELI	TIMOTHY	W
MIERKE JR	ALBERT	H
MILATOVIC	JOHN	
MILAVEC	BETTY	S
MILES	ARCHIE	
MILES	MARIE	H
MILLER	LARRY	E
MISLEY	JAMES	J
MISLEY	RAYMOND	F
MOCADLO	PAUL	E
MOG	DAVID	
MORAN	ROBERT	J
MULLEN	JAKE	
MYERS	ARVY	C
NEELD	DAVID	R
NELTON	JANET	
NEWMAN	DONALD	P
NEWTON	FRANCIS	
NICOL	ANDREW	
NOKAVAR	DENNIS	
OAKLEY	ADDIE	E
OERGEL	BETTY	F
OMAHEN	WITT	
OPALK	CAROLINE	V
OPALK	MARTIN	
OSBORNE	MILDRED	M
OSTROWSKI	JEROME	J
PAGE	RICHARD	N
PAGON	PHILIP	
PAKISH	FRANK	J
PALEK	FRANCIS	Z
PALEY	NORMAN	
PARKER	CLYDE	H
PARKER	HENRY	C
PARKER	WILLIAM	
PARKHURST	PHILIP	
PARMERTON	ROBERT	
PARMERTOR	JAMES	C
PARMERTOR	ROBERT	A
PATTON	PATRICIA E	
PAULETTO	ROBERT	
PAVELLA	DANIEL	
PAVELLA	THOMAS	O
PEARCE	AGNES	A
PECK	LOIS	E
PEGORANO	RANDOLPH	J
PERSELL	CHARLES	
PESEC	EDWARD	J
PETRECCA	JOSEPH	
PETRICH	DANIEL	J
PETSCHE	JOHN	
PHIPPS	EARL	E
PIERCE	MARTIN	J
PLANTSCH	MICHAEL	
POCKAR	JEROME	J
POLLOCK	JOSEPH	R
PORTER	GRACE	
POTTS	ROBERT	D
PRAYNER	EUGENE	
PREISS	LOU	COLE
PRICE	DONALD	F
PRUSNICK	LEONARD	A
PUSTARE	ROBERT	G
QUICK	WILLIAM	
RAIMONDO	ALFRED	J
RAMONA	JOSEPH	M
RAMSEY	RONALD	
REARDON	EDWARD	
REARDON	ROGER	J
REARDON JR	ROGER	S
REED	GARY	
REED	MARIE	E
REESE	LOUISE	B
REINHARD	GEORGE	M
REINKE	ROBERT	
REINKE	WILLIAM	H
RICHARDS	ANDREW	P
RISTAU	ALAN	A
ROBERTS	JULIUS	
ROBERTS	RUSSELL	
ROCKFORD	JAMES	M
RODERICK	ALICE	M
ROSS	JERRY	L
ROTH	CARL	J
RUCKER	CARTER	H
RYAN	ALAN	W
RYDMAN	ELIZABETH	
SAJOVEC	RICHARD	
SANDERS	CALVIN	M
SANDS	WILLIAM	T
SANEFSKI	EVELYN	
SANTORELLI	JAMES	
SAUVAGEOT	MARTHA	M
SCHAEFER	RUTH	E
SCHAUB	GEORGE	A
SCHILL	AUGUST	J
SCHOONOVER	JAMES	
SCHOTT	ISADORE	W
SCOTT	HAROLD	
SCOTT	PHILLIP	
SEGER	THOMAS	M
SELJAN	ROBERT	
SEMAN	JAMES	J
SENNISH	GEORGE	W
SEVEROVICH	THOMAS	G
SHANIUK	FRANK	

"Administration Building, Euclid Beach Park, Cleveland, Ohio"

LAST	FIRST	MIDDLE
SHAVER	LORETTA	
SHILLIDAY	KATHRYN	
SHIMP	ERIC	J
SISTEK	ELSIE	B
SIVIK	VIOLET	N
SLATER	LOUISE	M
SMITH	CHARLES	L
SMITH	IRENE	M
SMITH	JACK	D
SMITH	JACK	J
SMITH	MARY	C
SMITH	THOMAS	R
SMITH JR	JOHNNIE	
SMRDEL	STANLEY	J
SNIDER	THOMAS	A
SOJEBA	BETTY	
SOJEBA	HARRY	G
SOLLIE	GENE	
SOVICK	JOANNE	
SRSA	WELLA	D
STALLA	DONALD	W
STEELE	ARTHUR	W
STEFANCIC	WILLIAM	M
STEH	JOSEPH	S
STEPHENS	BRIAN	
STEPP	ROBERT	L
STEPP	ROGER	L
STIH	NORMAN	
STRAVOR	LEON	
STRUNA	EDWARD	L
STUCK	RICHARD	E
STUPICA	RAYMOND	J
SUKENIK	JAMES	
SUMPH	GEORGE	W
SUSTARIC	STEVE	
SWANEY	KENNETH	A
SWEENEY	ANGELA	M
SWEENEY	DANIEL	J
SWEENEY	JOHN	
TADYCH	VICTOR	V
TAGLIA	JOSEPH	
TATE	JOSEPHINE	R
TAYLOR	THOMAS	
THOMAS	WAYNE	A
THOMPSON	VAUGHN	
THUNHURST	EDYTHE	
TIMKO	JOHN	
TIMKO	RUTH	A
TIMPERIO	FILOMENA	
TINCK	WILLIAM	T
TKATCH	JAMES	T
TOMARO	VICTOR	P
TOMINO	NICK	
TOPOLY	BART	
TRAUB	NORMAN	L
TRAVERS	BENJAMIN	M
TREM	NICHOLAS	
TRIVISON	CHARLES	
TROUER	BENJAMIN	M
TROYER	BELLE	
TROYER	ISABELLE	M
TURK	JERRY	A
TURK	RAY	
TUSHAR	THOMAS	L
ULMAN	LILLIAN	H
VACCARIELLO	ANITA	
VACCARIELLO	URBAN	
VALENTINE	EDWARD	J
VALENTINO	PAT	A
VANDIEST	THOMAS	W
VARRECCHIO	ELVIRA	
VAUGHN	ELMER	L
VERH	EUGENE	C
VICIC	JOSEPH	J
VOGEL	ROBERT	B
VRANEKOVIC	JOSEPH	D
VRANEZA	CHRISTOPHER	
VRANEZA	SUSAN	
WADE	ROBERT	J
WALKER	HERBERT	LEE
WARD	ROBERT	J
WARD	ROBERT	T
WATERS	WILLIAM	C
WATSON	CHARLES	A
WATSON	ROBERT	D
WELLBAUM	ANDRAE	J
WERBER	DOUGLAS	M
WEST	NETTIE	
WILDER	JESS	L
WILEY SR	FRED	R
WILLIAMS	DORA	T
WILLIAMS	GLORIA	D
WILLIAMS	KENNETH	A
WILLIAMS	LEANNA	
WILLIAMS	LINDA	D
WILLIAMS	WALTER	D
WILSON	PAUL	
WILSON	RICHARD	G
WILSON	ROBERT	O
WILSON	WILLIAM	S
WISE	JAMES	R
WOHLGEMUTH	GARY	G
WOLF	ALAN	P
WOLFF	LESTER	W
WRETSCHKO	DAVID	A
WRIGHT	ETHEL	M
WURM	CHARLOTTE	
YAGER	FRED	H
YAKOS	JAMES	R
YOUNG	REGINALD	E
YOUNG	ROBERTA	
ZACK	ANTHONY	C
ZAGAR	STEVEN	
ZANDER	RICHARD	
ZANELLA JR	JOHN	
ZAPOL	RICHARD	
ZEITZ	RAYMOND	C
ZEZZO	LUCILLE	
ZGONETZ	KENNETH	P
ZIETZ	WILLIAM	S
ZIFKO	MARTIN	A
ZIPAY	JOHN	A
ZORN	JEFFREY	L
ZUST	FRANK	D
ZUST	JAMES	A

1965-66
EMPLOYEE ROSTER

LAST	FIRST	MIDDLE
ADAMS	MARY	C
ADKINS	ROY	J
AMEN	JOHN	H
ANDRY	JOSEPH	E
ASHDOWN JR	ROBERT	
BACHER	FRED	W
BAILEY	WILLIAM	H
BALDINI	RICHARD	A
BANASZEK JR	THEODORE	W
BARBATO	JERRY	M
BARBERIO	FRANK	L
BARRY	DOROTHY	M
BATES	EDNA	M
BEEBE	BEN	
BENNETT	ARVIA	P
BENNETT	CHARLES	D
BENSI	JOSEPH	A
BERGOCH	STANLEY	J
BERNSTEIN	GARY	M
BLACK	JOHN	
BLACKMAN	MARY	L
BLAUCH	DAVID	C
BLEAM	CLARENCE	M
BLEAM	LULA	E
BONJACK	BRIAN	T
BONNER	ADDISON	L
BOUNCE	JOSEPH	H
BOWERS	ROBERT	M
BOWSER	RICHARD	D
BRADISH	GARY	J
BRAZAUSKAS	ROSALIND	M
BRODNICK	FRANK	J
BRODNICK	JIM	
BRODNICK	ROBERT	D
BROOKS	GEORGE	
BRUNO	ROBERT	H
BRYANT	THOMAS	E
BUCKLEY	HAROLD	D
BUCKLEY	PATRICK	M
BUCKNER JR	ROY	W
BUDAN	KAREN	L
BURKE	FLORENCE	A
BUSCHER	GEORGE	E
CABOT	WILLIAM	J
CAMPBELL	FRED	
CARNEY	ROGER	C
CATALANO	FRANK	A
CHAL	THOMAS	M
CHEBO	ROBERT	T
CHERVON	LEONARD	L
CHILDS	ERNEST	
CHINN	DALLAS	D
CIPITI	THOMAS	F
CLARK	EVELYN	E
COLLINS	EDWARD	M
CONNOR	NORMAN	
COOK	ALICE	S
COWIE	THOMAS	F
CUNNINGHAM	DOUGLAS	J
CUNNINGHAM	RICHARD	A
CUNNINGHAM	STEPHEN	
DAVIS	JOHN	B
DAY	JENNIE	
DAY	JEREMIAH	S
DEADWYLER	MILTON	
DEBALSO	ANGELINA	
DELAMBO	FRANK	F
DENNIS	HERMAN	
DEUTSCH	GEORGE	
DEVNEY	JAMES	T
DIETRICH	ROBERT	A
DIETZ	GEORGE	N
DIETZEL	ROBERT	J
DIETZEL	THOMAS	J
DINUNZIO	RONALD	E
DORT	GEORGE	T
DOSSA	STEVE	
DRAHEIM	EDWARD	
DUCCA	MARLENE	
DUNCAN	EDNA	T
EDWARDSEN	ROBERT	
EISINGER	ALFRED	N
ELIE	CARRIE	
ETZLER	JAMES	R
FARMAN	DONALD	L
FEDELE	CHRIS	P
FERRITTO	WILLIAM	F
FIER	ADA	
FIER JR	FRANK	J
FINKLER	CLARENCE	M
FIRESTONE	LOUIS	H
FITZMAURICE	MATTHEW	G
FOLTZ	CALVIN	
FOSTER	ROGER	A
FREDLE	JOSEPH	J
FULTON	JOHN	E
GALASKA	RICHARD	C
GALLON JR	EDMUND	
GEDEON	LEO	
GEISE	HENRY	C
GEISE	MAE	R
GELLER	ELLIOTT	P
GEORGE	LEVI	
GERMANO	CARRIE	
GIANFAGNA	TERESA	M
GILLESPIE	JAMES	L
GIULIANO	MICHAEL	
GLAVAN	FRANK	
GOGOLIN	ROY	E
GOLDSTEIN	MELVYN	
GOLDSTEIN	SAM	
GORNIK	FRANK	
GORNIK	LEONARD	
GOVE	ROLAND	S
GRDOLNIK	NEIL	V
GRIFFIN JR	FLOYD	J
GROGAN	THOMAS	J
GRONERT	HERMAN	
GUENTHER	HIRAM	G
GUHDE	DONALD	J
HAAKE	THEODORA	F
HALL	CLIFTON	A
HANNAM	TIMOTHY	J
HANSHEW	MICHAEL	J
HARPSTER	HARRY	H
HARRISON	LORENZO	
HAWKINS	LESTER	E
HEITMAN	DANIEL	
HEITMAN	NORMA	M
HENRY	JAMES	G
HERBST	DENNIS	J
HERTENSTEIN	EDWARD	
HIGHLAND	WILLIAM	A

LAST	FIRST	MIDDLE
HOERNIG	MARY	
HOKAVAR	DENNIS	
HOPPEL	WILLIAM	R
HUBBARD	MARY	
HUDSON	EDWIN	P
HUER	WILLIAM	H
HUFFMAN	SARA	L
HUMPHREY	GERTRUDE	G
HYDE	GARY	L
IANNICELLI	HELEN	
IULIANI	CAROL	
JACKLITZ	ALBERT	
JACKSON	ANDREW	I
JANIS	WILLIAM	
JENKINS	MOSE	
JORDENS	HERBERT	J
JUDD	ESTIL	
JUDD	JULIANA	M
JUDD	NANCY	L
JURCEVIC	MARGARET	R
KASUNIC	LAWRENCE	
KEKIC	MARJORIE	H
KEPETS	DAWN	M
KIKOL	FRANK	
KOENIG	WILLIAM	R
KOPACZ	JOSEPH	C
KOPCAK	DAVID	E
KOPLOW	PHILIP	
KORACIN	JOHN	E
KOSEC	NICHOLAS	R
KOVACH	JOSEPH	
KRAMER	JANE	F
KRATZERT	HERMAN	L
KRAUSE	DOROTHY	L
KRAUSE	HARRY	
KRECZKO	THOMAS	S
KRISTY	KENNETH	H
KUBELAVIC	KATHRYN	
KUNUGI	ALLEN	
LANDEN	JOSEPH	
LARGE	HELEN	
LAZZARO	MILDRED	
LEECE	MARIE	E
LEPA	FRANK	R
LIPP	FRED	D
LOMBARDO	ANTHONY	G
LONGO	JOHN	G
LUCE	HOMER	H
LUPTON	ROBERT	
LUPTON	RUTH	M
MACNAB	FRANK	E
MAHONE	WILLIE	B
MAHONE JR	JOHN	
MANNING	DANIEL	

LAST	FIRST	MIDDLE
MANTEY	KENNETH	L
MARELLA	ROSS	
MARIANO	MIKE	
MARINELLI	TONY	
MARN	LOUIS	
MARSICH	JOHN	F
MARTENS	ALBERT	M
MARTENS	GOLDA	E
MARTENS	WALTER	J
MARTIN	DELBERT	C
MARTINO	FRANK	L
MATKO	MARY	
MATUNA	PETER	
MCDONALD	MARVIN	J
MCGINNIS	MARI	BARBARA
MCINTYRE	JAMES	G
MCKELVEY	DAVID	N
MCMASTER	ROBERT	R
MCNALLY	THOMAS	W
MEADE	JOHN	J
MELBER	EDWARD	F
MIERKE JR	ALBERT	H
MIHALIC	DAVID	P
MILATOVIC	JOHN	
MILES	MARIE	H
MILLER	LARRY	E
MILLS	FREDERICK	
MIMS	JIM	
MOCADLO	PAUL	E
MOORE	WALTER	W
MORRIS	DENNIS	
MORRIS	RICHARD	
MULLEN	JAKE	
MULLEN JR	SAM	
NEELD	DAVID	R
NELSON	MAY	
NERAD	DOROTHY	M
NEWMAN	DONALD	P
OLESKI	KAREN	J
OPALK	CAROLINE	V
OPALK	MARTIN	
OSBORNE	EDWARD	L
OSBORNE	MILDRED	M
PAGE	RICHARD	N
PARISI	PAUL	J
PARKER	CLYDE	H
PARKER	HENRY	C
PARKER	WILLIAM	
PARKHURST	PHILIP	A
PEARCE	AGNES	A
PECK	KENNETH	L
PECK	LOIS	E
PERKINS	JOHN	M
PERSELL	CHARLES	
PETRECCA	JOSEPH	D
PETRELLO	RAYMOND	P
PETRICH	DANIEL	J
PETTI	JEFFREY	M
PHIPPS	EARL	E
PHIPPS	LINDA	M
PIERCE	BARBARA	J
PIERCE	MARTIN	J
PIERCE	RONALD	J
POCHERVINA	ERNEST	G
POLLACK JR	JOSEPH	R
POLLUTRO	ROCCO	M
POROK	PETER	P
POTTS	ROBERT	D
PREGIZER	FRED	
PREISS	HERBERT	M
PREISS	LOU	C
PUFFENBARGER III	MARSHALL	B
PUSTARE	ROBERT	G
RAFFA	PATRICK	C
RAIMONDO	ALFRED	

LAST	FIRST	MIDDLE
REARDON JR	ROGER	J
REARDON SR	ROGER	
REED	GARY	C
REED	MARIE	E
REESE	LOUISE	B
REID JR	JAMES	
RICHARDS	ANDREW	P
RIEBE	ROBERT	R
ROCKFORD	JAMES	M
RODERICK	ALICE	M
ROSS	RICHARD	B
RUSNAK	GEORGE	A
RUTH	ROBERT	F
RYDMAN	ELIZABETH	
SAMS	JOSEPH	
SANDS	WILLIAM	T
SCHAEFER	RUTH	E
SCHAUB	GEORGE	A
SCHOONOVER	JAMES	
SCHOTT	ISADORE	W
SCHULZ	DONALD	A
SEBOCK	JOSEPH	M
SEGER	THOMAS	M
SENNISH	GEORGE	W
SHAFFER	FRANK	
SHAVER	LORETTA	
SHILLIDAY	KATHRYN	G
SIMONCIC	RUDOLPH	
SISTEK	ELSIE	B
SIVIK	VIOLET	
SKEDEL	KENNETH	J
SLATER	LOUISE	M
SMITH	JACK	D
SMITH	JAMES	F
SMITH JR	JOHNNIE	
SNYDER	RICHARD	L
SOJEBA	BETTY	L
SOJEBA	HARRY	G
SPENKO	MARGARET	
SPRING	TIMOTHY	L
STANLEY	NANCY	JO
STARZYNSKI	JEFF	B
STEARNS	HELEN	
STEELE	ARTHUR	W
STIH	JAMES	E
STOUT JR	FRED	
STRAPP	CATHERINE	A
STUBBS	HENRY	
STUCK	RICHARD	E
SUTTON	JUDITH	A
SWANEY	KENNETH	A
SWEENEY	ANGELA	M
SWISHER	GEOFFREY	A
TAGLIA	JOSEPH	
TATE	JOSEPHINE	R
THUNHURST	EDYTHE	
TIMKO	JOHN	
TIMKO	RUTH	A
TIMPERIO	FILOMENA	
TOMARIC JR	JOSEPH	E
TOMARO	VICTOR	P
TOMINO	NICK	
TRIVISON	CHARLES	
TROYER	ISABELLE	M
TUSHAR	THOMAS	L
VACCARIELLO	ANITA	
VACCARIELLO	ANTHONY	
VACCARIELLO	URBAN	
VALENTINE	EDWARD	J
VALENTINO	NICK	A
VALENTY	ANTHONY	
VERH	EUGENE	C
VITANTONIO	KENNETH	L
VOGEL	ROBERT	B
WARD	DANIEL	R
WARD	ROBERT	J
WARGO	DANIEL	R
WASSIL	MICHAEL	
WATERS	WILLIAM	
WEBER	ALAN	G
WELTON	RUTH	
WHITE	ALBERT	A
WHITE	ROY	H
WIKTOROWSKI	JOHN	A
WILEY SR	FRED	R
WILLIAMS	CARL	
WILLIAMS	CHARLIE	
WILLIAMS	DORA	T
WILLIAMS	JOHNNIE	
WILLIAMS	KENNETH	S
WILLIAMS	LEANNA	
WILLIAMS	PAULETTE	
WILLIAMS	VERNA	T
WILLIAMS	WALTER	D
WILLIAMS	WAYNE	
WILLIAMS	WILSON	
WILLIAMSON	EARL	
WILSON	RICHARD	G
WISE	JAMES	R
WOHLGEMUTH	GARY	G
YAGER	FRED	H
ZELZNICK	PHILLIP	G
ZORN	JEFFREY	L
ZUPANCIC	THOMAS	J

* INCOMPLETE

The Thriller.

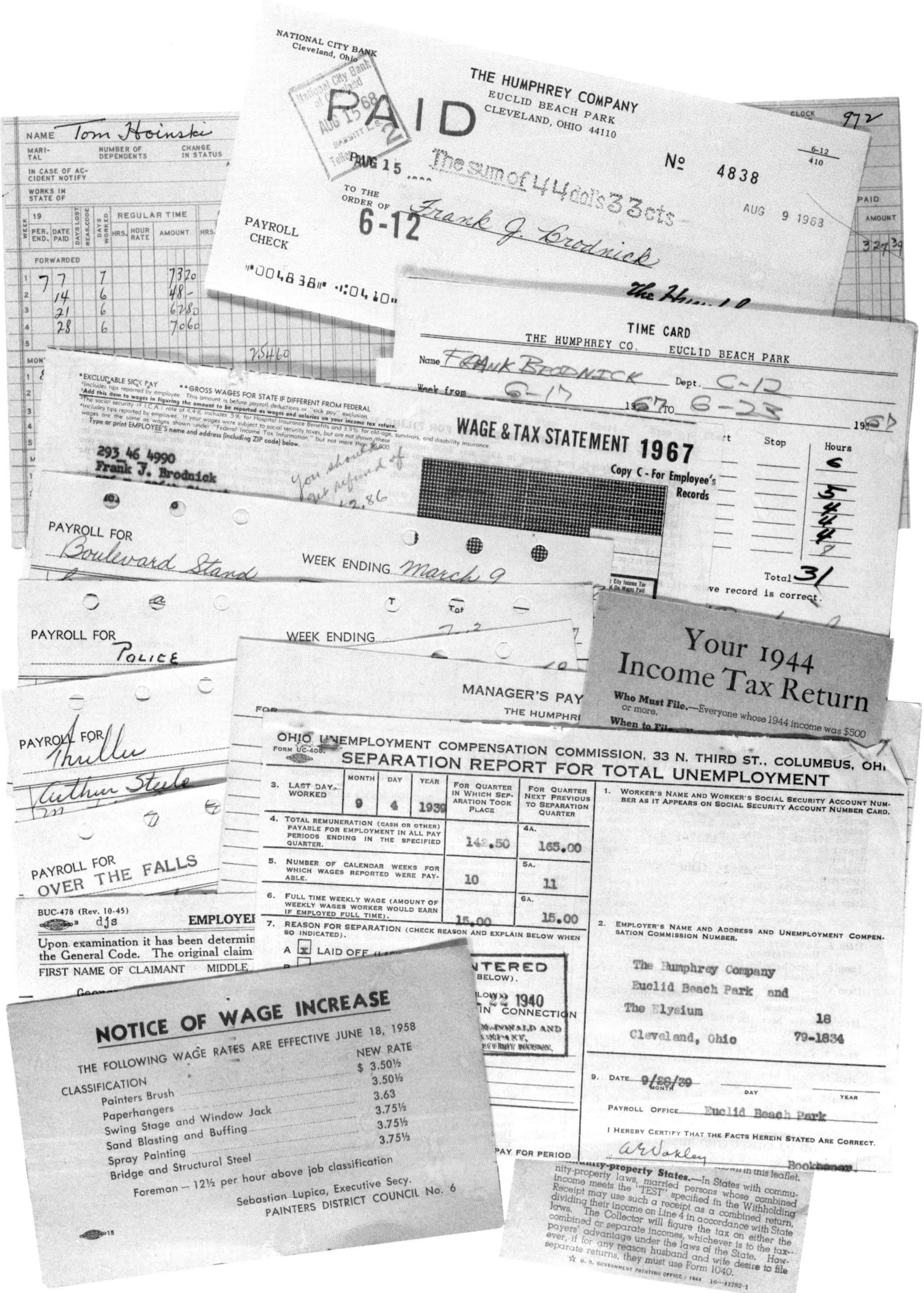
NATIONAL CITY BANK
Cleveland, Ohio
PAID
THE HUMPHREY COMPANY
EUCLID BEACH PARK
CLEVELAND, OHIO 44110
No 4838
AUG 9 1968
TO THE ORDER OF
6-12
PAYROLL CHECK
The sum of 44 dols 33 cts
Frank J. Brodnick
NAME Tom Hoinski
TIME CARD
THE HUMPHREY CO. EUCLID BEACH PARK
Name FRANK BRODNICK
Dept. C-12
Week from 6-17 TO 6-23
Stop
Hours
Total 31
WAGE & TAX STATEMENT 1967
Copy C - For Employee's Records
293 46 4990
Frank J. Brodnick
PAYROLL FOR Boulevard Stand
WEEK ENDING March 9
PAYROLL FOR Police
WEEK ENDING
PAYROLL FOR Thriller
Arthur Steele
PAYROLL FOR OVER THE FALLS
MANAGER'S PAY
THE HUMPHREY
Your 1944 Income Tax Return
Who Must File.—Everyone whose 1944 income was $500 or more.
OHIO UNEMPLOYMENT COMPENSATION COMMISSION, 33 N. THIRD ST., COLUMBUS, OHIO
SEPARATION REPORT FOR TOTAL UNEMPLOYMENT
3. LAST DAY WORKED 9 4 1939
4. TOTAL REMUNERATION (CASH OR OTHER) PAYABLE FOR EMPLOYMENT IN ALL PAY PERIODS ENDING IN THE SPECIFIED QUARTER. 142.50 165.00
5. NUMBER OF CALENDAR WEEKS FOR WHICH WAGES REPORTED WERE PAYABLE. 10 11
6. FULL TIME WEEKLY WAGE (AMOUNT OF WEEKLY WAGES WORKER WOULD EARN IF EMPLOYED FULL TIME). 15.00 15.00
7. REASON FOR SEPARATION (CHECK REASON AND EXPLAIN BELOW WHEN SO INDICATED).
LAID OFF
1. WORKER'S NAME AND WORKER'S SOCIAL SECURITY ACCOUNT NUMBER AS IT APPEARS ON SOCIAL SECURITY ACCOUNT NUMBER CARD.
2. EMPLOYER'S NAME AND ADDRESS AND UNEMPLOYMENT COMPENSATION COMMISSION NUMBER.
The Humphrey Company
Euclid Beach Park and
The Elysium 18
Cleveland, Ohio 79-1834
9. DATE 9/28/39
PAYROLL OFFICE Euclid Beach Park
I HEREBY CERTIFY THAT THE FACTS HEREIN STATED ARE CORRECT.
Bookkeeper
BUC-478 (Rev. 10-45)
EMPLOYER
Upon examination it has been determin
the General Code. The original claim
FIRST NAME OF CLAIMANT MIDDLE
NOTICE OF WAGE INCREASE
THE FOLLOWING WAGE RATES ARE EFFECTIVE JUNE 18, 1958
CLASSIFICATION NEW RATE
Painters Brush $ 3.50½
Paperhangers 3.50½
Swing Stage and Window Jack 3.63
Sand Blasting and Buffing 3.75½
Spray Painting 3.75½
Bridge and Structural Steel 3.75½
Foreman — 12½ per hour above job classification
Sebastian Lupica, Executive Secy.
PAINTERS DISTRICT COUNCIL No. 6
Community-property States.—In States with community-property laws, married persons whose combined income meets the "TEST" specified in the Withholding Receipt may use such a receipt as a combined return, dividing their income on Line 4 in accordance with State laws. The Collector will figure the tax on either the combined or separate incomes, whichever is to the taxpayers' advantage under the laws of the State. However, if for any reason husband and wife desire to file separate returns, they must use Form 1040.
U. S. GOVERNMENT PRINTING OFFICE : 1944

1917 - 1966
YEARLY EMPLOYEE WAGE TOTALS

Year	Amount	Year	Amount
1917	108,620.00	1942	306,695.81
1918	101,089.06	1943	307,270.13
1919	121,006.29	1944	360,867.10
1920	175,449.82	1945	383,212.70
1921	232,011.48	1946	419,359.55
1922	140,620.22	1947	342,329.50
1923	185,173.54	1948	350,500.50
1924	228,569.68	1949	337,583.55
1925	245,773.43	1950	389,074.40
1926	245,194.05	1951	354,765.95
1927	237,158.35	1952	411,690.70
1928	226,712.10	1953	410,156.65
1929	236,533.70	1954	419,774.65
1930	233,488.30	1955	431,569.55
1931	165,101.95	1956	397,240.95
1932	90,072.60	1957	432,334.90
1933	82,957.20	1958	460,461.00
1934	144,464.20	1959	427,391.05
1935	123,580.35	1960	426,541.00
1936	127,597.33	1961	420,647.04
1937	176,581.00	1962	423,925.27
1938	172,089.60	1963	444,277.13
1939	179,186.10	1964	447,408.64
1940	220,970.26	1965	402,735.19
1941	246,152.78	1966	370,426.74

CONCESSION OPERATORS

Barber Shop

1929 - 1932	Ed Begmer
1933 - 1939	Geo. Clements
1940 - 1941	C. E. Harvey
1942 -	Tom Collins

Beauty Shop

1945 - 1948	Helen Cheseth

Boat Rental

1950	Joe Longwell
1951 - 1953	Tony Svetina
1954 - 1957	Victor Coertnik

Cigar Stand

1926	J.M. Frese
1927 - 1930	E.A. Erving
1932 - 1941	Crosby's
1942 - 1944	S.L. Minor
1945	Harry Hell
1947 - 1957	Ralph Wagner

Photo Gallary

1925 - 1942	C.F. Riblet
1942 - 1957	Grant Photo Corp.

Pony Track

1937 - 1954	H.C. Dressler

Arcade

1924 - 1958	William Gent and Son

Euclid Beach Hotel

1925 - 1950	Mrs. W.S. Walker
1950 - 1953	Rhodema Walker

1918 - 1966
YEARLY TICKET TOTALS

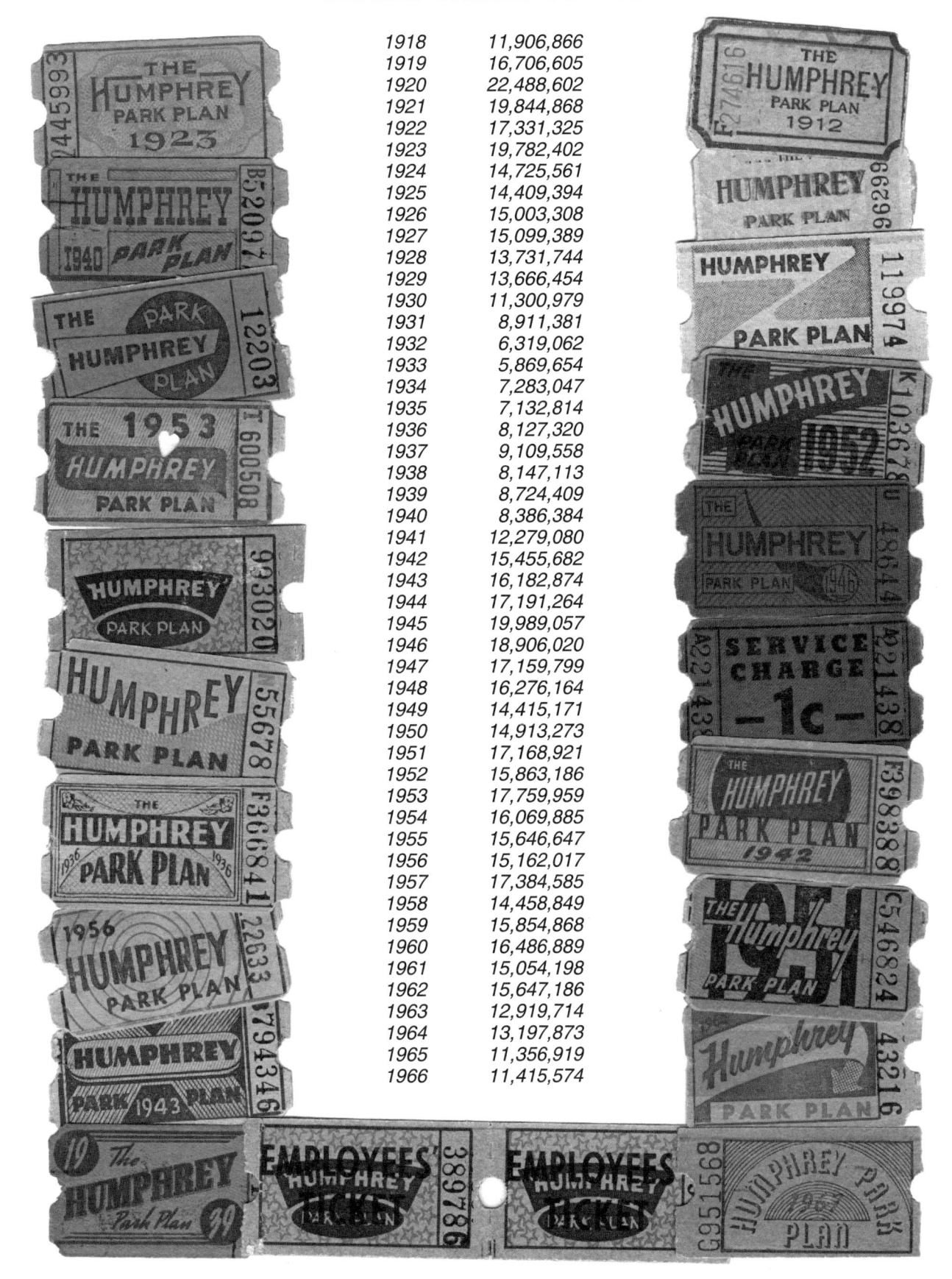

Year	Total
1918	11,906,866
1919	16,706,605
1920	22,488,602
1921	19,844,868
1922	17,331,325
1923	19,782,402
1924	14,725,561
1925	14,409,394
1926	15,003,308
1927	15,099,389
1928	13,731,744
1929	13,666,454
1930	11,300,979
1931	8,911,381
1932	6,319,062
1933	5,869,654
1934	7,283,047
1935	7,132,814
1936	8,127,320
1937	9,109,558
1938	8,147,113
1939	8,724,409
1940	8,386,384
1941	12,279,080
1942	15,455,682
1943	16,182,874
1944	17,191,264
1945	19,989,057
1946	18,906,020
1947	17,159,799
1948	16,276,164
1949	14,415,171
1950	14,913,273
1951	17,168,921
1952	15,863,186
1953	17,759,959
1954	16,069,885
1955	15,646,647
1956	15,162,017
1957	17,384,585
1958	14,458,849
1959	15,854,868
1960	16,486,889
1961	15,054,198
1962	15,647,186
1963	12,919,714
1964	13,197,873
1965	11,356,919
1966	11,415,574

JULY
DAILY TICKET COUNTS

	1954		1955		1956		1957		1958		1959
1	60020	1	72055	1	222911	1		1	81033	1	35717
2	62493	2	141997	2	-0-	2	111122	2	57703	2	65191
3	109324			3	102381	3	87653	3	51530	3	118653
		3	267636	4	374299	4	445543	4	432695	4	530098
4	291913	4	418807	5	44857	5	117180	5	115548		752441
5	368480	5	67075	6	79809	6	154520			5	115138
6	51336	6	76073	7	166861			6	172285	6	
7	44581	7	82218		997254	7	248351	7		7	108657
8	82019	8	81886	8	175215	8	-0-	8	78836	8	83132
9	68133	9	212893	9	-0-	9	100790	9	73543	9	220678
10	187412			10	77866	10	83542	10	194172	10	83987
		10	240774	11	80881	11	192287	11	64385	11	224198
11	223853	11	-0-	12	196298	12	87348	12	247681		835740
12	92619	12	85951	13	184658	13	253765			12	287767
13	78238	13	76799	14	226271			13	189982	13	
14	117796	14	62556		941189	14	252678	14		14	110947
15	214580	15	161281	15	244610	15	-0-	15	239229	15	206223
16	166376	16	217308	16	-0-	16	309667	16	197955	16	72898
17	169912			17	244342	17	199158	17	84776	17	65283
		17	232171	18	198892	18	82109	18	181190	18	78759
18	198186	18	-0-	19	64387	19	202377	19	141805		771877
19	270883	19	309345	20	50568	20	208146			19	213922
20	59958	20	182849	21	164888			20	164476	20	
21	181537	21	90355		957687	21	207854	21		21	226623
22	79325	22	67180	22	236990	22		22	186544	22	131823
23	58666	23	125219	23	-0-	23	287383	23	138070	23	59072
24	136218			24	305999	24	130669	24	66296	24	192615
		24	213625	25	108590	25	74960	25	142207	25	136197
25	192561	26	296171	26	68046	26	66746	26	117124		960752
26	272540	25	-0-	27	76837	27	128967			26	203228
27	64673	27	114415	28	122704			27	179438	27	
28	146055	28	65994		919166	28	217644	28		28	205995
29	62891	29	76787	29	209240	29		29	241783	29	150481
30	68301	30	117761	30		30	311374	30	239880	30	57774
31	85167			31	269216	31	245158	31	31443	31	69603
		31	202462								

	1960		1961		1962		1963		1964		1965
1.	183209	1.	170001	1.	274915	1		1	86771	1	40350
2.	165564			2.		2	122330	2	89783	2	15345
		2.	147555	3.	53573	3	64872	3	39770	3	134613
3.	197391	3.		4.	505126	4	45031	4	67683		
4.	529326	4.	446974	5.	82290	5	61883			4	195452
5.		5.	107800	6.	87790	6	154009	5	212613	5	45730
6.	189394	6.	102325	7.	184337			6		6	
7.	118196	7.	81713			7	129295	7	116172	7	42627
8.	103444	8.	140810	8.	116013	8		8	57148	8	58298
9.	168907			9.		9	44458	9	70456	9	40425
		9.	231751	10.	169673	10	166782	10	114680	10	148877
10.	216461	10.		11.	203857	11	165196	11	129415		
11.		11.	91075	12.	123557	12	91414			11	174039
12.	102755	12.	162792	13.	90166	13	218291	12	33518	12	
13.	112868	13.	162936	14.	192172			13	68875	13	87968
14.	210883	14.	64033			14	72977	14		14	9640
15.	229587	15.	221916	15.	196602	15		15	60149	15	56349
16.	273458			16.		16	82946	16	178120	16	129975
		16.	183030	17.	123978	17	105309	17	172742	17	214856
17.	242936	17.		18.	124654	18	159185	18	237862		
13.		18.	237828	19.	91274	19	147121			18	224705
19.	233654	19.	92157	20.	129164	20	99077	19	244699	19	
20.	134201	20.	80934	21.	160010			20		20	302563
21.	73447	21.	210643			21	183423	21	294981	21	92277
22.	74003	22.	44193	22.	179337	22		22	167074	22	61319
23.	114051			23.		23	292127	23	65518	23	49579
		23.	168147	24.	287010	24	146069	24	145779	24	149582
24.	226125	24.		25.	130490	25	119955	25	127436		
25.		25.	94702	26.	59679	26	150206			25	215984
26.	70890	26.	177344	27.	174633	27	128069	26	182786	26	
27.	223011	27.	133580	28.	113514			27		27	144642
28.	108225	28.	82353			28	121027	28	136540	28	55105
29.	76279	29.	138199	29.	197143	29		29	71994	29	62459
30.	131566			30.		30	203536	30	65559	30	177020
		30.	135961	31.	110105	31	153569	31	100738	31	174778
31.	230756	31.									

MONTHLY RIDE COUNTS

1942 RIDE COUNTS

	RIDE	APR	MAY	JUN	JUL	AUG	SEPT
1	DANCE HALL		73990	76850	75551	73770	34130
2	ROLLER RINK		8237	10348	12142	14468	6336
3	SUP. HOUSE		40957	53105	59118	62304	9113
4	AERO DIPS		16733	27406	38957	28568	6284
5	AM. DERBY		21530	28090	37720	58870	10835
6	AUTO TRAIN		3620	8449	22370	17940	5090
7	BUBBLE BOUNCE		12708	15293	20910	25605	8283
8	BUG		29277	46401	53882	55767	16699
9	CAROUSEL		6513	9260	16920	13210	6030
10	DIPPY WHIP		6468	17703	22339	21740	5025
11	DODGEM		61309	86535	93688	98848	37616
12	FLY. SCOOTERS		23045	32359	34874	39491	12926
13	" TURNS		46520	60803	69737	73523	25915
14	LAFF IN DARK		37030	47140	61869	65915	20035
15	OVER THE FALLS		19974	43364	53985	61306	20817
16	RACING COASTER		44192	64397	93382	87741	27769
17	ROCKET SHIPS		11986	20031	29594	26912	8195
18	THRILLER		105474	120820	131073	145566	55682
19	TRAIN		5802	14640	21750	23303	1629
20	PONIES		2945	2049	2168	2147	24

1943 RIDE COUNTS

	RIDE	APR	MAY	JUN	JULY	AUG	SEPT
1	DANCE HALL	19605	67615	74601	78120	66960	30202
2	ROLLER RINK	3687	9775	10027	12741	13444	7361
3	SUP. HOUSE	6738	39857	58542	54061	48654	13360
4	AERO DIPS	921	14942	33780	22971	20730	4226
5	AM. DERBY	3555	26725	45773	60893	49203	19910
6	BUBBLE BOUNCE	1317	15508	24480	25619	22339	4704
7	BUG	5230	25021	49107	59171	55242	12416
8	CAROUSEL	1136	9511	13995	19861	17653	4513
9	DODGEM	13667	58648	96207	88029	82701	25969
10	FLY. SCOOTERS	3153	18622	29803	34962	33342	8567
11	" TURNS	4844	36632	54985	47591	45416	12085
12	LAFF IN DARK	6518	37807	58168	61370	57677	15593
13	RACING COASTER	2064	55867	94846	79600	71806	17662
14	ROCKET SHIPS	292	10543	23132	32387	29074	6926
15	THRILLER	14853	105443	153720	108505	98066	23590
16	WHIP		7349	16937	23620	20284	4330
17	OVER THE FALLS		20025	44261	44167	41948	10499
18	AUTO TRAIN		5355	17370	14420	11540	2707
19	TRAIN		6885	13685	8705	10605	3485
20	PONIES		2388	14045	18644	21810	5137

EUCLID BEACH PARK

Recreational Amusement

THE HUMPHREY COMPANY
OWNERS AND OPERATORS

CLEVELAND, OHIO

Modern Tourist Camp

Nov. 22, 1943.

The Humphrey Company

Euclid Beach Park,

Cleveland, Ohio.

Gentlemen:-

The records for amusement taxes from April 15, 1943 to October 3, 1943, were checked and reconciled. All unused tickets were duly burned in my presence.

Samuel Klein

Deputy Collector

Bug

July 21

DAILY RECORD
OF
BRIGHT'S REGISTERING
TURN STILE

Flying Turns

July 8

DAILY RECORD
OF
BRIGHT'S REGISTERING
TURN STILE

Aero Dips

July 1
DAILY RECORD
OF
BRIGHT'S REGISTERING
TURN STILE

0 8 5 5 4
0 6 9 7 5

RACING COASTER
June 16, 1946
DAILY RECORD
OF

Roller Rink.
6/5/46
DAILY RECORD
OF
BRIGHT'S RE...

CLOSING

OPENING

Thriller

July 8

DAILY RECORD
OF
BRIGHT'S REGISTERING
TURN STILE

Over the Falls

July 17

DAILY RECORD
OF
BRIGHT'S REGISTER
TURN STILE

9 7 1 0 8
9 3 6 8 5

Laff in Dark

July 1

DAILY RECORD
OF

Dodgem

July 21

DAILY RECORD
OF
BRIGHT'S REGISTERING

CARROUSEL
May 31, 1946
DAILY RECORD
OF
BRIGHT'S REGISTERING
TURN STILE

1 4 2 1 2
1 2 5 1 2

BUBBLE BOUNCE
June 21, 1946
DAILY RECORD
OF
BRIGHT'S REGISTERING

FLYING SCOOTERS
June 16, 1946
DAILY RECORD

Rocket Ships

July 18

DAILY RECORD
OF
BRIGHT'S REGISTERING
TURN STILE

MINIATURE RAILWAY
June 26, 1946
DAILY RECORD
OF
BRIGHT'S REGISTERING
TURN STILE

0 7 7 1 6
0 6 4 6

MONTHLY RIDE COUNTS

1950 RIDE COUNTS

RIDE	APR	MAY	JUNE	JULY	AUG	SEPT
AUTO TRAIN	0	3969	10233	20847	22510	5658
KIDDIE RIDES	321	52053	78441	124813	119046	45456
PONIES	0	7290	15435	20980	21827	6015
AERO DIPS	0	11203	17452	41186	19678	5878
Amer. DERBY	717	15247	21238	49634	37435	16296
BUBBLE BOUNCE	538	11952	18299	24088	22367	5141
BUG	1171	29326	50838	65208	69816	19467
CARROUSEL	1286	13363	21837	40455	27362	5215
DODGEMS	3038	47328	75346	90250	96414	36230
FLYING SCOOTERS	904	17183	25821	33370	33942	11411
FLYING TURNS	1305	22253	27975	41905	39132	17619
HURRICANE	761	11137	11559	18804	16171	7447
LAFF IN DARK	1908	34510	50421	67072	60153	17854
Mini. TRAIN	786	16977	30719	60790	42805	11124
OVER FALLS	0	13888	30172	58871	37855	16034
RACING COASTER	0	29754	38130	70774	70300	23003
ROCKETS	0	6456	9661	22912	19013	3754
SUR. HOUSE	1682	24824	36357	44275	36447	5211
THRILLER	3683	53728	70234	87820	79295	29324
WHIP	0	6102	6499	16089	21325	5108

1951 RIDE COUNTS

RIDE	APR	MAY	JUNE	JULY	AUG	SEPT
AUTO TRAIN	1437	4577	10238	13621	13447	5781
KIDDIE RIDES	8510	57853	109426			
PONIES	0	22434	14858	1040	730	290
AERO DIPS	1095	11481	18651	26366	23860	8384
Amer. DERBY	2395	18742	29730	51219	44149	14353
BUBBLE BOUNCE	2220	14036	25519	26402	23348	6763
BUG	4441	30312	36801	71833	48476	13945
CAROUSEL	3311	12839	15296	25385	6337	1559
DODGEMS	11222	51807	74381	93496	100242	38075
FLY SCOOTERS	3955	18653	33448	37337	25443	8282
FLY TURNS	4498	24128	37582	48212	48057	22562
LAFF-IN-DARK	5892	33510	38995	64856	58309	23318
RAILWAY	2051	15895	26270	31678	20328	7315
OVER THE FALLS	0	16768	37072	42236	40355	14918
RACING COASTER	6433	32344	36154	80126	57257	24838
ROCKET SHIPS	916	5695	8329	20814	17131	6571
ROCK-O-PLANES	1471		29685	18598	13835	3347
SUR. HOUSE	4342	29082	42271	45948	32418	9421
THRILLER	11042	64516	94565	101332	109528	32447
WHIP	60	10585	9048	15839	17370	3400

YEAR	SKEE BALL GAMES	YEAR	SKEE BALL GAMES
1925	170569	1946	336295
1926	233534	1947	307402
1927	226572	1948	407003
1928	206385	1949	310373
1929	226300	1950	280953
1930	164350	1951	284668
1931	114835	1952	267776
1932	178400	1953	261855
1933	99808	1954	214049
1934	116651	1955	246099
1935	106610	1956	253779
1936	132852	1957	254652
1937	230487	1958	250563
1938	146837	1959	249041
1939	167883	1960	264954
1940	182426	1961	245883
1941	260695	1962	244749
1942	272730	1963	204724
1943	279732	1964	195334
1944	268778		
1945	303745		

EQUIPMENT AND BUILDINGS

1920
EVALUATION

Bowling Alley	17,705.56
Bath House	18,708.38
Dance Pavilion	39,575.42
Dance Platform	17,486.55

	EQUIPMENT	ORIGINAL COST	BUILDING
1902	Aerial Swing	13,810.76	
1902	World Theatre	7,734.04	13,671.51
1902	Log Cabin		17,248.09
1903	Flying Ponies	6,475.38	
1904	Roller Rink	5,303.34	17,988.83
1904	Rink Organ	11,048.62	
1907	Senic Railway	38,670.17	
1909	Aero Dips	19,887.52	
1910	Carousel	7,734.04	9,713.97
1912	Oak Pile Peir	15,468.07	
1912	Peir Approach	13,810.76	
1913	Racing Coaster	44,195.13	
1915	Whip	4,971.88	
1916	Auto Train	2,762.16	
1920	Skee Ball	5,524.31	Addition
1921	Dodgems	49,747.16	
1920's	Red Bug Boulevard	10,512.36	7,008.24
1920's	Zoomer	16,987.81	
1924	Thriller	90,805.75	
1924	Kiddie Ferris Wheel	1,958.10	
1926	Miniture Railway	12,294.58	10,794.48
1928	Bug	15,156.26	
1928	Dodgem Cars	15,289.59	Renewal
1930	Shooting Gallery	4,930.35	
1930	Flying Turns	44,736.18	10,000.00
1930	Sun Tan Links	2,667.90	
1930	Trainer Planes	1,701.89	
1931	Laff in the Dark	14,306.43	Remodeled
1931	Kiddie Bug	2,507.98	
1931	Kiddie Aiplanes	1,239.96	
1931	Kiddie Autos	1,382.43	
1933	Skeeball	10,730.10	Addition
1934	Kiddie Rodeo	1,903.60	
1935	Surprise House	17,368.53	
1935	Shooting Gallery	5,035.45	Addition
1936	Skee Ball	2,472.00	Addition
1937	Cuddle Up	1,321.19	
1937	Over the Falls	7,654.62	Remodeled
1938	Dippy Whip	8,751.80	
1938	Flying Scooters	12,434.99	

	EQUIPMENT	ORIGINAL COST	BUILDING
1939	Bubble Bounce	8,201.29	
1947	Bowlo	12,332.06	
1949	Kiddie Mill Chute	4,921.09	
1949	Hurricane	29,567.01	
1948	Miniture Railway	13,350.49	Renewal
1950	Kiddie Jeeps	5,741.96	
1951	Kiddie Hook & Ladder	2,763.56	
1951	Rock-o-Plane	9,880.25	
1951	Kiddie Autos	1,802.50	
1957	Kiddie Helicopter	9,832.35	
1957	Rotor	22,528.82	
1961	Giant Slide	4,415.00	
1962	Turnpike	24,844.16	
1963	Antique Autos	27,892.00	
1964	Coffee Break	14,579.04	
1964	Swinging Gyms	9,500.00	
1966	Tilt a Whirl	10,830.00	
1966	Ferris Wheel	10,220.00	

MISCELLANEOUS

1896 - 1903	Switchback Railway	
1900's	Ocean Wave	
1901	Baseball Bleachers	6,391.93
1903 - 1912	Figure Eight Coaster	
1904 - 1909	Whirl and Box Ball	2,485.94
1921 - 1936	Mill Chute	
1921 - 1966	American Racing Derby	
1920's	Witching Waves	
1928	Boulevard Stand	6,654.05
1928	Kiddie Swan Swing	
-1928	Caterpillar	
1960 - 1965	Commando Guns	
	Castle Inn	28,649.47
	Garage and Shop	15,048.62
-1965	French Frolic	
1966	Scrambler	

MISCELLANEOUS SOLD

1930	November	10 RED BUG CARS Palisades Park, N.J.
1938		THEATRE EQUIPMENT
1951	May	HURRICANE RIDE Festival Gardens London, England
1957	February	BUBBLE BOUNCE Park Amusements Inc.
1960	October	ROCK-O-PLANES
1960		DONATED ARCADE GAMES Stella Maris
1965	August	FRENCH FROLIC RIDE
1966		COFFE BREAK RIDE
1966	August	GIANT SLIDE Penzerard and Richman
1966	August	ORGAN Kap and Hachet
1966	March	AMERICAN RACING DERBY Cedar Point
1966	December	SWING AND GYMS
1967	May	34 ACRES (Humphrey Field) Visconi Co.
1967	August	16.33 ACRES (Grovewood) City of Cleveland

EVERYBODY HAS FUN
AT A
EUCLID BEACH PARK
PICNIC

YOUR ORGANIZATION CAN TOO! →

...STILL GREATER CLEVELAND'S

Euclid Beach has been Cleveland's major amusement park for over 50 years. The atmosphere at Euclid Beach Park is perfect for family outings and picnics of all sizes. Varied, wholesome entertainment is offered for people of all ages. There are no alcoholic beverages sold, nor are there any gambling devices of any sort on the premises. Only a large amusement park, with the staff and facilities of Euclid Beach, can offer the important ingredients that go into a successful picnic. Make your picnic an overwhelming success at Euclid Beach Park.

90 ACRES OF GROUNDS ON LAKE ERIE — WITH FACILITIES FOR YOUR PICNIC

PICNIC HEADQUARTERS
There are five Picnic Headquarters. The Log Cabin and Main Dance Hall are both equipped with platforms, public-address systems, and fine dance floors. For smaller groups there are the Annex, Lake Lunch, and Colonnade buildings. All are excellent facilities and are assigned on the basis of needs, to make your picnic a success.

BALL DIAMOND
There is a regulation-size ball diamond located adjacent to the Thriller. This ball diamond is suitable for either hard- or soft-ball games and has a limited seating capacity for picnic rooters. The diamond may be booked, at no charge, when you make your picnic arrangements.

RACE COURSES
Three fine race courses are available for your use — each with plenty of room for any races and games which you might plan as part of your picnic program.

GAME DIRECTOR
A professional game director, if desired, will plan your program and run your races and games. There is a small charge for this service. Prices are quoted on request.

PICNIC AREAS
Several picnic areas with large seating capacities are located throughout the Park. Outside picnic tables will seat about 1,500 people and all tables are located in tree-shaded areas. Additional picnic tables around the Main, Colonnade, and Lake Lunch Stands — all under cover — will seat another 2,500 people.

SHELTERED WALKS
Most of the facilities of the Park are connected by a continuous canopy and many of the rides, amusements, and concessions are covered, so that rain need not be a deterrent to picnic fun.

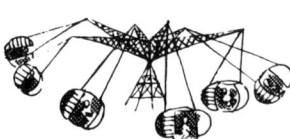

RIDES
There are thirty rides of which eleven are especially designed for small children. Roller coasters of varying speeds and heights are a park feature.

AMUSEMENTS
In addition to the rides there are other attractions. The Arcade, Miniature Golf, Bowl-O, Skee Ball, Shooting Gallery and Surprise House offer wholesome fun for the entire family and add immeasurably to the success of any picnic.

REFRESHMENTS
Humphrey popcorn, popcorn balls, and candy kisses are synonymous with Euclid Beach Park. In addition to these refreshments high-quality foods are served in the three park Lunch Stands and also in the restaurant overlooking the lake. A park feature is a quality wiener sandwich, always served piping hot. Popular soft drinks are served; also, frozen custard.

FREE PARKING
Free parking facilities for over 12,000 cars meet today's demands for ample parking. Entrance drives are attended by park personnel who direct cars to available parking spaces. Two public transportation systems use Euclid Beach as a terminal point.

SHADE TREES and PIER
Besides the rides, amusements, and other picnic facilities, Euclid Beach has always been particularly proud of the great number of shade trees on the park grounds, which add so much to the enjoyment of a picnic. Another feature of the park has been its pier which runs 550 feet out into Lake Erie and affords the fisherman the opportunity to try his luck. The Euclid Beach Pier is the only large private pier in the Northeastern Ohio Area.

ALWAYS POPULARLY PRICED...WITHIN

LEADING AMUSEMENT PARK

Many desirable features are to be found in the rides at Euclid Beach Park. Especially notable are: the sturdy construction, the large capacities, and the high moral and artistic standards of all scenes and illusions. Euclid Beach is one of America's few parks featuring four roller coasters. Consistently through the years Euclid Beach Park has put special emphasis on the maintenance of both the park facilities and rides.

FOR CHILDREN and ADULTS

AERO DIPS — a mild ride, especially good for introducing youngsters to roller coasters.

CARROUSEL — a big old-fashioned Merry-Go-'Round with hand-carved and hand-painted horses.

AUTO TRAIN — has been showing the camp grounds and the lake front to park patrons for over forty years.

MINIATURE RAILWAY — a little train pulled by a replica of a modern diesel engine, running over a scenic route.

JUST FOR KIDDIES

KIDDIE AUTOS — the most popular of the rides for the very small patrons.

KIDDIE BOATS — the beginner's boat ride.

KIDDIE CARROUSEL — a miniature Merry-Go-'Round.

KIDDIE HELICOPTER — an extremely popular modern children's ride, in which the youngster controls the flight of his plane.

These five kiddie rides are exact miniatures of adult rides:
KIDDIE MILL   KIDDIE BUG
KIDDIE WHIP   KIDDIE ROCKET SHIPS
          KIDDIE CARROUSEL

HOOK & LADDER — a fire engine ride, complete with siren and bell.

RODEO — western horses with guns and holsters for the use of the young riders in shooting at bad-men targets.

FOR CHILDREN, TEEN AGERS, and ADULTS

AMERICAN RACING DERBY — the modern version of the Merry-Go-'Round. Capacity: 128 passengers per trip.

BUG — a combination of the coaster and whip; fast and vigorous.

DIPPY WHIP — a mild, flat ride that has remained popular in amusement parks for over forty years.

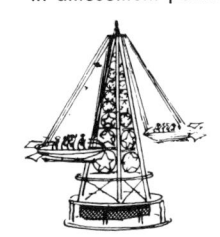

RACING COASTER — a fast-moving, roller-coaster race between two trains on parallel tracks. Capacity: 72 passengers on each trip.

ROCKET SHIPS — the oldest ride in Euclid Beach Park, an aerial swing, from a 101-foot tower.

ROCK-O-PLANE — a ferris-wheel type of ride in which the rider may somersault his car.

LAFF-IN-THE-DARK — a ride through a large, dark room filled with startling glimpses of funny and frightening creatures with appropriate sound effects. Capacity: 360 per hour.

SURPRISE HOUSE — Entertaining fun house full of illusions, moving stairways, and walkways. Fun for all. Capacity: over 300 per hour.

DODGEM — Rider-driven cars, encircled with bumpers for gay, care-free, reckless driving. Capacity: 100 per ride. Largest in the United States.

FLYING SCOOTERS — the modern circle swing in which the planes can be maneuvered to a height of 31 feet by manipulating a forward rudder.

FLYING TURNS — an exciting bob-sled ride in a twisting, spiraling, banking trough — one of very few of its kind in the world.

OVER-THE-FALLS — boats, propelled by water power through a covered course, finally climb to the top of a chute and plunge down the steepest hill in the park.

ROTOR — a cylinder that spins around until riders are held against its walls by centrifugal force and then drops its floor to leave them suspended high on the walls. Considered by many as entertaining to watch as to ride.

THRILLER — has the highest, steepest, fastest, longest track of all the coasters in the park. Capacity: 800 people per hour.

EASY REACH OF EVERY FAMILY BUDGET

Picnic

City of Cleveland EMPLOYEES

32nd Annual Picnic M.B.A. OF THE CLEVELAND TWIST DRILL CO. Euclid Beach Park June 23 1934

Richman's Fun For Fussy Tailors Picnic
EUCLID BEACH PARK
August 12th, 1967

Euclid Community Day
EUCLID BEACH PARK
Thursday, June 14th 1962
This Badge Admits You to

NEWS-HERALD COMMUNITY BARGAIN DAY
Euclid Beach Park
Wednesday, Aug. 19, 1964
FREE-RIDE COUPONS
28

Good for One Ride On
AERO DIPS
EUCLID BEACH PARK
Wednesday, Aug. 19, 1964

Good for One Ride On
CARROUSEL
EUCLID BEACH PARK
1964

Cleveland AFL-CIO LABOR DAY PICNIC
Sunday, Sept. 1, 1968
EUCLID BEACH PARK

NORTHEAST COMMUNITY MERCHANTS' 72nd PICNIC SILVER DOLLAR DAY
EUCLID BEACH PARK — WEDNESDAY, JULY 18
WIN BAG$ OF $ILVER DOLLARS
FREE DANCING IN MAIN BALLROOM 9:00 TO 11:30 P.M.
ENTERTAINMENT — BEAUTY CONTEST
WINNER MUST BE PRESENT AND CLAIM PRIZE AT TIME OF DRAWING OR ANOTHER NUMBER WILL BE DRAWN
SAVE THIS COUPON

LAKE COUNTY SCHOOLS DAY
"Bring Your Family and Friends"
SATURDAY, MAY 24, 1969
1 P.M. UNTIL CLOSING

"Come and Bring a Guest"
PLAIN DEALER GREEN THUMB CLUB

REPUBLICAN FAMILY OUTING
Euclid Beach Park
SUNDAY, SEPT. 8, 1968
6 P.M.

AN RACING DERBY
BEACH PARK
Aug. 19, 1964

One Ride On
ET SHIPS
BEACH PARK
Aug. 19, 1964

Our Guest — AUGUST 2
Northern Ohio
ilers Association

23rd Annual **BROWNHOIST PICNIC**
Under the Auspices of the
Mutual Benefit Association of BROWNHOIST EMPLOYEES
to be held at
Euclid Beach Park, Saturday August 10, 19

OD ON ES

JR. gineer

ODGEM
BEACH PARK
Aug. 19, 1964

67th Annual MBA PICNIC
CLEVELAND TWIST DRILL CO.
Euclid Beach Park
JUNE 14, 1969

Good for One Ride On
FLYING SCOOTERS
EUCLID BEACH PARK
Wednesday, Aug. 19, 1964

PICNICS

DATE 1927	ITEMS	FOLIO
7/11	B[r]ot Fwd	
12	Cleveland Elec Ill Co	CR 638
12	Pesco Products	"
15	Sons & Daughters of Liberty	"
	Master Builders	
16	Temple on the Heights	"
17	Newburgh F & A M	639
	Al Sirat Grotto	"
18	Elwell Parker Co	"
	Chicago Pneumatic Tool	"
22	Women's Benefit Assn	"
	Broadway M E Church	"
31	Refunds	JR 7718
24	Cleveland Electric Ill Co	C. 639
25	Woodward Temple Assn	"
28	Chicago Pneumatic Tool	640
29	Purchasing Agents Assn	"
	Cleve Federation of Musicians	"
30	Al Sirat Grotto	"
	Elwell-Parker Electric Co	"
	Cleve Cap Screw Co	"
	Bloomfield Co The	
	B[r]ot Fwd	
8/31		JR 214
10/17	Higbee Co The	CR 618
31		8

DATE 1928	ITEMS	FOLIO
6/12	Euclid Ave Temple	CR 634
16	Container Corp of Amer	635
17	The Temple	"
19	Newburgh F & A M	"
20	Iris Lodge	"
21	Rosedale School	"
	Cleveland Twist Drill	"
27	Cleveland Twist Drill Co	636
28	Euclid Ave Temple	"
30	Iris F & A M	"
30	Refund	7706
30	"	"
30	"	"
30	"	"
30	"	"
7/10	Kiwanis	CR 638
11	Temple The	"

DATE 1927	ITEMS	FOLIO
7/11	B[r]ot Fwd	
12	Cleveland Elec Ill Co	CR 638
12	Pesco Products	"
15	Sons & Daughters of Liberty	"
	Master Builders	
16	Temple on the Heights	"
17	Newburgh F & A M	639
	Al Sirat Grotto	"
18	Elwell Parker Co	"
	Chicago Pneumatic Tool	"
22	Women's Benefit Assn	"
	Broadway M E Church	"
31	Refunds	JR 7718
24	Cleveland Electric Ill Co	C. 639
25	Woodward Temple Assn	"
28	Chicago Pneumatic Tool	640
29	Purchasing Agents Assn	"
	Cleve Federation of Musicians	"
30	Al Sirat Grotto	"
	Elwell-Parker Electric Co	"
	Cleve Cap Screw Co	"
	Bloomfield Co The	
31	May Co The	CR 640
8/4	Cleveland Diesel	641
6	Bailey Co The	"
7	Taylors (J M B A)	"
8	Al Koran Shrine	144.. "
9	Richman Family	642
12	May Co The	"
15	Bigelow F & A M	"
16	Monarch Al. Co	"
19	Cleveland Diesel	643
	Richman Bros	"
7/31	Refunds	2220
8/31	" Diesel	2226
23	Graphite Bronze	CR 643
	Monarch Aluminum	"
25	Federal Reserve Bank	"
	Bailey Co The	"
26	Accurate Geo Mfg Co	"
30	Federal Reserve Bank	644

DATE		ITEMS	FOLIO	DATE		ITEMS	
1947		Bal Fwd		1948 10	31	Balance	
9	10	Nickle Plate Assn	CR648	11	12	Accurate Parts Co	CR 678
	11	Goodyear Tire & Rubber	646	1949 6	16	Euclid Ave Temple	693
	17	Woodward F&AM	"		18	Cleveland Twist Drill	"
	20	Addressograph Multigraph	"		21	Newburgh F&AM	"
		Kiwanis	"		23	The Temple	694
	25	Bigelow Lodge	"		24	Cleveland Twist Drill	"
10	28	Accurate Pts Mfg Co	648		25	Willard Storage Battery	"
	31	Balance	8		28	" " "	"
10	31	Balance		7	1	Euclid Ave Temple	695
1947	7	Woodward Temple	CR 649		5	Newburgh Lodge	"
1948 6	15	The Temple	664		11	Kiwanis	696
	17	Newburgh F&AM	"		12	The Temple	"
	19	The Temple	"		14	Al Sirat Grotto	"
		Cleveland Twist Drill Co	"			Aluminum Co of America	"
	22	Oheb Zedek	665		22	Pesco Products	697
	30	St Vincents	"		23	Woodward F&AM	"
	30	Refunds	VR			Cleveland Electric Ill Co	"
	30	"	VR		28	Al Sirat Grotto	698
	25	Cleveland Twist Drill	CR 665	May		The	"
	26	Willard Storage Battery	"		29	Woodward Lodge	"
	28	Euclid Ave Temple	"		30	Tyler Co WS	"
	30	Picnic Tickets	VR			Elwell Parker Co	"
7	6	Newburgh Lodge	CR 666	8	2	Clev Electric Ill Co	699
	12	Kiwanis	667			Clev Cap Screw Co	"
	14	Pesco Products	669		5	Al Koran Shrine	"
	23	Willard Storage Battery Co	668		6	White Sewing Mach Co	"
		Woodward F&AM	"		11	Elwell Parker	700
	28	Cleveland Elec Ill Co	"		13	Al Koran Shrine	"
		Clev Fed of Musicians	"			Richman Family	"
	30	Oheb Zedek	669		17	Clev Hts Picnic	VR
		Woodward F&AM	"		17	Brooklyn Heights	VR
	31	Cleveland Cap Screw	"		19	Bailey Co The	CR 700
		Elwell Parker Co	"			Cleveland Diesel	"
		Tyler Co WS	"		20	Richman Bros Co	"
8	5	Taylors F&BA	670		23	Federal Res Bank	"
		Elwell Parker Co	"		24	Bailey Co	701
6	30		VR			Cleveland Diesel Eng	"
8	31	Goodyear T&R Co			27	Irish Picnic	"
10	12		676			Bigelow F&AM	"
	31		8	9	3	Fed Res Bk of Cleveland	702
				8	31	White Sewing Mach	"
				10	27	Kiwanis	702
					30	Clev Graphite Bronze Co	702

DATE	ITEMS	FOLIO	DATE	ITEMS	FOLIO
1949			1950		
10 30	Balance Fwd		9 16	Brot Fwd	
31		8249	8 31	Returns	
			10 31		8
1950	Euclid Ave Temple	721	1951		
6 15			6 14	Heights Temple	751
16	White Motor Co	"	16	Cleveland Twist Drill	752
17	Cleveland Twist Drill Co	"	21	Newburg F & A M	"
20	Temple The	"	22	Clev Press (Purple Tickets)	"
	Euclid Ave Temple	"	29	Euclid Ave Temple	753
23	Cleveland Twist Drill Co	722	7 7	Cleveland Twist Drill Co	756
	Newburg F & A M	"	9	Kiwanis	"
7 10	Newburg F & A M	723	10	Newburgh Lodge	"
	Kiwanis	"	11	The Temple	"
	Temple The	"	16	Cleve Fed of Physicians	757
13	Al Sirat Grotto	724	18	" " "	"
22	Cleveland Electric Ill Co	"	19	Taylors' T M B A	"
24	Al Sirat Grotto	"	21	Cleveland Elec Ill Co	"
27	Pesco Products Co	725	27	The May Co	758
	May Co The	"	28	Cleveland Cap Screw Co	759
28	Howard Lodge	"	8 1	Hickok Electric	760
29	Tyler Co W S	"		Cleveland Electric Ill Co	"
	Cleveland Cap Screw	"		Cleveland Graphite Bronze	"
8 3	Tyler Co W S	726			
	Cleveland Elec Ill Co	726	8 1	Brot Fwd	
	Taylor Sons Co Wm	"	3	Taylors T M B A	760
4	Al Koran Shrine	"	3	Al Koran Shrine	"
	May Co The	"	7	Kiwanis	761
5	Aluminium Co of Amer	"	11	Richman Family	"
6	Columbus Citizen	"	17	Bigelow F & A M	762
7	Wm Taylor Son Co Wm	"	18	Elwell Parker	"
11	Bigelow Lodge	727	31	White Sewing Mach	2548
	Al Koran	"	22	Federal Reserve Bank	762
16	Bigelow Lodge	"	23	Bigelow F & A M	"
18	Lynite Club	"	25	Elwell Parker Elec Co	763
	Cleveland Diesel	"	30	Federal Reserve Bank	"
19	Elwell-Parker	728	9 1	Cleveland Diesel	765
22	Federal Res. Bank	"	8	Addressograph Multigraph	"
25	Cleveland Diesel	"		Cleveland Diesel	"
27	Many Societies	"		County Engineers	"
31	Return Sales	768		May Co	"
9 9	Elwell Parker	730	11	Richman Bros Co	766
9	Federal Reserve Clev.	"	13	Nickel Plate Ass'n	"
13	Addressograph Mult.	731	14	Kiwanis	"
16	Kiwanis	"	28	Thornton Co	767
			10 10	Al Koran	768
			31		8 2573

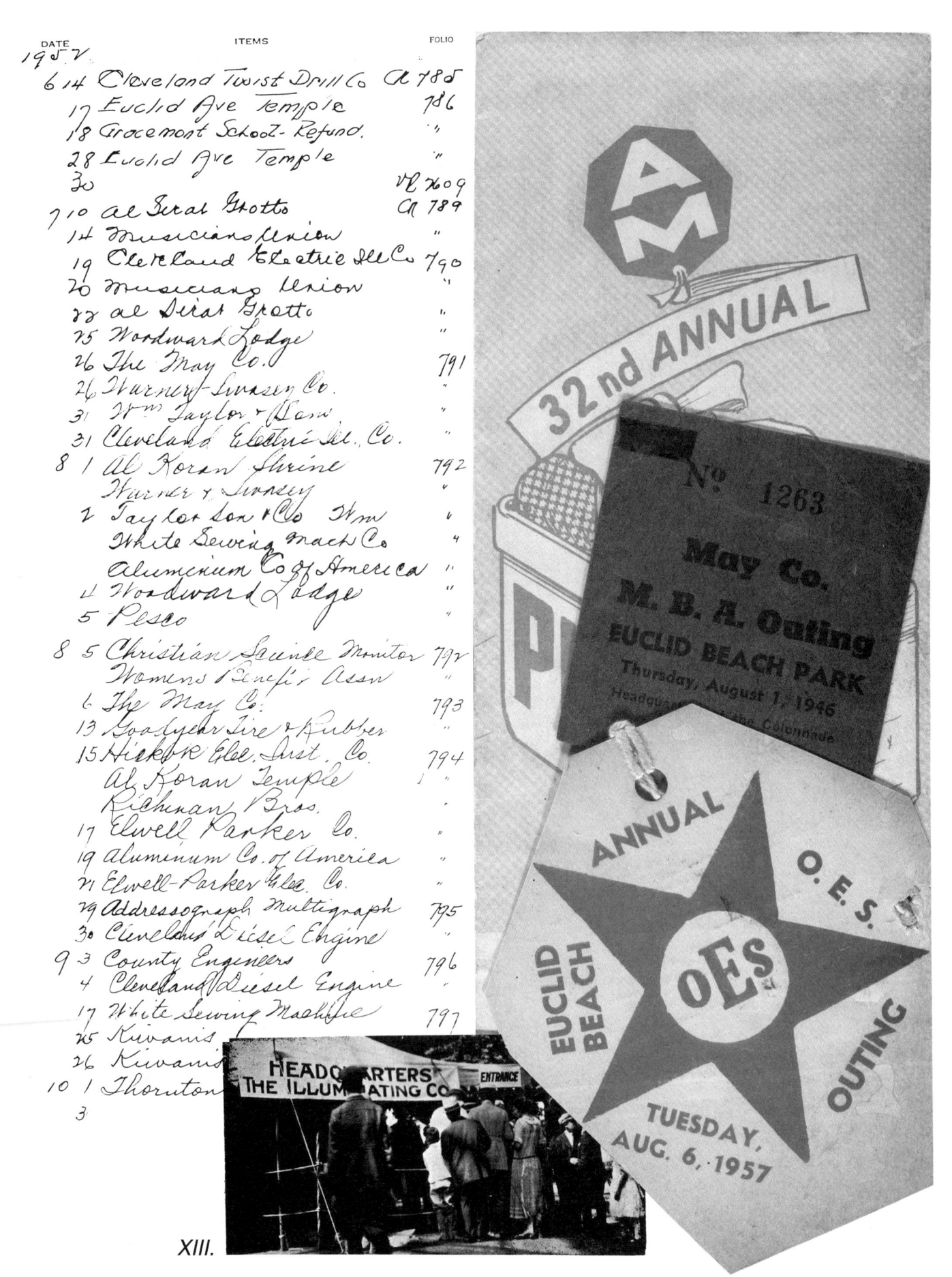

DATE	ITEMS	FOLIO
1957		
6 14	Cleveland Twist Drill Co	Ck 785
17	Euclid Ave Temple	786
18	Gracemont School - Refund	"
28	Euclid Ave Temple	"
30		VP 7609
7 10	Al Sirat Grotto	Ck 789
14	Musicians Union	"
19	Cleveland Electric Ill Co	790
20	Musicians Union	"
22	Al Sirat Grotto	"
25	Woodward Lodge	"
26	The May Co.	791
26	Warner-Swasey Co.	"
31	Wm Taylor & Son	"
31	Cleveland Electric Ill. Co.	"
8 1	Al Koran Shrine	792
	Warner & Swasey	"
2	Taylor Son & Co Wm	"
	White Sewing Mach Co	"
	Aluminum Co of America	"
4	Woodward Lodge	"
5	Pesco	"
8 5	Christian Science Monitor	792
	Womens Benefit Assn	"
6	The May Co.	793
13	Goodyear Tire & Rubber	"
15	Hickok Elec. Inst. Co.	794
	Al Koran Temple	"
	Richman Bros.	"
17	Elwell Parker Co.	"
19	Aluminum Co. of America	"
21	Elwell-Parker Elec. Co.	"
29	Addressograph Multigraph	795
30	Cleveland Diesel Engine	"
9 3	County Engineers	796
4	Cleveland Diesel Engine	"
17	White Sewing Machine	797
25	Kiwanis	"
26	Kiwanis	"
10 1	Thornton	"
3		

XIII.

SCHOOL DAYS

THE RIEHL PRINTING COMPANY
Printers · Lithographers

748 EAST 82nd STREET, CLEVELAND 3, OHIO
EXpress 1-4555

The Humphrey Company
Euclid Beach Park
Cleveland, Ohio 44110

80494 ✓ 4,000 strips - 4 coupons each - Dental Health Awards Day - May 21, 1967

80591 ✓ 10M Time slips - not padded - wrapped 500 per package

80592 ✓ 5,000 Time Cards - Form 246 - one change

80535 ✓ 16,700 School Badges - Bedford Schools Day, May 6, 1967 - 2 sides

80536 ✓ 5,900 School Badges - Chagrin Falls   "   "   "   "   "

80537 ✓ 8,200 School Badges - Fairview Park   "   "   "   "   "

80538 ✓ 3,200 School Badges - Richmond Heights Schools Day, May 6th   "

80539 ✓ 27,500 School Badges - South Euclid and Lyndhurst Schools Day - May 6, 1967 - 2 sides

80541 ✓ 10,100 School Badges - Brecksville Schools Day, May 13, 1967 - 2 sides

80543 ✓ 9,500 School Badges - Rocky River Schools Day,   "   - 2 sides

80585 ✓ 15,000 copies - To Homeroom Teachers: etc.

80494 — ✓ 4,000 strips - 4 coupons each - Dental Health Awards Day - May 21, 1967

80591 — ✓ 10M Time slips - not padded - wrapped 500 per package

80592 — ✓ 5,000 Time Cards - Form 246 - one change

80535 — ✓ 16,700 School Badges - Bedford Schools Day, May 6, 1967 - 2 sides

80536 — ✓ 5,900 School Badges - Chagrin Falls " " " . " "

80537 — ✓ 8,200 School Badges - Fairview Park " " " " "

80538 — ✓ 3,200 School Badges - Richmond Heights Schools Day, May 6th "

80539 — ✓ 27,500 School Badges - South Euclid and Lyndhurst Schools Day - May 6, 1967 - 2 sides

80541 — ✓ 10,100 School Badges - Brecksville Schools Day, May 13, 1967 - 2 sides

80543 — ✓ 9,500 School Badges - Rocky River Schools Day, " - 2 sides

80585 — ✓ 15,000 copies - To Homeroom Teachers: etc.

80548 — ✓ 500 School Badges - Bratenahl Schools Day, Friday, June 9, 1967 - 2 sides

80549 — ✓ 5,500 School Badges - Brooklyn Schools Day, Friday, June 9, 1967 - 2 sides

80551 — ✓ 3,300 School Badges - Independence Schools Day, Friday, June 9, 1967 2 sides

80553 — ✓ 12,800 School Badges - Mayfield School District - Friday, June 9, 1967 2 sides

80556 — ✓ 6,800 School Badges - Warrensville Hts. Schools Day, Friday, June 9, 1967 - 2 sides

80557 — ✓ 16,500 School Badges - East Cleveland Schools Outing, Wednesday, June 14, 1967 - 2 sides

80604 — ✓ 30,000 School Badges - Catholic Elementary Schools Day, Sunday, June 4, 1967 - 2 sides

80605 — ✓ 66,800 School Badges - Catholic Elementary Schools Day, Tuesday, June 13, 1967 - 2 sides

80558 ✓— 89,000 School Badges - Cleveland Elementary Schools Day, Tuesday, June 20, 1967 - 2 sides

80559 ✓— 90,600 School Badges - Cleveland Elementary Schools Day, Friday, June 23, 1967 - 2 sides

80606 ✓— 46,000 School Badges - Catholic High Schools Day, Thursday, June 15, 1967 - 2 sides

80620 ✓— 2,500 Identification Badges - punched - Lutheran Children' Day

80621 ✓— 225 Posters - Lutheran Children's Day - Thursday, June 29, 1967 2 colors

80607 ✓— 120 School Posters - Catholic High Schools Day, Thursday, June 15, 1967

81216	✓	9,000 Badges	Strongsville Schools Day - Saturday, May 11th - 2 sides
81217	✓	2,300 Badges	Cuyahoga Heights Schools Day - Friday, June 7th "
81218	✓	3,300 Badges	Independence Schools Day - Friday, June 7th - 2 sides
81219	✓	6,400 Badges	Olmsted Falls Schools Day - Friday, June 7th - 2 sides
81220	✓	13,000 Badges	Mayfield School District - Friday, June 7th - 2 sides
81221	✓	500 Badges	Bratenahl Schools Day - Friday, June 7th - 2 sides
81222	✓	16,000 Badges	Maple Heights Schools Day - Friday, June 7th - 2 sides
81223	✓	3,200 Badges	Richmond Heights Schools Day, Sunday, June 9th - 2 sides
81224	✓	7,400 Badges	North Royalton Schools Day, Sunday, June 9th - 2 sides
81225	✓	22,000 Badges	Ashtabula Schools Day, Sunday, June 9th - 2 sides
81226	✓	10 Posters	Strongsville Schools Day - Saturday, May 11th - 2 colors
81227	✓	12 Posters	Cuyahoga Heights Schools Day - Friday, June 7th - 2 colors
81228	✓	6 Posters	Independence Schools Day - Friday, June 7th - 2 colors
81229	✓	10 Posters	Olmsted Falls Day - Friday, June 7th - 2 colors
81230	✓	22 Posters	Mayfield School District - Friday, June 7th - 2 colors
81231	✓	2 Posters	Bratenahl Schools Day - Friday, June 7th - 2 colors
81232	✓	20 Posters	Maple Heights Schools Day - Friday, June 7th - 2 colors
81233	✓	8 Posters	Richmond Heights Schools Day - Sunday, June 9th - 2 colors
81234	✓	12 Posters	North Royalton Schools Day - Sunday, June 9th - 2 colors
81235	✓	40 Posters	Ashtabula Schools Day - Sunday, June 9th - 2 colors
81240	✓	106,000 School Badges	Lake County Schools Day - Saturday, May 25th - 2 sides
81241	✓	180 School Posters	Lake County Schools Day, Sat., May 25th - 2 colors
81242	✓	17,000 School Badges	East Cleveland Schools Day, Wednesday, June 12, 1968 - 2 sides
81243	✓	25 School Posters	East Cleveland Schools Outing, Wednesday, June 12, 1968 - 2 colors
81239	✓	100 5¢ Government stamped Post Cards printed	If interested in employment return this card etc.

XIV. Another Humphrey invention.

SPECIAL SHOWS

7-16-64	*Lou Adler*
7-22-64	*Four Seasons*
8-21-64	*Beach Boys*
6-25-65	*Gary Lewis and the Playboys*
7-30-65	*Jan and Dean*
8-16-65	*Jan Berry*
6-19-66	*Joe Keaper*
7-29-66	*Lovin' Spoonful*
8-21-66	*Bobby Shapiro*

MAPS...

CLEVELAND TO EUCLID BEACH
(11 MILES)

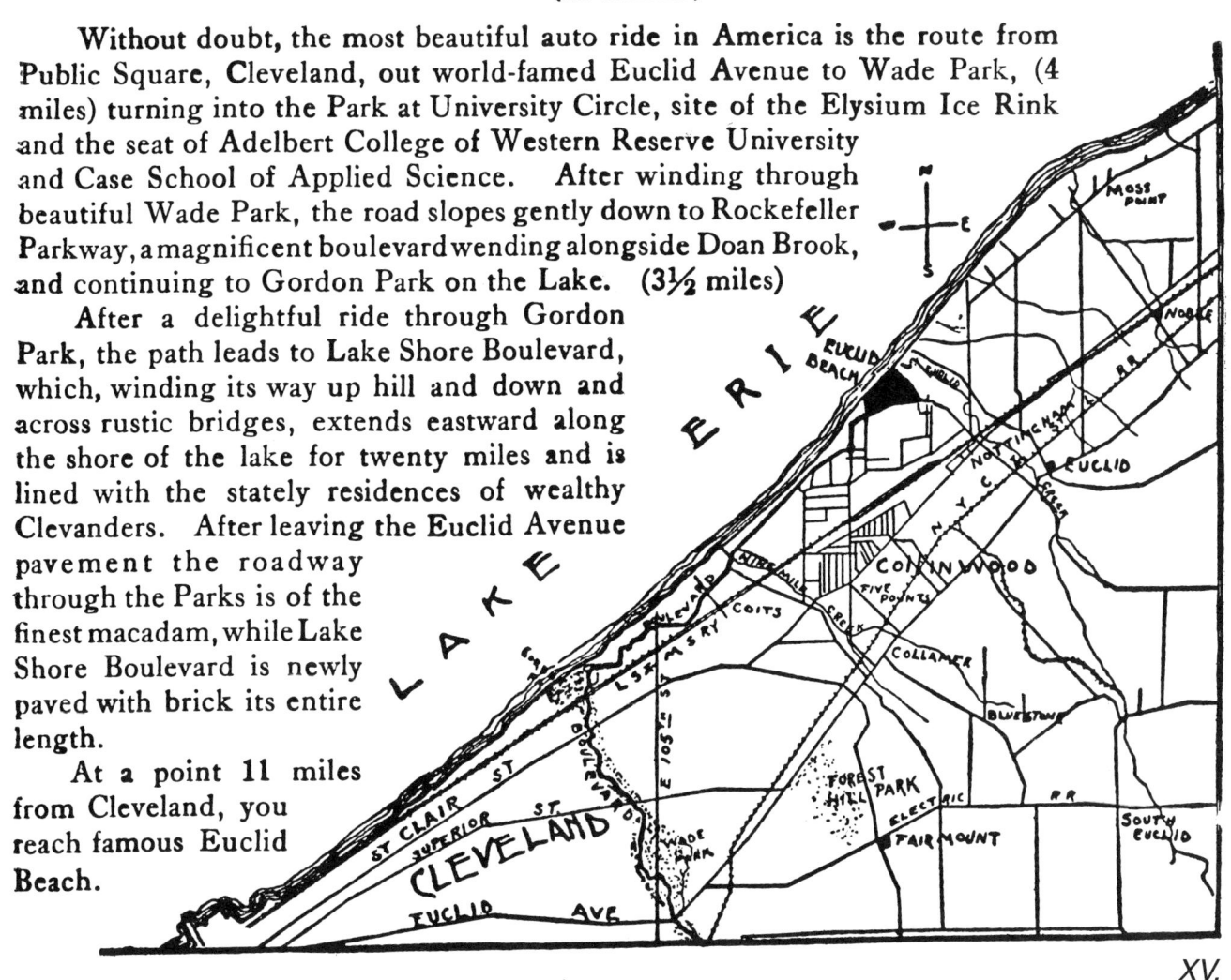

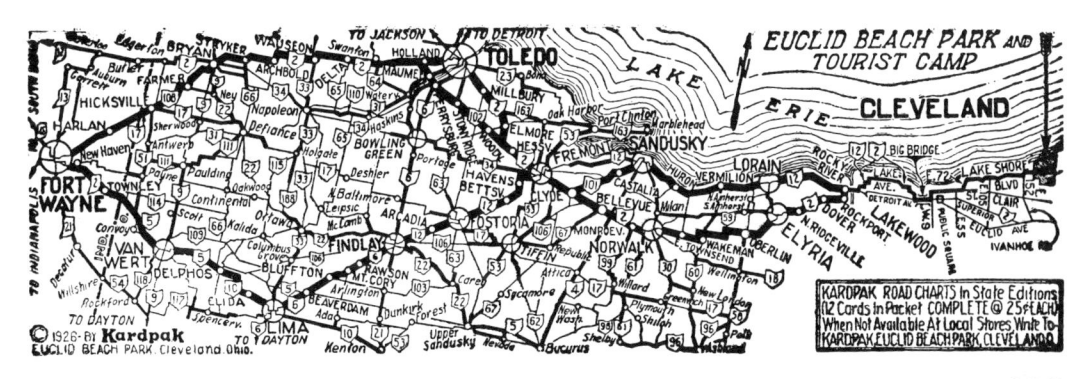

Without doubt, the most beautiful auto ride in America is the route from Public Square, Cleveland, out world-famed Euclid Avenue to Wade Park, (4 miles) turning into the Park at University Circle, site of the Elysium Ice Rink and the seat of Adelbert College of Western Reserve University and Case School of Applied Science. After winding through beautiful Wade Park, the road slopes gently down to Rockefeller Parkway, a magnificent boulevard wending alongside Doan Brook, and continuing to Gordon Park on the Lake. (3½ miles)

After a delightful ride through Gordon Park, the path leads to Lake Shore Boulevard, which, winding its way up hill and down and across rustic bridges, extends eastward along the shore of the lake for twenty miles and is lined with the stately residences of wealthy Clevanders. After leaving the Euclid Avenue pavement the roadway through the Parks is of the finest macadam, while Lake Shore Boulevard is newly paved with brick its entire length.

At a point 11 miles from Cleveland, you reach famous Euclid Beach.

XV.

XVI.

173

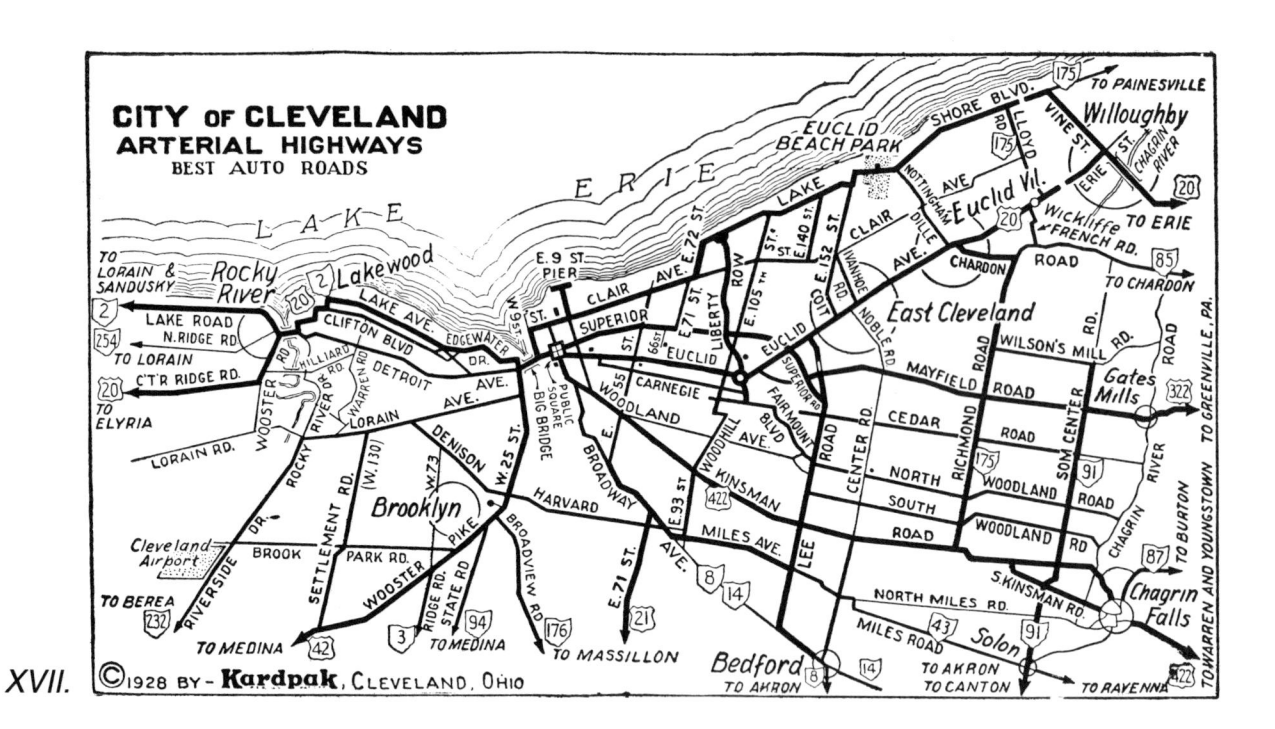

CITY OF CLEVELAND ARTERIAL HIGHWAYS
BEST AUTO ROADS

XVII. © 1928 BY — Kardpak, Cleveland, Ohio

XVIII.

XIX.

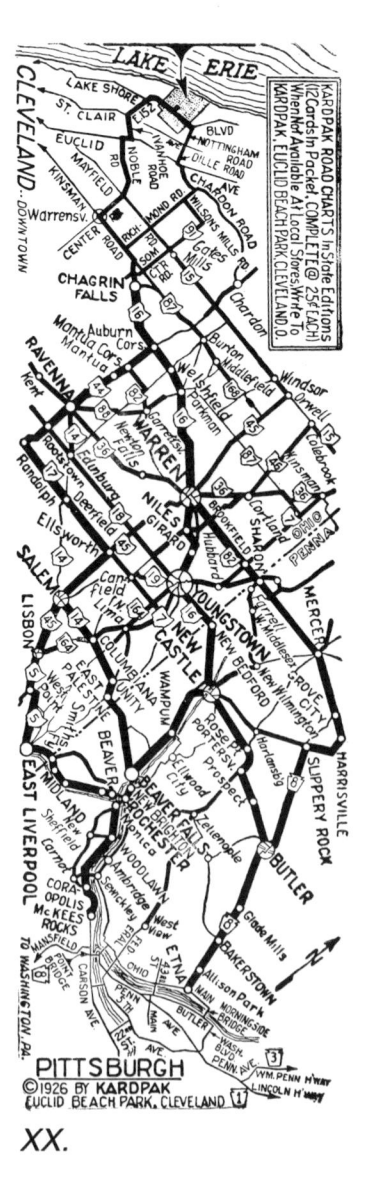

XX.

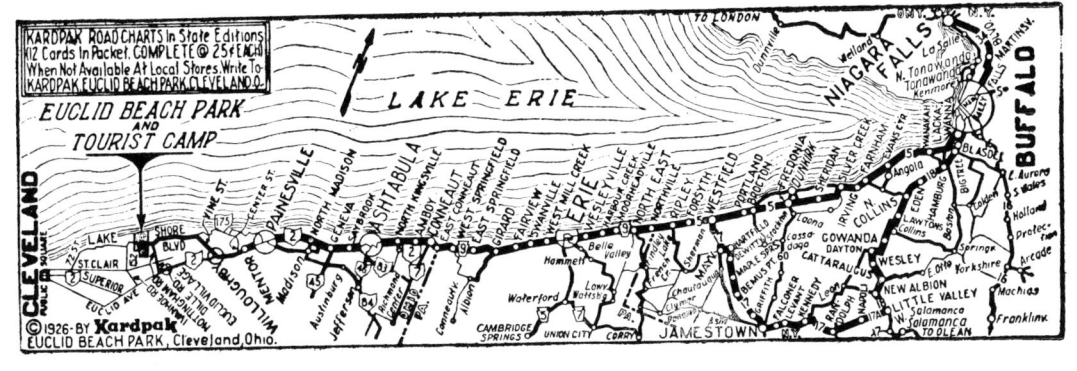

XXI.

... AND MILAGE

XXII.

XXIII.

XXIV.

		1 WEINERS	2 WEINER ROLLS	3 MUSTARD	4 HORSE RADISH	5 COFFEE	
		lbs.	DOZ.	Gal.	Gal.	lbs.	
1	1931		27786			5120	1
2	1932	14825	10558			6661	2
3	1933	14392	9117			6497	3
4	1934	14672	12896			6776	4
5	1935	15463	11766	399	175	6612	5
6	1936	20325	15292	318	235	6490	6
7	1937	25746	19798	400	327	6212	7
8	1938	23536	21248	390	235	7602	8
9	1939	28812	22665	495	293	7391	9
10	1940	30965	26287	530	308	7505	10
11	1941	48064	42902	864	554	7426	11
12	1942	58467	49186	938	652	7864	12
13	1943	50912	43385	804	533	6657	13
14	1944	55860	44885	891	470	9901	14
15	1945	54550	45348	876	531	11120	15
16	1946	46482	34938	697	427	12621	16
17	1947	45578	35891	715	530	8540	17
18	1948	48440	39995	791	575	10452	18
19	1949		34725	727	331	8968	19

	FINE SUGAR lbs.	TABLE SALT lbs.	JUMBO PEANUTS lbs.	POPCORN BOXES 000's	BALLOONS GROSS
1931	101953	2060	21015	297	
1932	95951	2304	16846	212	117
1933	99850	2722	22557	416	126
1934	85825	1924	30480	170	211
1935	91700		14930	184	171
1936	122820	2000	13079	158	247
1937	109950	2210	11350	196	304
1938	124250	1196	18570	173	149
1939	124610	1352	18710	182	166
1940	125490	1560	15665	173	120
1941	154700	2522	21060	249	183
1942	118417	2060	23300	271	RATIONED
1943	133732	1244	19160	334	RATIONED
1944	152129	2759	20900	436	RATIONED
1945	117945	2205	24890	243	RATIONED
1946	123400	2192	19490	686	RATIONED
1947	105665	2251	18040	320	RATIONED
1948	126603	2413	20610	373	458
1949		2382	18120	318	462

	BOILED HAM lbs.	PIES 9"	DOUGHNUTS DOZ.	WHIP FREEZE Gal.	BREAD LOAFS
1931					
1932	5697	5420	1881	7250	5018
1933	6323	4937	2227	7138	4655
1934	5391	5080	2693	8688	5289
1935	4171	5120	2684	9331	4664
1936	6412	5487	2496	10970	5114
1937	5775	6534	2542	11950	5521
1938	5573	6641	3017	11232	4905
1939	5971	7017		11607	5234
1940	5485	8379	2643	11280	4856
1941	5875	10491	2739	14979	6550
1942	7601	9701	3664	20109	7224
1943	5379	8507	9551	17580	5809
1944	7904	8946	8438	16530	5793
1945	8329	10540	9367	20148	5056
1946	7850	11390	10383	18108	4422
1947	4319	7867	4737	15390	3501
1948	2658	10969	3892	15729	3025
1949	2670	7190	2956		2730

	NAPKINS 000	PAPER PLATES 000	TOILET PAPER CASES	BROOMS #8 DOZEN	FLOOR WAX lbs.
1931	232	72	2250	126	
1932	158	98	2150	73	180
1933	147	112	2443	120	100
1934	268	91	2011	141	200
1935	265	103	2551	190	350
1936	220	109	2540	190	578
1937	433	118	2780	204	625
1938	419	125	2625	156	700
1939	505	134	2310	204	970
1940	575	136	2240	213	540
1941	837	417	2796	212	420
1942	888	155	2654	167	416
1943	765	146	3847	163	444
1944	654	161	3911	264	259
1945	996	146	3838	322	113
1946	780	133	3876	234	220
1947	325	88	3186	198	180
1948	825	133	3692	201	180
1949	612	121	3908	171	
1950					

Mat-Rubber for Fun House

I4 x 2' 6 ply #100 Nu Kraft   .65

B. F. Goodrich Co.

```
1941    Coffee Mugs

        Brown flaggon 8-0z.    Dz    2.25
        White    "       "           2.00
          "      mug     "           1.65
        Brown    "                   2.00

                    Bailey-Walker

'45     #350 Sterling China Co.,     2.37
        L.E. Samson, Cleveland, Ohio
        Plus freight & Carton
```

Bumpers, rubber for Dodgem Cars

    C.A. Stillman Jr.
    Goodyear Tire
    13 & Chester
Through Wm. Bingham Co.

ROOT BEER

25 lb. Sugar
12 " Cane
45 oz. Concentrate Extract
Water to 40 gal.

Article: Balloons   #12 Gas
Description: Imprint with our die
Size or No.

DATE	Order No. or Quoted by	FIRM	QUANTITY	UNIT	LIST	DISCOUNT	NET	TERMS	F.O.B.	Freight or Express	UNIT COST DELIVERED
/48		Barr Rubber Co	500	Gr			6.20			'48	458
/49		Inventory	42				5.60			'49	462

Article: Cheese
Description: 5 lb. Brick

DATE	Order No. or Quoted by	FIRM	QUANTITY	UN
Jan /36		Inventory	0	
May		D. J. Saegren	180	
June		" "	210	
July		" "		

Orange   19 to 1 strength   1942

    Orange   1 Gal
    Sugar   14 Lb.
    Corn   4-1/4 lb.
    Water to 20 gal.

Skee Ball Parts
For Wurlitzer & large alleys

    Julius M. Seidel
    741 East 42 St.,
    Brookly, N. Y.

Weiner Roll Slicers

    Chas. F. Warrick
    16251 Hamilton Ave.,
    Detroit, Mich.

Article: Pop   9-Oz bottle 24 to case
Description:

DATE	Order No. or Quoted by	FIRM	QUANTITY	UNIT	LIST	DISCOUNT	NET	TERMS	F.O.B.	Freight or Express	UNIT COST DELIVERED
May	30/42	Pepsi Cola Co	2466	case			.45				
Jun			2715								
Jul			5249								

Pop Corn Cartons

V.V. Sales Co.,
3259 Broadway, Chicago 13, Ill

Have a stock printed box
or will make up from our dies.
Sample on file.   Under V letter file

China Souvenirs,

    Acme Craftware, Inc.,
    Box 55 Wellsville, Ohio.

FLYINH SCOOTERS
Bisch-Rocco Amusement Co.
5441 South Cottage Grove,
Chicago, Ill.

Card 1: Cups-Dixie

Article	Cups-Dixie							Size or No. 2500		
Description	No. 185 9 Oz. Blvd. Stand only									
DATE	Order No. or Quoted by	FIRM	QUANTITY	UNIT	LIST	DISCOUNT	NET	TERMS	F.O.B.	Freight or Express	UNIT COST DELIVERED
Jan	/43	Union Paper	10000								
	/44	Inventory	9000		Late in season during				'43	1000	
		Union Paper	20000		Per M		5.65				
Jan	/45	Inventory	5000						'44	24 M	
			37500				5.65		'45	30 M	

Card 2: Bags

Article	Bags									
Description	3/4-lb. 25 lb. Glassine Printed							Size or No. $3\frac{1}{2} \times 2\frac{1}{4} \times 6\frac{3}{8}$		
	For Single Pop Corn Balls 25# Stock									
DATE	Order No. or Quoted by	FIRM	QUANTITY	UNIT	LIST	DISCOUNT	NET	TERMS	F.O.B.	Freight or Express	UNIT COST DELIVERED
Jan	/43	Inventory									
Jan	/44	Inventory	45	M			1.45		'43	45	
Dec		Moser Bag	100						'44	90	

Card 3: Coffee

COFFEE

1 lb. Coffee makes 2.6 gal. of brew
6.5 oz. per cup — 51 cups per lb.
or 255 cups per Urn @ 5 lb. to Urn

1½ lb. Sugar to 1 lb. Coffee

Card 4: Binding

Binding for Flying Scooter Cars
2" Airplane cloth per yd 05
Meister Bro's
2044 Euclid Ave.
City.

Card 5: Baskets

Baskets
Wire Baskets for Candy Kisses

#117-C   Dz   13.50

Massillon Wire Basket Co.

Card 6: Spring bumpers

Spring bumpers Laffindark front car

2 x 5/16" x 4'4-½" long
Onyx spring steel .60-.70 carbon
Crucible Steel of America (Payne)
Shaped by Perfection

Card 7: Pop Corn Shelled

Article	Pop Corn Shelled at Beach 1945			Size or No.	
DATE	Order No. or Quoted by	FIRM	QUANTITY	UNIT	
Jan	/45		1600	lb	
Feb			3700		
Mar			5200		
Apr			4900		
May			15200		
Jun			19600		
Jul			19600		

Card 8: Dodgem Steering Wheel Hubs

Dodgem Steering Wheel Hubs #45003

American Hard Rubber Co.
11 Mercer St., New York.
or Akron, Ohio.

Front wheels-cotton,
Devine Bro's. Co.,
Utica, N.Y.

Card 9: Spring seats

Spring seats for Coaster Cars.

Trenton Spring Products Co.,
Trenton, N.J.

'45 96500

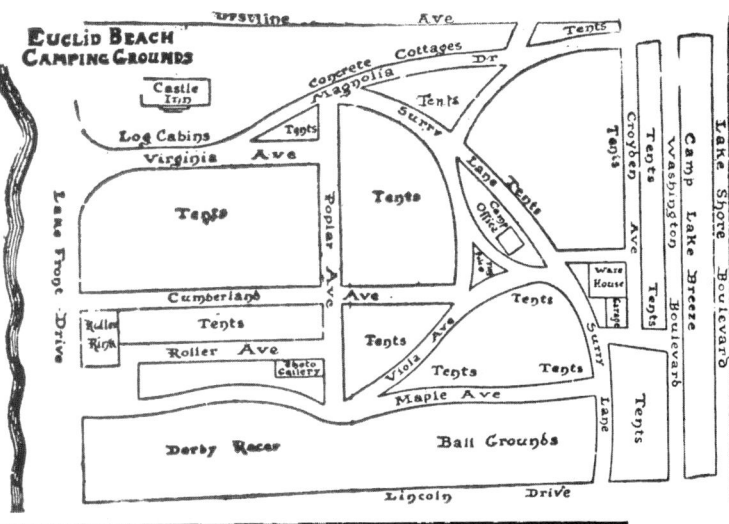

XXIV.

| SCHEDULE OF RATES | **Euclid Beach Park Camp** | ON LAKE ERIE |
| (Subject to Change) | | Cleveland |

Owned and Operated by THE HUMPHREY COMPANY, Cleveland, Ohio
(Address All Communications as Above)
Telephone IV. 1-7575

FOUR-ROOM COTTAGE TENTS
PERMANENT ROOFS

Furnished with two Double Beds, Camp Rest Chairs, Chiffonier, Clothes Tree, Dining Table and Chairs, Gas Stove for cooking — Free Gas and Electricity, Icebox, Wash Stand, and Dish Cupboard.

$4.00 per day $25.00 per week

EXTRA EQUIPMENT PRICES

PER DAY OR WEEK; Cots, 50c; Linens and Blankets 50c per bed. Dishes and cooking utensils, 25c per person.

Extra Cars 50c day — $2.00 per week.

Guests must furnish own towels.

Trailer Accommodations Community Showers and Laundry with Hot and Cold Water.

We will accept reservations upon receipt of a deposit equal to one-fourth of the total rental for the time desired. Checking out time 12:00 Noon. Boats, Bait and Fishing Tackle can be rented at the pier.
EUCLID BEACH PARK furnishes the finest in clean and wholesome amusements for young and old.

NO DOGS ALLOWED

181

XXVI. Rare Photoengravings of short-lived Outdoor Dance Patio. ca. 1926-1927.

XXVII. Photo by Hollywood Lighting Systems. June 16 1926.

Euclid Beach Park Official Program

1895   EUCLID BEACH PARK   1995

LAUGHING SAL